Environmental Quality and Safety
Vol. 5

Georg Thieme Publishers Stuttgart
Academic Press New York · San Francisco · London
A Subsidiary of Harcourt Brace Jovanovich, Publishers

Environmental Quality and Safety

Global Aspects of Chemistry, Toxicology and Technology as Applied to the Environment

Vol. 5

90 figures
61 tables

Contributors:

A. Azar, Newark, Delaware, USA
M. Beroza, Beltsville, Maryland, USA
B. G. Blaylock, Oak Ridge, Tennessee, USA
M. A. Bombace, Roma, Italy
E. A. Bondietti, Oak Ridge, Tennessee, USA
R. van den Bosch, Berkeley, California, USA
Ph. Bourdeau, Ispra, Italien
G. F. Clemente, Roma, Italien
J. A. Cooper, Richland, Washington, USA
F. Coulston, Albany, N. Y., USA
M. L. Dhar,
J. W. Dobrowolski,
W. J. Dongherty, New Mexico
A. Eisenberg, Yavne, Israel
Y. Feige, Jerusalem, Israel
C. W. Francis, Oak Ridge, Tennessee, USA
F. Geiss, Ispra, Italien
L. Golberg, Albany, N. Y., USA
F. D. Griffith, Newark, Delaware, USA
V. W. Henry III, Newark, Delaware, USA
R. J. Van Hook, Oak Ridge, Tennessee, USA
J. W. Huckabee, Oak Ridge, Tennessee, USA
J. B. Hunn, LaCrosse, Wisconsin, USA
H. Ishikura, Tokyo, Japan

J. D. Jansen, The Hagne, Netherlands
W. Klein, St. Augustin, Germany
J. Klopfer, Yavne, Israel
F. Korte, Attaching bei Freising, Germany
C. M. Menzie, Washington D.C., USA
R. L. Metcalf, Urbana Champaign, Illinois, USA
T. Misato, Saitama, Japan
E. M. Mrak, Davis, California, USA
W. P. Müller, Attaching bei Freising, Germany
R. W. Perkins, Richland, Washington, USA
Y. Proulov, Yavne, Israel
L. A. Rancitelli, Richland, Washington, USA
D. E. Reichle, Oak Ridge, Tennessee, USA
L. C. Rossi, Roma, Italy
L. Salonen, Helsinki, Finland
J. W. Sarver, Newark, Delaware, USA
R. D. Snee, Newark, Delaware, USA
F. H. Sweeton, Oak Ridge, Tennessee, USA
F. Takahashi, Kyoto, Japan
K. Trammel, Geneva, New York, USA
H. A. Vaajakorpi, Helsinki, Finland
J. Weisgerber, St. Augustin, Germany
J. P. Witherspoon, Oak Ridge, Tennessee, USA
J. Yamamoto, Tokyo, Japan

1976
Georg Thieme Publishers Stuttgart
Academic Press New York · San Francisco · London
A Subsidiary of Harcourt Brace Jovanovich, Publishers

© 1976 Georg Thieme Verlag, D-7000 Stuttgart 1, Herdweg 63, P.O.B. 732 – Printed in Germany by Maurersche Buchdruckerei, Geislingen an der Steige

For Georg Thieme Verlag ISBN 3 13 514501 8
For Academic Press ISBN 0-12-227005-3
Library of Congress Catalog Card Number: 70-145669

Table of Contents

Chemical Control of the Sea Lamprey the Addition of a Chemical to the Environment *)

By **Calvin M. Menzie** (1) and **Joseph B. Hunn** (2)

Abstract

Construction of the Welland Canal enabled shipping to by-pass Niagara Falls and enter the upper Great Lakes and also eliminated the barrier to the entry to the lakes by the sea lamprey (*Petromyzon marinus* Linnaeus). Within forty years the commercial fisheries of the Great Lakes was almost eliminated by this parasitic cyclostome.

A search for selective chemical control of the sea lamprey was undertaken in the 1950's and culminated with the discovery of TFM (3-Trifluoromethyl-4-nitrophenol).

At the request of the International Great Lakes Fishery Commission, the Bureau of Sport Fisheries and Wildlife undertook to assess the hazard of TFM to the aquatic ecosystem, to humans as well as to fish and wildlife. Studies were undertaken in Bureau laboratories as well as by contracts with university and private laboratories. Results of these studies to-date indicate that this material is not subject to biomagnification and does not pose a hazard to man or to the environment.

1. Office of Environmental Quality, Bureau of Sport Fisheries and Wildlife, USDI, Washington, D.C.
2. Fish Control Laboratory, Bureau of Sport Fisheries and Wildlife, La-Crosse, Wisconsin

---

*) This paper was presented at the Third International Symposium on Chemical and Toxicological Aspects of Environmental Quality, Tokyo, Japan, November 19–22, 1973

Introduction

Some time before 1920, the sea lamprey gained entry into the Upper Great Lakes by way of the Welland Canal. Before the appearance of this marine predator, the annual commercial production of lake trout, *Salvelinus namaycush* (Walbaum), from the Upper Great Lakes was about 15 million pounds. About 1941, production of lake trout began to decline and within 20 years the yield dropped to less than 300,000 pounds. Lakes Huron and Michigan, which had previously produced 5 and 6 million pounds respectively, produced less than 1,000 pounds in 1961. The annual loss to the commercial fishermen of the three lakes was estimated at between $ 7.3 and $ 8 million.

In the late 40's, studies were undertaken with the goal of controlling the sea lamprey. This could be accomplished by preventing the lamprey from spawning; by preventing the return to the lakes of adult lamprey; or by killing the larval lamprey. It was determined that the first two methods could not be adequately implemented. Most of the research has centered, therefore, on the latter approach (Schnick, 1972).

The adult stage of the sea lamprey lasts about one to one and one-half years. At that time, having become sexually mature, they migrate upstream where they spawn and die. While a female may produce as many as 100,000 eggs, the average is about 60,000. Within about 13 days, the eggs

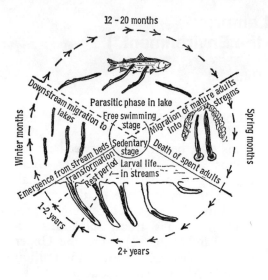

12 - 20 months

Fig. 1

hatch and the larvae burrow into the stream bottoms where they subsist on detritis and benthic organisms. This non-parasitic stage may last from 3 to 7 years or longer. These larvae, or ammocetes, then undergo a metamorphosis and become free-swimming parasitic adults. At this time they migrate down stream and enter the lakes where they begin their predacious existence (Fig. 1).

Surveys are used to locate streams and sections thereof that are infested with lamprey larvae. Tributary to Lakes Superior, Huron and Michigan are more than 200 streams and rivers which contained larvae.

Investigations begun in the mid 1950's led to the lampricide TFM (3-trifluoromethyl-4-nitrophenol). In 1958, applications of this material were undertaken in streams tributary to Lake Superior. Results were dramatic: 80% reduction of sea lamprey populations by 1962 and 90% reduction by 1966. Application was extended to Lakes Michigan and Huron in the 1960's with comparable results and in 1972 to Lake Ontario.

Application of the chemical is conducted by trained teams under controlled conditions. The stream to be treated is first surveyed to determine the presence, relative abundance and distribution of lamprey larvae. Then a team of men with a mobile laboratory conducts streamside bioassays to determine the concentration and exposure required to kill the larvae without excessive damage to resident fish. TFM, in a liquid formulation, is metered into the stream according to the determined concentration and exposure. To minimize the amount of toxicant used, applications are made when stream and weather conditions are optimal.

In order to obtain a registration for this use pattern of TFM, the Bureau of Sport Fisheries and Wildlife, as the agent of the Great Lakes Fisheries Commission, undertook a program of research to evaluate the effects and influences of TFM on plants, invertebrates, fish, birds, and mammals. These studies included short-term and range-finding assays and long-term (2 years) studies. They were designed to detect and determine acute and chronic toxicities, carcinogenicity, teratogenicity, adverse reproductive effects and biotransformation or degradation by mammals, fish, plants, microorganisms, and sunlight or ultraviolet light (Fig. 2).

Results and Discussion

The entire theme of our studies is to assess the hazard to humans and the environment of the continued use of TFM to control the sea lamprey. To lay the basis for all subsequent studies,

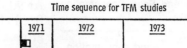

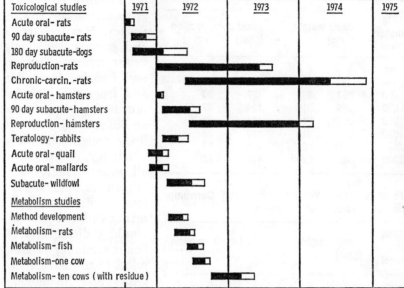

Fig. 2

acute and subacute toxicity of TFM was determined. Preliminary studies were used to determine palatability and any severe toxic responses which could interfere with or prohibit lengthy feeding studies. Experimental design and methods were the same for the rat, hamster and dog studies. Preceding studies assessed the hazard of TFM use to humans. The assessment of environmental hazard of TFM use included birds, fish, invertebrates and plants native or representative of the Great Lakes fauna and flora.

Rat Studies

Acute oral administration of TFM consisted of a 14-day observation period on at least 6 dosage levels. Male and female Sprague-Dawley albino rats, with a minimum of 5 of each sex at each level, were observed at 1 hour, 4 hours, and then daily for 28 days.

Subacute dietary administration was conducted to determine effects if any of TFM on rats and to define the maximum tolerated dose (MTD) to be used in subsequent chronic studies. Parameters measured were body weight, feed consumption, blood and urine clinical data, and gross histologic changes of internal organ tissues.

Table 1　Distribution of Animals and TFM Diets Sublethal Studies

	No. Animals	Sex	Mg TFM per kg diet
1	20	M	0
2	10	M	500
3	10	M	900
4	10	M	1620
5	10	M	2916
6	10	M	5248
7	20	F	0
8	10	F	500
9	10	F	900
10	10	F	1620
11	10	F	2916
12	10	F	5248

Table 2   Body Weights and Feed Consumption of Rats on Test 13 Weeks

TFM dietary level	Body weight (gms)		Food consumption (gms)	
ppm	M	F	M	F
0	353	214	127	88
500	355	216	124	87
900	360	216	124	81
1620	355	218	122	86
2916	352	205	127	87
5248	340	210	118	88

Table 2a   Body Weights and Feed Consumption of Rats on Test 26 Weeks

TFM dietary level	Body weight (gms)		Feed consumption (gms)	
ppm	M	F	M	F
0	453	263	154	120
300	460	266	155	122
1250	458	268	154	124
5000	432	259	149	117

Sprague-Dawley strain weanling albino rats, seventy each of males and females, 45—55 grams, were used. Animals were individually housed. Water and feed was provided *ad libitum* throughout the study. Six test diets were used: the control (zero level) and five levels of TFM in the basal. The animals were randomly distributed into 6 male and 6 female groups: 20 animals per control diet and 10 animals per test diet (Table 1).

At 90 days on test, animals were sacrificed (with CO_2) and examined for gross alterations of internal organs and tissues. Heart, liver, spleen, kidneys, and gonads were weighed to 3 significant figures. Tissues from all males and females from the control groups and from the high level groups were prepared with hematoxylin and eosin and examined histologically.

Body weight, feed consumption, feed efficiency, and organ weight values for all groups, males and females, were within normal ranges for the age and strain animal used. Test and control rats exhibited no differences which would indicate any biologically significant tendency or trend which may be related to administration of TFM. Gross and histologic observations of abnormalities were of low incidence and degree; not specific for any groups, test or control; and not considered related to test administration of the compound (Tables 2, 3, 4 and 5). Survival rates in all Fla groups of the chronic studies were good: one death in each of group 4 (5000 ppm-male), group 8 (300 ppm-female), group 10 (5000 ppm-female). Overall survival at six months on test was 99% (Table 2b). At 10 weeks, animals generally had hematologic values within clinically normal ranges. Similarly urinalysis data generally showed no clinically significant abnormalities.

At 26 weeks, only a few differences in body weight were observed in males at

Table 2b   Body weight, feed consumption, mortality: Fla rats after 40 weeks on test

TFM (ppm)	No. animals	Body weight (gms)		Feed consumption (gms)		No. animals died	
		M	F	M	F	M	F
0	75	494	282	159	111	1	1
300	75	497	285	154	106	1	1
1250	75	493	287	144	103	1	2
5000	75	467	277	149	105	3	1

Table 3 Hematology Data: Rats After 90 Days on Test

TFM dietary level (ppm)	RBC (10^6 cells/mm³)		WBC (10^3 cells/mm³)		Hgb (mg%)		Hematocrit (% vol)	
	M	F	M	F	M	F	M	F
0	8.23	8.12	27.7	18.2	14.5	14.3	49	47
500	8.16	8.01	23.0	20.1	14.6	14.9	48	46
900	8.09	7.89	21.9	20.6	14.7	14.3	50	50
1620	8.43	7.84	22.4	12.4	14.2	14.5	48	46
2916	8.13	7.84	22.8	12.7	14.6	14.3	49	48
5248	8.33	7.79	25.0	14.5	15.3	14.3	52	47

the high test level. No notable differences were observed among female groups. Feed consumption was relatively uniform among males and females (Table 2a).

In general, after 26 weeks on test Fla rat hematologic values were within normal ranges (Table 3a). Several hematologic abnormalities were observed but are of doubtful significance (Table 6). Urinalysis data showed normal response values.

Hamster Studies

Weanling Syrian hamsters (Mesocricetus auratus) were used as test animals. Seventy male and seventy female weanlings were placed on test diets. Test animals were held in individual cages in a temperature-humidity controlled room. Water and feed was ad libitum. Animals were observed twice daily for abnormal appearance

or behavior and appropriate notations regarding room temperature and feed and water level. Six test diets were used: the control (zero level) and five levels of TFM in the basal. Animals were randomly distributed into 6 male and 6 female groups: 20 animals per control diet and 10 animals per test diet. Daily observations showed most animals to be in good health throughout the study. There were no indications of any illness or abnormal behavior associated with administration of the test compound.

Body weight and feed consumption values were within normal values and showed no biologically significant differences among groups (Table 7). No differences in hematology (Table 8) and blood chemistry (Table 9) were detected among test and control groups. Similarly, no differences in the urinalysis was measured among groups (Table 10). No gross or histologic alterations were observed which could be

Table 3a Hematology Data: Fla Rats After 26 Weeks on Test

TFM dietary level (ppm)	RBC (10^6 cells/mm³)		WBC (10^3 cells/mm³)		Hgb (mg%)	
	M	F	M	F	M	F
0	7.49	6.89	21.0	17.1	15.4	13.5
300	7.89	7.03	18.4	16.2	16.3	13.7
1250	7.82	7.31	16.7	14.3	16.5	13.7
5000	7.80	7.13	16.4	12.5	13.7	13.3

Table 4   Blood Chemistry Data: Rats After 90 Days

| TFM dietary level (ppm) | Blood Chemistry (International units) | | | |
| | SGOT | | SAP | |
	M	F	M	F
0	208	171	196	183
500	161	142	206	179
900	166	154	242	194
1620	148	141	254	185
2916	142	143	231	241
5248	146	158	207	207

related to administration of the test compound. Few differences in terminal weight of organs could be related to consumption of the test material. The only exception was a slight increase in kidney weight to body weight ratio in females at dietary levels of 2916 ppm and 5248 ppm.

Of all animals on test, only one died after 16 weeks on test. At this time, F_0 females were mated 1:1 with males of the corresponding diet group. Weanlings from these matings were used in the chronic studies.

Average numbers born, per cent survival, 4-day weight, and weaning weights were normal. No difference among the groups were observed.

Table 6   Hematology Observations: Fla Rats After 26 Weeks

Group	TFM dietary level (ppm)	No. Animals Affected	Observation
1	0	1	low red cell parameters, slightly elevated WBC of normal differential
1	0	1	elevated WBC of normal differential
3	1250	1	possible leucocytosis
7	0	1	elevated WBC with increase in neutrophils with a high band cell count
7	0	2	low borderline ranges for RBC count
10	5000	1	low borderline in RBC parameters

Through 24 weeks of test, all weights for Fla animals appeared normal and no differences among groups was observed. Feed consumption in the high level TFM group males was slightly less than in the control group. Otherwise feed consumption appeared normal for all groups.

After 26 weeks on test, ten animals were randomly selected from each group for clinical examination. Hema-

Table 5   Urinalysis Data: Rats After 90 Days

| TFM dietary level | pH | | Protein (mg%) | | Bilirubin Ketones Glucose | Blood | |
ppm	M	F	M	F	M & F	M	F
0	6.0—7.0	6.0—6.4	Neg.	Neg.-30	Neg.	Neg.	Neg.
500	6.0—7.0	6.0—6.3	Trace	Neg.-30	Neg.	Trace (in one)	Neg.
900	6.0—7.0	6.0—7.3	Trace	Neg.-trace	Neg.	Neg.	Neg.
1620	6.0—7.0	6.0—6.2	Neg.	Neg.-trace	Neg.	Neg.	Neg.
2916	6.0—7.0	6.0—8.0	Neg.	Neg.-30	Neg.	Neg.	Neg.
5248	6.0—7.0	6.0—6.3	Neg.	Neg.-30	Neg.	Neg.	Neg.

Table 7 Body Weights, Feed Consumption, Survival: Hamsters on Test 13 Weeks

TFM dietary level (ppm)	Body weight (gms)		Feed consumed (gms)		Survival (%)	
	M	F	M	F	M	F
0	118	127	57	58	70	80
500	125	121	55	58	80	100
900	121	119	49	54	90	90
1620	120	123	56	50	90	70
2916	123	111	58	64	80	80
5248	123	115	49	61	90	60

Table 7a Body weight, feed consumption, mortality: Fla hamsters after 26 weeks on test

TFM (ppm)	No. Animals	Body weight (gms)		Feed consumption (gms)		No. animals died	
		M	F	M	F	M	F
0	75	139	140	85	69	6	1
300	75	142	148	77	78	1	2
1250	75	139	142	77	76	1	1
5000	75	142	142	78	79	1	1

tology data are generally normal for all groups of animals. A few high and borderline high WBC's are seen sporadically in the male groups (Table 8a). Urinalysis data also appear to be normal for all groups.

Survival, in general, has been good. In group 1 (0 ppm TFM), 6 of 75 male animals died; and in group 6 (300 ppm TFM), 2 of 75 female animals died. All other groups had only one mortality per 75 animals (Table 7a).

Dog Studies

Beagle dogs, 16 each of males and females were used. Animals were individually housed. Water and feed was provided *ad libitum* throughout the study. Four test diets were used: the control (zero level) and three levels of TFM in the basal. The animals were randomly distributed into 4 male and 4 female groups: 8 animals per control diet and 8 animals per test diet.

Table 8 Hematology Data: Hamsters on Test 90 Days

TFM (ppm)	RBC (10⁶ cells/mm³)		WBC (10³ cells/mm³)		Hgb (mg%)		Hemat. (% volume)	
	M	F	M	F	M	F	M	F
0	21.6	8.4	9.93	9.28	21.1	18.6	57	53
500	10.7	9.4	9.24	9.11	20.7	19.7	56	54
900	12.2	9.4	9.63	9.18	21.1	18.6	58	52
1620	10.1	8.2	9.58	8.87	19.9	20.2	56	56
2916	12.6	19.3	9.25	9.77	19.9	19.9	55	56
5348	10.7	10.6	9.55	9.69	20.5	19.7	58	56

Table 8a   Hematology Data: Fla Hamsters on Test 26 Weeks

| TFM | RBC (10^6 cells/mm³) | | WBC (10^3 cells/mm) | | Hgb (mg %) | |
(ppm)	M	F	M	F	M	F
0	8.8	8.5	18.1	10.4	17.9	16.6
300	8.9	9.0	20.9	11.7	18.1	17.1
1250	9.3	8.9	18.3	12.8	17.7	17.3
5000	8.7	8.7	17.1	15.7	17.2	16.8

Table 9   Blood Chemistry: Hamsters on Test 90 Days

| TFM | SGOT (International Units) | | SAP (International Units) | |
(ppm)	M	F	M	F
0	264	111	103	123
500	206	66	119	141
900	165	132	117	134
1620	154	177	116	143
2916	146	84	100	167
5248	163	122	102	130

Through a 26-week feeding study of TFM, body weights for all dogs appeared about normal. Animals on highest dietary levels of TFM ate somewhat less feed and had slightly smaller body weights. Through week 26, cumulative weight gains were 9.70, 9.78, 10.1 and 8.49 kg for males and 9.64, 9.70, 9.24 and 8.62 kg for females for control, low, mid, and high dietary levels (Table 11). At 6 months, all hematology and blood chemistry values were within normal ranges. Differences observed did not appear to be significant (Table 12). Similarly, urinalysis values were normal and no differences were observed among the groups.

Dairy Cow Studies

A single encapsulated oral dose of 11.0 grams of ^{14}C-TFM (specific activity of 84.5 μc/mmol), based on a theoretical water consumption of 15 gallons per head per day at a dose level 10 times use level (20 ppm), was given to test animals.

Nine hours post treatment, TFM and related radioactive compounds in urine reached a high level of 870 ppm. Nineteen hours post-treatment, the level had dropped to 112 ppm; 26-hour pooled urine and feces showed a level of 58 ppm. The level of TFM and related compounds in the first post-treatment milking was 0.550 to 0.995 ppm; in the second milking, 0.197 to

Table 10   Urinalysis Data: Hamsters on Test 90 Days

| Dietary Level (ppm) | pH | | Protein (mg/100 ml) | | Glucose & Ketones | | Bilirubin | |
	M	F	M	F	M	F	M	F
0	6.0—8.0	6.0—8.0	trace 30	trace-30	Neg.	Neg.	Neg.	Neg.
500	6.0—8.0	6.0—8.0	30	trace-30	Neg.	Neg.	Neg.	Neg.
900	6.5—8.0	7.0—8.0	trace-30	trace-100	Neg.	Neg.	Neg.	Neg.
1620	7.0—8.0	7.0—8.0	trace-30	trace-30	Neg.	Neg.	Neg.	Neg.
2916	6.5—7.5	6.0—7.5	trace-30	trace-30	Neg.	Neg.	Neg.	Neg.-trace
5248	6.5—8.0	6.0—7.0	trace-30	trace-30	Neg.	Neg.	Neg.	Neg.

Table 11 Body Weight and Feed Consumption: Beagle Dogs After 6 Months on Test

TFM (ppm)	Body weight (gms)		Feed consumption (kg)	
	M	F	M	F
0	9.70	9.64	63.08	62.50
300	9.78	9.70	64.10	64.06
1250	10.1	9.24	62.28	63.49
5000	8.49	8.62	58.22	54.65

Table 13 Blood Chemistry Data: Beagle Dogs After 6 Months on Test

TFM (ppm)	SGOT (International units)		SGPT (International units)		SAP (International units)	
	M	F	M	F	M	F
0	16	34	50	41	52	72
300	16	157	47	42	61	86
1250	68	35	40	25	58	73
5000	76	41	38	36	70	70

0.199 ppm. Tissue levels of TFM and related compounds at 26 hours post-treatment were less than 10 ppb in fat, 10 to 19 ppb in muscle, 162 to 166 ppb in liver, and 702 to 721 ppb in kidney.

Bird Toxicity

Mallard drakes (Anas platyrhynchos), male and female ringbilled gulls (Larus delawarensis), and female California quail (Lophortyx californicus) were fed TFM. The LD_{50} was determined to be 308 (range 237–400), 250 (range 170–368), and 546 (range 313–950 mg/kg, respectively (Hudson, 1972).

Fish Toxicity

Studies of the toxicity of TFM to different species of warmwater fishes showed that the differential toxic effects of this chemical varied widely with fish species and diminished as conductivity and alkalinity of the water increased. Smallmouth bass and centrarchids were most tolerant of TFM; bullheads, walleye, white suckers, and yellow perch were most susceptible (Table 15) (Applegate and King, 1961).

In water of comparable hardness, toxicity of TFM to fish in flow-through assays was similar to that of static assays. Continuous exposure in hard water indicated an LC_{50} of TFM for coho salmon of 10.5 ppm; for lake trout, 16.9 ppm; and for brook trout, 19.6 ppm (Marking 1972).

Invertebrate Toxicity

The realationship of aquatic invertebrates as fish food organisms is so important that we must understand the effects of TFM use on these organisms. In laboratory studies with aquatic in-

Table 12 Hematology Data: Beagle Dogs After 6 Months on Test

TFM (ppm)	RBC (10^6 cells/mm³)		WBC (10^3 cells/mm³)		WBC Differential									
					N		L		M		E		B	
	M	F	M	F	M	F	M	F	M	F	M	F	M	F
0	6.54	6.77	13.3	12.4	54	51	43	41	3	6	1	2	0	0
300	6.88	6.65	13.3	11.8	62	58	36	32	1	8	1	2	0	0
1250	6.42	6.70	12.9	13.0	59	59	35	33	5	6	1	3	0	0
5000	6.53	6.61	14.4	16.1	59	59	31	33	8	5	3	2	0	0

Table 14   Samples From Au Gres River

| Sample | TFM residues (μg/g) | | | |
	Pre-treatment	During treatment	Post-treatment 24 hr.	96 hr.
Insects	<0.01	2.58	0.18	0.04
Crayfish	<0.01	1.14	0.22	0.02
Fish	<0.01	9.10	4.12	0.05
Snails	<0.01	15.26	0.58	0.37
Soil	<0.01	0.62	0.11	<0.01
Plants	<0.01	4.83	0.27	0.06

vertebrates, the hydras, turbellarians, blackflies were most severely affected with 100% mortality at 8 ppm TFM. Mortality of other invertebrates varied from 0 to over 90% (Smith, 1967). In five streams tributary to Lake Superior and four streams tributary to Lake Michigan, most groups of aquatic organisms were not adversely affected by TFM treatments. Although the total number of invertebrates was smaller 1 week after treatment than before treatment, some recovery was observed by 6 weeks and numbers returned to pretreatment levels by 1 year after treatment (Torblaa, 1968).
Chironomid larvae, after 8 hr. exposure at 10 ppb, contained 20 ppm TFM.

This fell to less than 10 ppm 96 hr. after withdrawal (Kawatski, 1972). Residues in scud (Gammarus pseudolimnaeus), crayfish (Orconectes nais), mayfly nymphs (Hexagenia bilineata), damselfly naiads (Ischnura verticalis), sow bugs (Asellus brevicaudus), and water fleas (Daphnia magna), reached equilibrium after 7 days exposure at 26 ppm and ranged from 2 to 50 times the water concentration. When scud were transferred to clean water, residues declined by 41% in 3 days; by 90% in 8 days; and by 98% in 14 days (Sanders, 1972).
Clams were exposed to ^{14}C-labeled TFM. Residue levels in the gill plateaued after 7 hours at 12° C in water of pH 7.2. Residue concentration did not exceed exposure concentration. When the clams were removed to toxicant-free water, elimination of half the radioactive material occurred in 6 hours and 95% after 24 hours (Johnson, 1972).

Plant Residues

When unialgal cultures of green, blue-green, and diatom species were ex-

Table 15   TFM Toxicity to Larval Lampreys and Fish

| Test fish | Av. TFM dose required to kill | | Differential toxicity (ratio) |
	100% larval lampreys (ppm)	25% Test fish (ppm)	
Largemouth bass (Micropterus salmoides)	6.3	31.5	5.0
Smallmouth bass (Micropterus dolomieui)	4.5	38.2	8.5
Bluegill (Lepomis macrochirus)	6.3	32.0	5.1
Walleye (Stizostedion vitreum vitreum)	6.7	8.3	1.3
Yellow perch (Perca flavescens)	6.3	18.6	2.2
White sucker (Catostomus commersoni)	5.7	8.3	1.5
Yellow bullhead (Ictalurus natalis)	6.0	10.3	1.7
Blacknose shiner (Notropis heterolepis)	6.3	19.9	3.2
Golden shiner (Notemigonus crysoleucas)	6.0	23.4	3.9
Fathead minnows (Pimephales promelas)	6.0	25.3	4.2
Rainbow trout (Salmo gairdneri)	5.3	19.3	3.6

posed to 5—8 ppm TFM, 50% inhibition of growth occurred. At field levels, TFM would be algastatic rather than algicidal (Maki *et al.*, in press).

Environmental Residues

Samples of insects, fish, snails, soil, and plants were collected from two stations on the AuGres River. Analyses were conducted on pre- and post-treatment collected samples (Table 14). Ninety-six hours post-treatment, residues were 0.06 ppm or less except in snails (Allen & Sills, 1972). Muscle tissue of adult trout collected from Lake Superior also were analyzed for TFM. None was detected. Detection level was 0.01 μg/g (Allen & Sills, in press).

Degradation Studies

In natural waters that are in contact with bottom sediments, TFM disappeared within 2 weeks. Some fluoride ion is liberated in the bottom sediment. This is decreased by the addition of phenol (Magadanz and Kempe, 1968; Sutton, 1970). Although stable under aerobic conditions, TFM was reduced, presumably through bacterial action, to the aminophenol (RTFM) under low oxygen tensions (Bothwell *et al.*, in press).

Culture techniques were used to obtain TFM-degrading bacteria from lake and river muds. Although one mixed culture of bacteria was able to degrade the TFM in a solution of 100 ppm within seven days, a pure culture having the characteristics of a *Pseudomonas* sp. was unable to fully degrade the entire amount of TFM present in the media (Sutton and Kempe, 1970). After ip administration of TFM to

male Holtzman rats, analysis of urine indicated that some TFM was reduced to the aminophenol (RTFM). Both TFM and the reduced form were excreted as polar, acid labile compounds. Treatment with β-glucuronidase released both TFM and RTFM, indicating the presence of some glucuronides (Lech, 1971).

In vitro studies were conducted with rainbow trout and labeled TFM. Trout liver and kidney contained enzymes capable of reducing TFM to RTFM, of acetylating RTFM to the N-acetyl RTFM, and of conjugating TFM with glucuronic acid. When rainbow trout were exposed *in vivo* to ^{14}C-labeled TFM, only the glucuronide conjugate was detected. This was excreted largely in bile. Highest levels of the glucuronide, as well as TFM, were found in the liver. Although TFM has been observed in trout blood, brain, heart, muscle, and liver, the glucuronide has been found only in the blood and liver (Lech, 1972a and b; Lech and Costrini, 1972) (Fig. 3).

In initial studies with sea lamprey exposed to TFM, sea lampreys rapidly took up the chemical. Blood and tissue levels of TFM were higher than in trout under similar exposure conditions. However, no metabolites were found in the exposed lampreys (Lech, 1972).

The mechanism of action of TFM in lamprey is not understood. Initially, anoxia was thought to be the cause of death in TFM exposed sea lampreys. Electrocardiograms were used to compare changes resulting from anoxia induced by stopping the aeration of water perfusing test lampreys and changes resulting from TFM exposure. The recordings showed significant differences. Also, all lampreys killed with TFM were found to have red blood, indi-

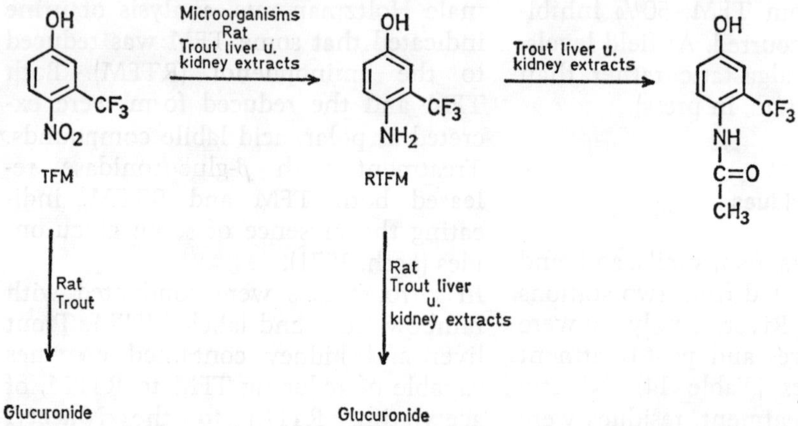

OH

CF₃

NO₂

TFM

→ Microorganisms
Rat
Trout liver u.
kidney extracts →

OH

CF₃

NH₂

RTFM

→ Trout liver u.
kidney extracts →

OH

CF₃

NH
|
C=O
|
CH₃

| Rat
Trout ↓

| Rat
Trout liver
u.
kidney extracts ↓

Glucuronide

Glucuronide

Fig. 3

cating the absence of methemoglobin (Agris, 1967).

Summary

Through the years, a general protocol for assessing hazard to humans of a chemical has evolved. Individual use patterns have periodically dictated some modifications. Short term (90 day) toxicological studies have been conducted with rats, dogs, and hamsters. Effects on growth, feed in-take, organs and tissues, blood and enzyme parameters, and finally lethality have been observed. Long term studies also are being conducted with rats and hamsters to determine adverse effects on reproduction through observation of multiple successive generations continuously exposed to TFM.

Survival rates of all animals has been excellent. Generally body weight, feed consumption, and organ weights were within normal ranges. Female hamsters at dietary levels of 2916 ppm and 5248 ppm exhibited a slight increase in ratio of kidney to body weight. Some depression of body weight and feed consumption was observed with beagle dogs and rats on highest dietary levels. Urinalysis and hematological values were within clinically normal ranges. Gross and histologic observations were of low incidence and degree, not specific for test or control groups, and not considered related to administration of TFM.

From F₀ hamsters, average numbers born, per cent survival, and 4-day and weaning weights were normal. At 7 weeks, too, all weights for Fla animals appeared normal.

Studies were also conducted to assess the hazard of TFM to fish, wildlife and other aquatic life. Toxicity of TFM to fish is related to water quality especially pH. Consequently, streamside analyses are conducted prior to addition of TFM to the waterway. Concentrations and amounts of TFM used are thereby kept to a minimum and incidental damage to aquatic organisms is minimized.

Residue analyses indicate that TFM is not subject to biomagnification; and that tissue levels in fish, high after initial exposure, decline rapidly and approach the limits of detection by the end of 4 days. Residues in fish from Lake Superior were not detected. The detection level is 0.01 ppm.

The fate of TFM in fish, mammals, and in the environment has been studied. Although in vitro the glucuronide and N-acetyl conjugates of RTFM have been observed with fish liver and kidney, the only conjugate observed *in vivo* was TFM glucuronide. This was excreted largely in the bile. Rats *in vivo* produced the glucuronide of TFM and RTFM. Bacteria apparently are capable of a very limited removal of fluoride from TFM. Although stable aerobically, TFM undergoes reduction anaerobically to the aminophenol (RTFM).

From the studies with TFM conducted thus far, we feel that there is no evidence of hazard to man nor to the environment.

References

1 Agris, P. F. (1967): Comparative Effects of a Lampricide and of Anoxia on the Sea Lamprey. Journal of the Fisheries Research Board of Canada, vol. 24, No. 8, p. 1819 to 1822.

2 Allen, J. L., and J. B. Sills (1972): Residues of TFM in Fish and Environmental Samples. Appendix XII e, 2pp. *In* Great Lakes Fishery Commission, Report of Interim Meeting, Ann Arbor, Michigan, December 5–6, 1972.

3 Allen, J. L., and J. B. Sills: (In press). GLC Determination of 3-Trifluormethyl-4-nitrophenol (TFM) Residue in Fish. Journal of the Association of Official Analytical Chemists.

4 Applegate, V. C., and E. L. King, Jr. (1961): Comparative Toxicity of 3-Trifluormethyl-4-nitrophenol (TFM) to Larval Lampreys and Eleven Species of Fishes. Transactions of the American Fisheries Society, vol. 90, No. 4, p. 342–345.

5 Bothwell, M. L., A. M. Beeton, and J. J. Lech: (In press). Degradation of the Lampricide (3-trifluormethyl-4-nitrophenol) by Bottom Sediments. Journal of the Fisheries Research Board of Canada.

6 Hudson, R. H. (1972): TFM Field Formulation. Appendix XII 1, 3pp. *In* Great Lakes Fisheries Commission, Report of Interim Meeting, Ann Arbor, Michigan, December 5–6, 1972.

7 Johnson, H. E. (1972): The Toxicity of TFM to Stream Invertebrates, Algae and Macrophytes. Appendix XII g, 2pp. *In* Great Lakes Fishery Commission, Report of Interim Meeting, Ann Arbor, Michigan, December 5–6, 1972.

8 Kawatski, J. A., and M. M. Ledvina: (In press). Acute Toxicities of TFM (3-trifluormethyl-4-nitrophenol) and Bay 73 (2',-dichloro-4'-nitrosalicylanilide) to the Aquatic Midge *Chironomus tentans*. U.S. Fish and Wildlife Service, Investigations in Fish Control.

9 Lech, J. J. (1971): Metabolism of 3-Trifluoromethyl-4-nitrophenol in the rat. Toxicology and Applied Pharmacology, vol. 20, p. 216 to 226.

10 Lech, J. J. (1972): Isolation and Identification of TFM Glucuronide in Bile of TFM Exposed Rainbow Trout. Federation Proceedings, vol. 31, No. 2, p. 606.

11 Lech, J. J. (1973): Isolation and Identification of 3-trifluoromethyl-4-nitrophenol Glucuronide from Bile of Rainbow Trout Exposed to 3-trifluoromethyl-4-nitrophenol. Toxicology and Applied Pharmacology, vol. 24, No. 1, p. 114–124.

12 Lech, J. J., and N. V. Costrini (1972): In vitro and in vivo Metabolism of 3-trifluoromethyl-4-nitrophenol (TFM) in Rainbow Trout. Comparative and General Pharmacology, vol. 3, No. 10, p. 160–166.

13 Magadanz, H. E., and L. L. Kempe (1968): The removal of 3-Trifluoromethyl-4-nitrophenol from natural water by bottom sediments. Presented at ACS Student Affiliate Regional Convention, Indianapolis, Indiana.

14 Maki, A. W., L. D. Geissel, and H. E. Johnson: (In press). Toxicity of TFM (lampricide) to 10 Species of Algae. U.S. Fish and Wildlife Service, Investigations in Fish Control.

15 Marking, L. L. (1972): Toxicity of TFM to Fish. Appendix XII d, 2 pp. *In* Great Lakes Fishery Commission, Report of Interim Meeting, Ann Arbor, Michigan, December 5–6, 1972.

16 Marking, L. L., and L. E. Olson: (In press). Toxicity of TFM (lampricide) to Non-target Fish in Static Toxicity Tests. U.S. Fish and Wildlife Service. Investigations in Fish Control.

17 Sanders, H. O., and D. F. Walsh: (In press). Toxicity and Residue Dynamics of the Lampricide TFM (3-trifluormethyl-4-nitrophenol) in Aquatic Invertebrates. U.S. Fish and Wildlife Service. Investigations in Fish Control.

18 Schnick, R. A. (1972): A Review of Literature on TFM (3-Trifluoromethyl-4-nitrophenol) as a lamprey larvicide. U.S. Bureau of Sport Fisheries and Wildlife, *Investigations in Fish Control*, No. 44, 31 pp.
19 Smith, A. J. (1967): The Effect of the Lamprey Larvicide, 3-trifluoromethyl-4-nitro-phenol, on Selected Aquatic Invertebrates. Transactions of the American Fisheries Society, vol. 96, No. 4, p. 410–413.
20 Sutton, P., and L. L. Kempe (1970): The Removal of TFM from Natural Water by River Muds. Report for Humanities 499. The University of Michigan, 19 pp.
21 Torblaa, R. L. (1968): Effects of Lamprey Larvicides on Invertebrates in Streams. U.S. Fish and Wildlife Service, Special Scientific Report — Fisheries, No. 572: 1–13.

Acknowledgement

The work with rats, hamsters and dogs was carried out under contract to the Bureau of Sport Fisheries and Wildlife by WARF, Inc., Madison, Wisconsin.

ECDIN, An EC Data Bank for Environmental Chemicals *)

By **F. Geiss** and **Ph. Bourdeau**

European Communities, Joint Research Centre, I-21020 Ispra (Italy)

Summary

All environmentally relevant information is computer stored for each organic chemical manufactured in quantities above 1.000 tons per year (whether it is considered as potentially toxic or not), known toxic chemicals (independently of their production figures) and certain classes of inorganic chemicals. The main data fields ("attributes" for each chemical) are: 1. substance identification and physicochemical data; 2. production and use data; 3. handling, transportation, and disposal; 4. environmental involvement (breakdown under biotic and abiotic conditions, average concentrations in environmental matrices); 5. biological effects and toxicity; 6. protection standards, administrative and regulatory data. Structure and substructure codification permits screening for substances with similar structures and coupling to GC-MS-computer analysis units with full or approximative structure outputs.

A pilot project has been started permitting queries via direct and inverted file. It will include all available information on a limited number of compounds (some 5.000) from different chemical and use classes. The further software development foresees automatic indexing, automatic encoding of formatted data, automatic query formulation and search on a probabilistic basis for natural-language source data. Easy data flow is ensured by a networking between the central unit at the EC Research Centre in Ispra (Italy) and specialised national data centres (ECDIN = European Chemicals Data and Information Network).

A. A Definition

Environmental chemicals are substances which occur in the environment as a result of human activity and which may be present in quantities capable of harming man, other living beings and the environment. They include chemical elements and compounds organic or inorganic in nature and of synthetic or natural origin. Human activity may be affected directly of indirectly, intentionally of unintentionally. The harmful effect — if there is any — of these substances may be acute or chronic, and may also occur by way of accumulation, chemical changes, or synergisms.

B. Why a data bank for environmental chemicals?

The world's total production of organic chemicals (i.e. fully synthetic products, excluding lubricating oils) amounted to 7 million tons in 1950, it increased 63 million tons in 1970 and is estimated to be 100 million tons in 1974 and 250 million tons in 1987 (1) (see Fig. 1).

*) This paper was presented at the Third International Symposium on Chemical and Toxicological Aspects of Environmental Quality, Tokyo, Japan, November 19–22, 1973

About one third of the manufactured chemicals is released into the environment. Finally, a certain quantity of any chemical which is not reclaimed or recycled, will, even after chemical modification and biological degradation end up mostly somewhere in the environment and especially in the hydrosphere. Hence the total environmental charge by synthetic chemicals in 1973 lies certainly between 60 and 100 millions tons.

	1950	1970	1985	Release in Env. 1970
Grand total 10^6 t	7	63	250	20

Organic Chemicals — World Production

Manufactured 10^6 t		natural sources	10^6 t
Solvents	10	methane	1600
Detergents	1,5	terpene type hydrocarbons	170
Pesticides	1		
Gaseous base chemicals	1	lubricating and industrial oils	2-5
Miscellaneous	7		

Release of Organic Compounds in the Environment, 1970 (Diff. 1971)

Fig. 1 World Production of Organic Chemicals (after Iliff).

The production figures of synthetical chemicals are relatively low compared to the natural production of chemicals (e.g. terpenes), but we have to take into account that the latter are generally easily degradable and are in a certain equilibrium with nature; however, they can react with man-made chemicals.

The great majority of these anthropogenic environmental chemicals goes into the ecosphere without any registration before marketing or without more than superficial toxicity testing, often because they are, a priori, regarded as "harmless", or because extensive testing is too expensive. In practice, regulations, registration and control of chemicals released into the environment in most countries is limited to drugs, food additives, and pesticides, eventually to cosmetic ingredients.

There are altogether some 15.000 to 20.000 chemicals manufactured in quantities above some 500 kg/year (drugs excepted), only some hundreds of which are submitted to registration procedures.

With a certain regularity, again and again, chemicals once regarded as harmless, are revealed to be dangerous for man and animals. The most spectacular case was that of the polychlorinated biphenyls (PCPs), which owe their detection in the environment (by chance) to the fact that they constantly disturbed gas chromatographic determination of DDT in a detecting device which is specifically sensitive for halogenated aromatic hydrocarbons. Once the PCBs were discovered, people concerned had no clear picture as to where they came from. The PCBs have been used as an additive to rubber, paints, plastics, adhesives, insecticides and printing ink, and as a heat transfer and insulating fluid. The lack of any detailed information as to their sources, uses and quantities produced, and a complete file listing all their known properties, complicates any systematic attempt to map their flows through the environment and to determine the amount already present. This case illustrates some of the deficiencies in the present information systems on toxic environmental chemicals.

The exemplary value of the PCB's case is the fact that large quantities of a potentially hazardous compound were released into the environment without any public knowledge of its sources and effects. Which one will be next?

To assess the environmental impact of an anthropogenic chemical therefore the storage of data on production, properties, use pattern, disposal, persistence, dispersion tendency, conversion under biotic and abiotic conditions, biological consequences, and structure-activity relationship is necessary.

The basic principle of our data bank ECDIN is to store relevant information pertaining to any individual chemical compound produced in sizable quantities e.g. 1000 kg per year, regardless of the form in which it is used, or its intended function, or its presumed degree of toxicity or "harmlessness".

An essential stimulation for the EC project ECDIN came from the SCOPE*-Committee's recommendation to built up an "International Registry on Chemical Compounds." The build-up of such a registry has also been recommended by the UN Conference on Environment and Man, held in Stockholm in 1972.

C. Structure of ECDIN

ECDIN is conceived essentially for the retrieval and manipulation of *hard data* and not (or only indirectly) for the retrieval of documents. It is more research-orientated. All information pertaining to a single compound or a formulation is recorded on its computer file containing variable length fields in semi-hierarchical order. Retrieval is

made by direct or inverted file access (see below). The provisional structure of the direct file is shown in the following table.

Its final structure is still submitted to evolutive changes.

List of attributes for each compound to be stored

1. Substance identification

1.1.   ECDIN system number
1.2.   Chemical name, trivial name(s), trade names (+ country) (d, e, f, i, n)
1.3.   CAS number
1.4.   Molecular formula
1.5.   Molecular weight
1.6.   Wiswesser line notation
1.7.   Other structure codes
1.8.   Substructure fragment code

2. Physico-chemical data

2.1.   Melting point
2.2.   Boiling point
2.3.   Vapour pressure
2.4.   Density
2.5.   Solubilities
2.6.   Organoleptic properties (smell, taste, colour etc.)
2.7.   Analytical methods
2.8.   Mass spectrum
2.9.   Infrared spectrum
2.10. Other spectroscopic data
2.11. . . . . .
2.12. Chromatographic data

3. Production and use data

3.1.   Manufacturer(s) and location of production plant(s)
3.2.   Approx. production figures (decade ranges)

---

*) *Scientific Committee on Problems of Environment, in the International Council of Scientifice Unions (ICSU)*

3.3. By-products and contaminants (pointer to 1.)

3.4. Waste products of manufacture and modes of release into the environment

3.5. Import to EC region (from/to)

3.6. Export from EC countries

3.7. Intra-Community commercial exchange

3.8. Use, modes of application and corresponding quantities (approximate figures)

3.9. Locations and quantities of use (per hydrographic basin)

3.10. Concentration in a "formulation" (blend)

4. Handling, transportation and disposal

4.1. Danger class

4.2. Transport class, transport routes, shipment methods

4.3. Safety, transport and storage rules

4.4. Counter measures in case of accident (environmental, medical)

4.5. Corrosiveness, explosiveness

4.6. Cleansing and emergency services

4.7. Contamination control

4.8. Disposal of waste

4.9. Known accidents (type, place, quantities involved, danger, measure, lessons learnt)

5. Environmental involvement

5.1. Breakdown under biotic conditions, rate

5.2. Breakdown under abiotic conditions, rate

5.3. Other reactions of environmental relevance, persistance

5.4. Dispersion pathways

5.5. Yearly average concentrations in selected matrices per region or country

6. Data on biological effects and toxicity

6.1. Effects on man

6.2. Effects on animals

6.3. Effects on lower organisms

6.4. Effects on plants

6.5. Carcinogenicity data

6.6. Teratogenicity data

6.7. Mutagenicity data

6.8. Allergenicity data

6.9. Synergistic effects

6.10. Antagonistic effects

6.11. Toxicological and biochemical effects on isolated biochemical systems

6.12. Absorption, metabolism, excretion

6.13. Therapeutic measures in case of intoxication

6.14. Other data of pharmacological relevance

7. Protection standards, administrative and regulatory data

Specified for each:
— B/Lux, DK, D, D (Länder), GB, F, EIR, J, NL, USA, others
EC, WHO
other international, intergovernmental
or interorganizational harmonizations
— recommendation
— voluntary regulation
— legal enforcement

7.1. Basic protection level, target

7.2. (No effect level)

7.3. Environmental quality objectives, target

7.4. Product standards

7.5. Process standards

7.6. Operation standards

7.7. Emission standards

7.8. "Immission" standards (environmental quality standards)

7.9. Maximum daily intake (MDI)

7.10. Occupational health standards
7.11. Public health standards
7.12. Food and drug standards
7.13. Animal feed standards
7.14. Waste disposal standards
7.15. Other regulations and recommendations
7.16. Cases of infringement and conflict

Not too much effort will be made to reach a highly rational classification of the fields in the file, where overlapping and semantic difficulties are inevitable, especially for the toxicological data fields. These difficulties are overcome by a well elaborated thesaurus for the inverted file search.

At this time the field of chemical structure codification is rapidly evolving. Instead of making a premature choice of the final system to be adopted for the time being we rely on transcoding from one system to another in order to have easy access to data sets with different structure codes. The importance of chemical structure codification lies in the necessity of substructure screening.

The quantity and rate of production of a chemical is evidently at the basis of any estimation of potential environmental burden. It is obvious that producers will not be very willing to disclose precise data on the production and use of chemicals, unless they are legally enforced to do so. Nevertheless, in general it will be possible to obtain approximate production rather than sales figures. One could record production figures in the decade range. Even then approximative figures could permit a rather good impact statement if they are sufficiently differentiated by type and region of use.

The *dispersion tendency* indicates in which sectors outside its application field a chemical or its degradation product can emerge.

D. Users and Uses

A system like ECDIN must meet a variety of needs. By effecting the consolidation into a single system of all data pertaining to the problems of environmental toxicology, it provides for availability of data to governments at all levels, agencies, industry and scientists. These data should be accessible immediately upon request and cover a large number of separate but interrelated disciplines. In addition, it allows the use of current research results as an early warning system for environmental hazards, it provides a screening device to establish priorities for toxicity testing and it stimulates research, basic and applied, directed towards the hazards of toxic chemicals. Hence, the ECDIN data bank has to fulfill the demands of the following user categories (list not exhaustive):

— governmental authorities at all levels, international organizations and institutions involved in environmental protection (establishment of quality objectives, regulatory activities etc.) and control
— police and other services (countermeasures in case of accidental contamination, inspections, law enforcements)
— water management authorities
— industry (decisions on product development and marketing, research priorities etc.)
— designers of monitoring networks
— experimental scientists (research priorities for toxicity tests, identification and bibliography, screening of pollutants in environmental samples)

— epidemiologists identifying environmental disease factors

— ecologists, chemists and "modellists", who study the dispersion mechanisms of pollutants.

At the actual development of ECDIN several thousand types and combinations of queries are possible. The following is a random sample of possible questions.

1. Complete or partial information on a particular compound? (trivial).

2. All compounds with certain structure elements (e.g. dichlorinated aliphatic ethers) produced in a certain country or region?

3. All environmental chemicals which are structurally related to one discovered to be toxic (screening). Purpose: toxicity tests with these compounds.

4. In a water or air sample a chemical compound has been approximately identified. Which compounds of similar structure are stored in ECDIN? Are they recognized as toxic? Is it worthwhile to fully identify it? What is its use? Quantities? (early warning system)

5. How many tons of a certain pesticide have been produced in, imported to certain countries, and used in regions of a hydrographic-basin?

6. Are there seasonal restrictions for the application of insecticide 'B' in fruit cultures?

7. How much of the insecticide 'A' is a) produced in, b) imported in Italy? How much of the production is exported?

8. What are the "environmental quality standards" for a given chemical in different countries?

9. How much corrosive chlorinated acids was transported over Belgian roads in 1973?

10. All chlorinated aromatics containing two nitro groups made in France, used in Italy, having LD_{50} in the range of 10 to 100 mg/kg for rats, but not toxic to fish.

11. Which commercial formulations contain the tenside 'C'? In what concentrations? In what solvents?

12. Which factories located in the hydrographic system of the Rhone produce compound 'D'? Final or intermediate product? What kind of containers are used for shipping?

13. Are there efficient and economic methods to destroy the toxic compound 'E' or to convert it to innocuous ones? References.

14. Display all stored substances containing the ring system of benzimidazol. Which of them have been tested for mutagenicity? Display results of testing.

15. Is anything known about bacteriostatic properties of compound 'F'?

16. Are there reports on accidental water pollution with dimethylaniline or similar compounds (aryldialkylamines)? Is this substance sedimented? Solubility in water at pH 4? Skin absorption from aqueous solution?

17. Display pathway and metabolic conversion of 1,2 dioxyanthrachione in mammals.

Let us consider case 4 in more detail: The chemical analysis of organic pollutants in environmental samples is characterized by a great number of compounds to be separated and identified (Fig. 2) (2). They can be of both natural and anthropogenic origin.

The analytical method of choice is gas chromatography combined with mass spectrometry (Fig. 3). However the manual identification of the peaks in the chromatrograms as e.g. shown in

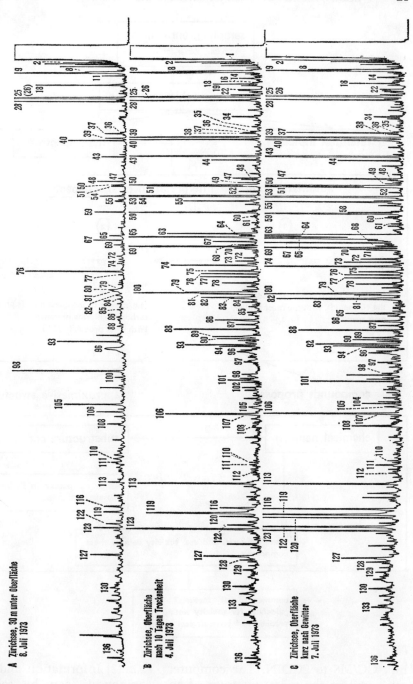

Fig. 2

A Zürichsee, 30 m unter Oberfläche
8. Juli 1973

B Zürichsee, Oberfläche
nach 10 Tagen Trockenheit
4. Juli 1973

C Zürichsee, Oberfläche
kurz nach Gewitter
7. Juli 1973

Fig. 2 is tedious, time consuming and requires highly skilled and experienced personnel. Fortunately, the techniques of automatic computer interpretation of mass spectra are now advanced sufficiently to simplify this task. They give a new dimension to the organic analysis of environmental samples and

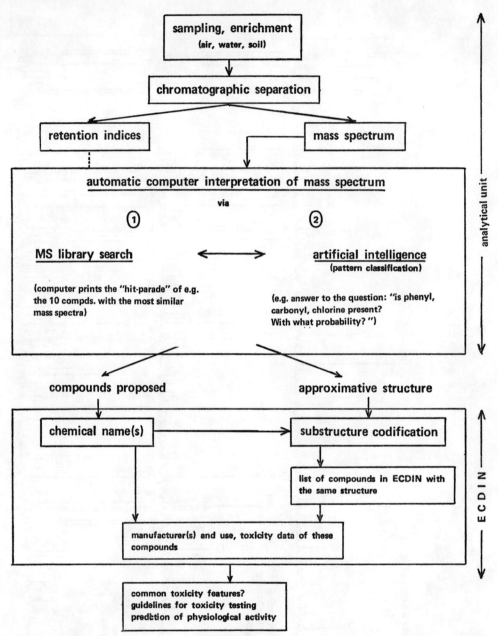

Fig. 3

link GC/MS to ECDIN. The computer can compare the mass spectrum of an unknown compound to spectra in the library and then make structure proposals. Fig. 4 and 5 show such a printout obtained with the method of Clerc *et al.*(3) Information and data on these compounds can be searched for in ECDIN directly via their name, or after breaking them down into substructure, information can be obtained on their chemicals "cousins". MS spec-

MS-SEARCH   VERSION FEB 73   BMAG1/SET100   20/03/73 PAGE 67
SAMPLE COMPOUND NO 422
MAXIMUM S   27600

NUMBER		REL. SIMI-LARITY FACTOR	MOLECULAR FORMULA	NAME
1	422	0	100.00	C4.H6.CL2	1,4-DICHLOROBUTENE-2
2	8949	0	96.58	C4.H6.CL2	1,4-DICHLOROBUT-2-ENE
3	8948	0	94.50	C4.H6.CL2	1,2-DICHLOROBUT-3-ENE
4	2389	0	93.62	C4.H6.CL2	3,4-DICHLORO-1-BUTENE
5	2391	0	92.71	C4.H6.CL2	1,4-DICHLORO-2-BUTENE
6	423	0	91.71	C4.H6.CL2	2,3-DICHLOROBUTENE-2
7	7227	0	89.26	C4.H6.CL2	1,4-DICHLOROBUT-2-ENE
8	2414	0	86.41	C4.H8.CL2	1,3-DICHLORO-2-METHYL-PROPANE
9	418	0	85.33	C3.H4.CL2	1,3-DICHLOROPROPENE-1
10	303	0	84.84	C3.H6.CL2	1,2-DICHLOROPROPANE
11	8958	0	84.70	C4.H6.CL4	1,2,3,4-TETRACHLOROBUTANE
12	3372	0	84.02	C11.H13.O.CL	3-CHLOROALLYL 2,4-DIMETHYLPHENYL ETHER
13	6498	0	83.26	C23.H21.N.O.	2-PHENYLIMINO-3,3-DIMETHYL-4,4-DIPHENYL OXETANE
14	2937	0	83.16	C4.H7.CL3	TRICHLOROBUTANE
15	399	0	82.58	C3.H4.CL2	3,3-DICHLOROPROPENE-1
16	200	0	82.55	C4.H4.CL2	1,4-DICHLORO-2-BUTYNE
17	5503	0	82.55	C3.H6.CL2	1,2-DICHLOROPROPANE
18	2253	0	82.19	C3.H6.CL2	1,2-DICHLORO-PROPANE
19	825	0	82.14	C8.H7.CL	BETA-CHLOROSTYRENE
20	3227	0	81.80	C10.H11.O.CL	3-CHLOROALLYL-O-METHYLPHENYL ETHER

F. Erni u. J.T. Clerc, 1973

Fig. 4

tra interpretation by artificial intelligence (4) will for a while propose only structure fragments of an unknown compound. Hence, the entrance into ECDIN is straight forward as well.

To summarize: A compound with an "approximative" structure (from MS) can be checked against compounds or a series of similar compounds and their environmental relevance, screened out of ECDIN, to decide whether it is worthwhile to fully identify the substance or to neglect it because of probable harmlessness.

E. Operational Features

ECDIN has now been launched as a 4-year pilot project. During this period the input will be restricted to a limited number of compounds (certain chemical families and use classes) but not to the number of obtainable data in all fields. The retrieval system SIMAS at the Joint Research Centre at Ispra will be used during the pilot phase. Its operational characteristics are

— Time-sharing system on IBM 370/165
— Access by inverted file
— Access language: Acronyms
— Access to numerical data through data ranges

Though it does not have the most convenient imaginable accessibility for the user (it has been developed for other purposes) it is fully sufficient for the pilot study and can range up to 60,000 compounds (if necessary even to an order of magnitude more) and 30,000(!) retrieval key words.

After two years the SIMAS system will be substituted by a more sophisticated retrieval system under development

MS-SEARCH   VERSION FEB 73   BMAG1/SET100   20/03/73 PAGE 93
SAMPLE COMPOUND NO 8409
MAXIMUM S   25900

NUMBER		REL. SIMI-LARITY FACTOR	MOLECULAR FORMULA	NAME
1	8409	0	100.00	C15.H14.N2.O2	2-ETHOXYCARBONYL-AZOBENZENE
2	8408	0	86.82	C13.H10.N2.O2	4-CARBOXY-AZOBENZENE
3	3539	0	86.54	C14.H12.O2	BENZYL BENZOATE
4	8416	0	85.49	C25.H20.N4	P,P≤-DI≥BENZENEAZO↑DIPHENYLMETHANE
5	8499	0	85.13	C20.H15.N.O2	N-PHENYLDIBENZAMIDE
6	2792	0	85.08	C8.H6.O3	PHTHALALDEHYDIC ACID
7	8596	0	84.94	C14.H14.N2.O	N-NITROSO-DIBENZYLAMINE
8	6491	0	84.68	C14.H10.N2.O3.S	4,6-DIPHENYL-1,2,3,5-OXATHIADIAZINE-2,2-DIOXIDE
9	8500	0	83.99	C21.H17.N.O2	N-BENZYLDIBENZAMIDE
10	3396	0	83.98	C13.H10.O2	1-(2-FURYL)-3-PHENYL-2-PROPENE-3-ONE
11	8414	0	83.92	C18.H14.N4	P-BENZENEAZO-AZOBENZENE
12	3017	0	83.40	C9.H10.O3	2-PHENOXYPROPIONIC ACID
13	9058	0	83.37	C19.H18.O7	CARBETHYL SALICYLATE
14	348	0	82.94	C7.H7.N	4-VINYLPYRIDINE
15	8413	0	82.93	C13.H9.N2.O2.CL	4-CHLORO-2≤-CARBOXY-AZOBENZENE
16	8415	0	82.63	C24.H18.N4	BIS≥P-AZOBENZENE↑
17	1183	0	82.53	C18.H20.O2	3,5-DIMETHYLBENZYL-3,5-DIMETHYLBENZOATE
18	3315	0	82.50	C12.H14.O2	ALLYL-BETA-PHENYLPROPIONATE
19	8401	0	82.31	C12.H9.N2.I	4-IODO-AZOBENZENE
20	7694	0	82.29	C11.H9.N.O	1-PHENYL-2-PYRIDONE

F. Erni u. J.T. Clerc, 1973

Fig. 5

performing automatic indexing, automatic encoding of formatted data, automatic query formulation and search on a probabilistic basis for natural language source data.

Acknowledgement:

We are very grateful to Dr. C. Levinthal, Columbia University, New York, for giving us his data base on chlorinated aromatic hydrocarbons.

References

1 Iliff, N. A.: 2nd Internat. Symposium "Chemical and Toxicological Aspects of Environmental Quality", München, 27./28. 5. 1971.

2 Grob, K. and G.: "Organische Stoffe in Zürichs Wasser", "Neue Zürcher Zeitung", 10. 2. 1973, Beilage Forschung und Technik.

3 Clerc, J. T., F. Erni, C. Jost, J. Meili, P. Nägeli, und R. Schwarzenbach: Z. Anal. Chem. 264, 192 (1973).

4 a. Franzen, J., and H. Hillig: 6th Internat. Mass Spectrometry Conference, Edinburgh, 14. 9. 1973, to be published in "Advances in Mass Spectrometry", Vol. 6, Pergamon Press Ltd., Oxford, Spring 1974;
   b. Isenhour, T. L., and J. B. Justice: idem.

Technology Assessment on the Use of Pesticides *)

By **Hidetsugu Ishikura**

Former Science Councillor, Science and Technology Agency, Tokyo, Japan

Abstract

The extensive use of pesticides in Japanese agriculture was studied as a problem-oriented case of technology assessment with an aim to contribute to the establishment of appropriate procedures of assessing real and potential impacts which modern technologies have or may have on health, industries, economy, society and on the environment and to obtain clues to the development of safer use of pesticides.

Direct and indirect impacts, both real and potential, favourable and adverse, were intended to identify and evaluate systematically and comprehensively as their cause and effect sequences were studied. Adverse impacts were divided into tradable and untradable categories; untradable adverse impacts were related mostly to human health. As to the evaluation of impacts, it was suggested that the size of area and population affected, irreversibility and controllability of the impact be considered. It was recognized as urgent and requisite to develop and establish a more efficient and satisfactory method of testing the safety of pesticides and their metabolites over a wider spectrum of organisms and with respect to newer aspects of toxicology as mutagenesis, teratogenesis and cancerogenesis.

*) This paper was presented at the Third International Symposium on Chemical and Toxicological Aspects of Environmental Quality, Tokyo, Japan, November 19–22, 1973

1. Needs of Technology Assesment on the Use of Pesticides in Agriculture

In the early 1940s, science and technology entered a new epoch of development and a number of new technologies emerged. During the 1950s, industrial production in many developed countries grew remarkably by applying these new technologies. The gross national production increasd by 1.6 fold in the United States of America, 1.6 fold in Great Britain, 2.5 fold in France and in the Federal Republic of Germany and 2.9 fold in Japan in the said decade.

Since the early 1960s, however, some of the new technologies began to be recognized as affecting and deteriorating societal systems and human environments. Critiques for the application of new technologies to industrial and societal systems without advanced deliberation on their potential effects on these systems became strong enough to seek new approaches to the problems of science and technology.

Development of synthetic pesticides and their use in agriculture was one of the postwar innovative technologies as is indicated by the year of discovery of major pesticides. The potent insecticidal property of DDT was demonstrated in 1938; 2,4-D was discovered in 1941, BHC in 1942, dithiocarbamates in 1943, parathion in 1944 and aldrin in 1948. Consumption of pesticides increased tremendously in the 1950s in many advanced countries. In the United States of America, total sales of pesticides soared from 200 to 350 million dollars and in Japan from 2 to 24 billion Yen in the said decade.

The introduction and increased use of modern pesticides in agriculture contributed very much to the improvement of agricultural production by increasing yield, upgrading the quality of product, and safeguarding production from the damage caused by pests.

However, as the use of pesticides in agriculture became extensive, many unanticipated adverse effects, really or suspectedly caused by the use of pesticides, were recognized. Farmers who were directly exposed to pesticides during application were affected healthwise; cattle and fish pastured or cultured in the treated or neighbouring environments were damaged by contaminated feeds or by polluted air and water; pest organisms, particularly insect pests, became more difficult to control due to the development of resistance to pesticides or the destruction of natural enemies by pesticides. Furthermore, widespread contamination of foods, produced with the use of pesticides, with pesticide residues warned the general public of the potential health hazard of the spreading contamination of the environment with pesticide residues and their degraded chemicals. The movement of these chemical compounds in the ecosystem were demonstrated or suspected of causing drastic effects on many living organisms and their function in nature. Silent Spring written by Miss Rachel Carson (1962) pointed out these adverse effects in detail and drew the attention of the public to the abuse of pesticides.

In the later 1960s, apprehension in regard to the real and potential adverse impacts, which new technologies have or may have on societal system and human environment, became serious enough to encourage a search for measures to identify, recognize and prevent or mitigate such adverse impacts in advance of the decision-making or the application of these new technologies. In 1966, Representative Emilio Dadario, Chairman of the Subcommittee on Science, Research and Development of the House Committee on Science and Astronautics of the United States of America, introduced a new concept of technology assessment, intending to systematically identify, recognize, analyze, and evaluate real and potential impacts of new technologies on social, economic, environmental and political systems.

The importance of this concept was recognized by the Advisory Council of Science and Technology of the Government of Japan. In the Council's report on the fundamental policy of science and technology for the 1970s, it was recommended that, in order to help science and technology serve social development and well-being in a more rational way, technology assessment be introduced into the planning and administration of scientific and technological programs. However, since knowledge on the procedures for assessing various impacts of new technologies are very meager, this study was conducted to learn the use of pesticides in Japanese agriculture as a case of technology which already has so many problems in the societal and environmental systems.

2. Aim and Scope of the Study

The use of pesticides in Japanese agriculture increased tremendously during the past two decades because of the recognition of their usefulness by farmers in food production increase encouraged by the government. By 1972, some 15,000 tons of DDT, 40,000 tons of BHC, 3,500 tons of aldrin tnd endrin, 27,000 tons of carbamyl and other carbamate insecticides, 4,400 tons of dicofol, 3,200 tons of chlorobenzilate and 32,000 tons of organophosphorus insecticides were consumed for insect control. Organomercurial fungicides equivalent to 2,300 tons of metal mercury were employed for seed and plant treatments. Among herbicides, sodium pentachlorphenate was used in large quantities for rice weed control before it was replaced by other herbicides of low fish-toxicity.

As the consumption of pesticides increased, many adverse impacts of pesticide use became apparent and the need for reviewing the use of pesticides in Japanese agriculture in respect to usefulness, safety, and the possible effects on environment was more and more strongly felt. Thus an initial concept of technology assessment on the use of pesticides existed before this study was launched.

However, since technology assessment is a new field of soft science, no appropriate methodology has been developed for identifying, recognizing and assessing real and potential impacts which a given technology has or may have on societal and environmental systems. Furthermore, a methodology developed in one country for a given technology may not be adoptable in other countries for other technologies because societal and environmental systems may differ. This study, therefore, aimed to contribute to the development of methodology for problem-oriented assessment, taking the use of pesticides in Japanese agriculture as the case for study. In addition, since it was already prescribed for the technology of Japan that by the year 2000 safer pesticides and safer pest control

Block Diagram of Technology Assessment

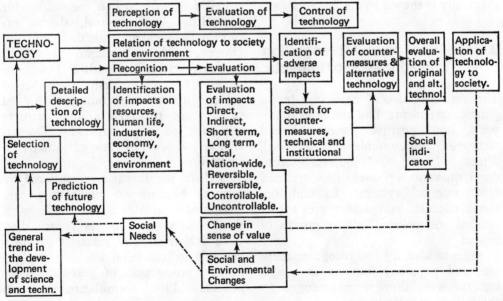

Fig. 1

measures are essential for crop protection of that age, it was intended to find initial clues to develop such pest control technology in this study.

Pest control with pesticides involves the development of more potent pesticide compounds, their formulation for safe and easy application, equipment for efficient and effective application as well as a means of predicting the occurrence of pests and determining the best time for the application of pesticides. Institutions for guaranteeing the good quality of pesticide products, for fostering wise and safe use of pesticides, for restricting, if required, the use of specified pesticides under particular conditions, for regulating the registration, distribution, transportation and application of pesticides with safety, and for checking the residues of pesticides in and on foods are required for safe and effective use of pesticides in agriculture. This study covered all the technologies and institutions associated with the use of pesticides.

The sequence of procedures adopted in this study is shown in Fig. 1.

3. Identification and Recognition and Intensification of Impacts

The extensive use of pesticides in Japanese agriculture had many real impacts, both positive (favorable) and negative (unfavorable or adverse), direct (primary) and indirect (secondary), on various phases of societal and environmental systems. In addition, some potential impacts were apprehended and pointed out by groups of people.

In order to identify and recognize these real and potential impacts in a systematic way, they were arranged in the following categories.

Human categories:
Health, mentality and life.
Industrial categories:
agriculture and other bioindustries, manufacturing industries, and service.
Societal categories:
family, local community, societal organization and nation.
Environmental categories:
soil, water, air and organisms.

The major impacts identified with each category, with an emphasis on adverse ones, are briefly described in the following. *Health:* Increased production of foods of high quality achieved by the effective control of pests with pesticides and associated modern agricultural techniques contributed very much to the improvement of nutrition and health of farmers and the general public. Increase in income of farmers by the increase of farm production and by earning through side jobs enabled the farmer to receive medical care if required. Decrease of mosquitos, flies, gnats and other disease spreading pests, which indirectly resulted from the use of pesticides for agricultural pest control, almost exterminated local endemic diseases.

However, while pesticides of high mammalian toxicity such as parathion were in use, a considerable number of toxicity instances including some fatal cases were observed among farmers who applied these pesticides. Dermatitis and eczema of skin, conjunctivitis and corneitis of eye, bronchitis and catarrh of the throat were suffered by farmers because of inadequate pretesting of pesticides for their effects on these organs and tissues before they were registered and released for use. Hazards to industrial workers engaged in the production of pesticide compounds and their formulation were also reported but in limited numbers. Irritating fumes from soil-treated fields annoyed non-agrarian residents in semiurban areas where vegetable growing is practiced and profitable. Spray and dust fumes drifting from treated fields and orchards were occassionally reported to have affected or annoyed neighboring residents and passers-by.

The detection of residues of persistent chlorinated insecticides and mercury in rice, vegetables and milk shocked the general public by suspected unsafety of foods. Finding mercury to be the metal causing Minamata Disease amplified their fear, and plant treatment with organomercurial fungicides was replaced by other fungicides in 1966; the use of organomercurial fungicides for plant treatment was withdrawn absolutely in 1968. Detection of residues of BHC isomers, particularly the beta isomer which is more stable in nature than others, in cow milk as well as in mother's milk at a rather high level forced withdrawal of this pesticide in 1971.

The experimentally demonstrated potential cancerogenic and teratogenic properties of some pesticides both abroad and in domestic use shocked some groups of people. 2,4,5-T employed in the reforestation program of the National Forest Agency, and suspected of having trace amounts of dioxine, was deleted as the result of the protest of the forest workers.

Human mentality: Liberation from the fear of food shortage for the general public encouraged them in their economic and social activities and contributed to a rapid economic and social development of the nation. Liberation of farmers from hard farm labor made it possible for them to obtain more advanced knowledges of their farming and to plan and practice their farming in a more rational way. Superstitions, which once prevailed among farmers, faded with the scientific solution of their problems including the menace by pests.

Farmers were gradually accustomed to think more about farm economy because of the increased input of capital for pest control.

On the other hand, farmers were burdened with new needs of learing safe and effective use of pesticides many of which are toxic to human beings. Rural communities in which very toxic pesticides are obtainable rather easily are confronted with the abuse of pesticides for suicides and homocides. Great numbers of people living both in rural and urban areas became very concerned about the potential dangers which pesticides may have for their health.

Impacts on agriculture: The successful control of pests with pesticides increased agricultural production to a great extent when it was combined with the introduction of high yielding varieties, heavy fertilization, and dense planting. The quality of products was also upgraded by the improved pest control and, together with the increased yield, very much increased farm income. Much farm labor was saved by the introduction of herbicides in weed control and the surplus labor was utilized in more intensive and profitable sectors of agriculture or in side jobs to increase the income of the farm family. Free choice of crops and varieties became feasible without paying attention to their resistance to pests. More intensive utilization of fields became practicable by discontinuing rotation and fallowing which had been employed to suppress the buildup of pest populations. Direct planting of crops which had so far been transplanted also became feasible by the successful control of weeds with pesticides. Introduction of power sprayers and dusters for efficient pest control stimulated the mechanization of other farm practices.

On the other hand, yields of some crops were decreased and their quality degraded due to injuries caused by careless use of pesticides. Destruction of natural enemies of insect pests caused resurgence of pests including those which had so far been of minor importance. Development of resistance to pesticides were observed with a number of insect pests and, because of these facts, more frequent application of pesticides at higher dosages or switching to newer pesticides which were usually more costly were required in order to maintain the efficiency of control. Because of the decrease of pollinators in orchards, farmers had to work for artificial pollination or protect and release the pollinating bees. They had to refrain from planting crops such as cucumber and potato which take up the residues of chlorinated insecticides in fields contaminated with these residues. Fumigation of soil with fumigants for controlling soil-borne diseases and nematodes disturbed the microfloral activities in the soil and subsequent abnormal growth of crops in treated fields occurred.

Impacts on silkworm rearing and bee keeping: Many major pests of mulberry shrubs and infectious diseases of silkworm were effectively controlled by pesticides and relevant chemicals. However, silkworm rearing was frequently threatened with loss caused by feeding mulberry leaves contaminated with pesticides drifting from near-by treated fields. Although tolerance of various pesticides to silkworms and the time required for the disappearance of pesticide residues on mulberry leaves

have been investigated in details, silkworm-rearing farmers are still suffering from accidental loss of their harvest.

Bee-keeping is a small industry in Japan and accordingly no serious dispute arose between bee-keeping farmers and others. In areas where aerial application of pesticides are scheduled, bee-keepers are advised to remove their bee-hives outside of the scheduled areas, at least 7 kilometers away from the areas in advance of the application. Honey produced in the country was demonstrated to contain pesticide residues, although in a very minute amount.

Impacts on cattle and poultry raising: Insect pests of cattle and poultry were effectively controlled by pesticides; general sanitary conditions of barns and sheds were much improved. A mosquito species breeding in rice fields and transmitting encephalitis-virus, which causes abortive black babies of swine, was believed to be controlled by rice insect control with pesticides.

However, fatal and affected cases of cattle and poultry were found occasionally during the pest control season resulting from feeding contaminated feeds or exposure to spray and dust drift. Dairy production was seriously affected by the detection of BHC residues, particularly of the beta-isomer which is more persistent in the environment and whose toxicity to warm-blooded animals including human beings is not known. Since this residue was believed to have been stored in the straw of rice plants treated with BHC for control of rice insect pests, this detection cancelled the use of BHC in agriculture in 1972.

Impacts on fish and shell-fish culture: Some compounds employed as pesticides are highly toxic to fish and shell-fish and are used in large quantities in flooded rice fields during the rice growing season. As a consequence, both natural and cultured stocks of fish and shell-fish in inland and coastal waters were seriously damaged frequently and this damage developed into a big problem between farmers and fishermen. In order to settle the problem, an amendment was made to the Agricultural Chemicals Registration Act in 1968 to enforce a stringent restriction on the use of pesticides which are highly toxic to aquatic organisms.

Impact on manufacturing industries: The growing demand for new pesticides, encouraged by food production campaign which the government promoted, spurred the development of the pesticide industry. As was exemplified by the findings of fenitrothion in 1956 and blasticidin in 1958, the Japanese pesticide industry acquired the potential in ten years after World War II to develop new and safer pesticide compounds. Pesticide industries had to increase research and development facilities and corps of scientists of various specialities which was a heavy financial burden to the industry. The successive development of safer pesticides by different companies induced excessive competition in the pesticide market and the pesticide industry lost its profitability.

Pest control equipment manufacturing industries also grew as a result of encouragement by the government program to augment rural pest control. The industry had to develop with the rapid change in pesticide formulation and the scale of application by developing suitable equipment.

Relevant services: More than 70 per cent of the pesticides consumed in Japanese agriculture have been distributed through the nation-wide network of Agricultural Purchase Cooperatives to the ultimate consumers. This business of pesticide sale strengthened the economic potential of agricultural cooperatives very much, but the cooperatives had to recruit technical staffs to advise farmers on the safe and effective use of pesticides.

Coping with an increase in the size of cooperative control of pests, agricultural aviation service emerged as a new service. This service contributed very much to the development of pest control with pesticides in respect to the saving of labor; however, the large scale operation of pest control conducted by aerial application agitated the public against the use of pesticides and the business has to stand stagnant recently.

Impacts on societal systems and functions: The saving of farm labor by the introduction of pesticides and associated technologies caused a movement of the younger generation from the rural community so that the size of the farm family and the rural community dwindled in many areas. While highly toxic pesticides had to be employed and cooperative operation under supervision of authorized technical officers was practiced, a sense of unity was strengthened among farmers. However, the development of safer pesticides diminished cooperative pest control and the sense of unity decreased. Partly due to the decrease of population, particularly of an active young generation, and partly due to the decreased sense of unity, rural societal functions became less active.

In order to encourage the use of pesticides to achieve food production increase, both the central and local governments required new organizations and institutions to approve and register pesticides, to secure production and distribution, to educate farmers on the use of pesticides, to restrict the use of specified pesticides, to test safety of pesticides, to monitor the residue of pesticides on and in foods as well as in the environment, and to seek measures for preventing or mitigating adverse effects caused by the use of pesticides. These administrative needs required a large manpower and a great amount of financial expenditure to be borne both by the central and local governments.

The research efforts of pest control had been concentrated during the past two decades into the development of new pesticides and their effective use in pest control. Although effective control measures have been established against many important pests by this concentrated effort, the general pattern of pest control was distorted to a considerable extent.

Impacts on environmental systems and functions: Residues of BHC and other persistent organochlorinated insecticides and metallic mercury derived from organomercurial fungicides extensively contaminated the crop fields; products from these fields were found to be contaminated with these residues through absorption by plants. Residues of chlorinated insecticides were further conveyed into animal products. Excessive vegetative growth was observed with potato, sweet potato and other crops planted in fumigated fields; this was attributed to the retarded nitrification of ammonium due to reduced microfloral activity in fumigated soil. Edaphonic population seems to have been affected in many ways by the pesticide residues in soil, but available evidence is very scarce.

Since almost half of the pesticides used in Japanese agriculture were applied for rice pest control in flooded fields, the aquatic environment was extensively and seriously polluted with these pesticides. Sodium pentachlorphenate employed for rice weed control seriously affected fish and shell-fish populations. Reduction of freshwater snail was believed to be related to the decrease of Bilharziosis to the causal organism of which the snail serves as intermediate host. Swarms of firefly, the larvae of which prey on aquatic snails, and designated as a natural monument because of their aesthetic beauty in summer evenings, became almost extinct. Fish and other aquatic organisms inhabiting rice growing areas were found to accumulate residues of BHC in their bodies, but ecological significance of this accumulation is not yet studied in detail.

Pollution of air by pesticide drift and vapor generated from treated fields has been investigated very little except in and around treated fields and potential adverse impacts are not well defined.

4. Cause and Effect Sequence and Intensification of Impacts

Direct impacts of the use of pesticides observed on human, industrial, societal and environmental systems caused further impacts on the same or other aspects of these systems. Because of this cause and effect sequence, some impacts were intensified when the use of pesticides was coupled with other technologies.

Persistent pesticides were preferred by farmers because of their extended effectiveness against pests. With these pesticides the farmers could save both cost and labor through less frequent application. However, residues of these pesticides were more liable to remain on and in the product, accumulate in the soil more quickly, be absorbed by the crop, and produce products contaminated with residues; this induced fear of contamination of foods among the public and disturbed the marketing of agricultural commodities including uncontaminated products.

As the holdings of respective farmers are small, cooperative operation for applying pesticides was encouraged in order to achieve more satisfactory control. In fact, the operation increased the efficacy of control, reduced the cost and labor of operation, stimulated the cooperation in other farming practices and strengthened the communal sense and tie among farmers. However, the cooperative operation over extensive acreage caused serious contamination of the environment, killed fish and shell-fish, upset prey and predator balance, and encouraged the development of pest resistance to pesticides.

Along with the transportation of pesticides and their residues in the environment by physical forces such as wind and water flow, the impact of the pesticide use spreads into different aspects of societal and environmental systems. Potential hazards to human health are caused by direct exposure to pesticides during their manufacture, transportation, handling and application, by intake of residue in and on foods, and by contact with environmental components contaminated with pesticides and their residues. Foods are contaminated by direct application of pesticides and by absorbing pesticides and their residues, and by bioconcentration in the course of the food web. Some impacts of the use of pesticides were intensified by coupling the use of pesticides with other technologies. Success in controlling the rice stem borer

with pesticides not only saved the loss caused by the said pest but also made it feasible to transplant early and fertilize the crop heavily. The remarkable increase in yield achieved after the establishment of the measures for controlling the rice stem borer should be attributed more to early transplanting and heavier fertilizing.

To save labor and expenditure for application in controlling the rice stem borer with BHC, a hand-broadcasting granular formulation was manufactured by an advanced formulation technique. This formulation gave a better control than dust or spray and mitigated to some extent the adverse effect of BHC on spiders and other predators and parasites of rice leaf- and planthoppers and accordingly reduced their resurgence. However, since this formulation required two-fold as much pesticide ingredient per unit acreage, the introduction of this formulation accelerated the contamination of rice field environments with BHC residues. BHC absorbed by rice plants was accumulated in the straw and dairy cattle fed with contaminated straw were found to produce milk contaminated with BHC residues. The general public, informed of this contamination, strongly opposed the use of BHC and this insecticide was completely withdrawn from agricultural and forestry uses in 1972.

5. Evaluation and Compensation of Impacts

In order to evaluate the importance of impacts, whether they may be favorable or adverse, it is adequate to consider the size of areas and of population actually or supposed to be affected, extent of irreversibility of effect, and the controllability of the impacts.

Since pesticides are easily disseminated in the environment through physical forces such as wind and water flow, through foodweb and moving of agricultural products from rural to urban areas, the area directly or indirectly affected by the use of pesticides is nation-wide or even global. The topography of Japan is mostly mountainous and agricultural fields are distributed along rivers and coast line. The percentage of areas directly affected by the use of pesticides is 16 per cent of the whole territory of the nation. However, since a large proportion of the population inhabits lowland areas, the percentage of population directly or indirectly affected by the use of pesticides is very high. Rural population has been on a quick decline during the past two decades due to the outflow of the younger generation and accordingly the size of population affected by the direct impacts of the use of pesticides is dwindling. On the other hand, urban population which is potentially affected by various indirect impacts is increasing and, therefore, those indirect impacts are assuming more and more importance.

Among groups of people who are affected by direct impacts of the use of pesticides, people engaged in fishery are second in number. Use of pesticides highly toxic to fish and shell-fish was accordingly opposed strongly and the restricted use of those pesticides needed enforcement.

It is quite difficult to identify whether the effect of impact is reversible or not unless time permits some evidences. Switching from persistent to less persistent insecticides and from toxic to less toxic herbicides to fish and shell-fish in rice insect and weed control seems to be restoring some aquatic organisms which had been almost extinct.

There is definite indication that residues of persistent chlorinated insecticides in soil are decreasing since the use of these insecticides was withdrawn.

Most of the direct adverse impacts of the use of pesticides seem to be controllable by the proper choice of pesticides and the method of application. Indirect adverse impacts caused by the spreading of pesticides in the environment are mostly difficult to control unless the use of pesticides is restricted or substituted by other methods of pest control.

Although many adverse impacts of the use of pesticides were observed on various human, industrial, societal and environmental aspects, most of them may be compensated for by the favorable impacts. Adverse impacts are most easily compensated for by favorable impacts within the same individual or within the same group of people. Farmers easily and profitably compensate the adverse impacts with favorable impacts and this is why the consumption of pesticides increased very rapidly in Japanese agriculture. Compensation becomes more and more difficult between individuals or groups of individuals having less mutual economic interests and communal sense. Damage of cattle, poultry, fish and shell-fish could be compensated for by advantages to be obtained in agriculture by compensating the economic loss of farmers. Compensation of this kind is less difficult in local communities, but is extremely difficult or almost impossible on the nation-wide scale. When fish and shell-fish kill caused by sodium pentachlorphenate for rice weed control became nation-wide, the central government had to set stringent restriction on the use of this herbicide. Compensation of adverse impacts on matters of aesthetic significance and on human health by economically favorable impacts is almost impossible. Diverse thinking on the philosophy of life and sense of values are two big barriers in compensating negative impacts concerned with different groups of people.

6. Removal of Adverse Impacts

In order to make the adverse impacts of the use of pesticides as small as possible, measures to remove or lessen respective adverse impacts should be sought. However, they are so varied in nature and in extent, it would be practically impossible to treat all the impacts at this stage and priority should be considered with respective impacts. It is logical to put the first priority for removing or lessening adverse impacts on those related to human health and those which are difficult to compensate by other favorable impacts. Thanks to the rapid advance in the development of pesticides of more safety, there is much possibility to reduce acute health hazards caused by direct exposure to pesticides. Switching from persistent to less persistent and biodegradable pesticides helps decrease the contamination of foods with pesticide residues and reduces potential hazards due to chronic toxicity. However, in order to guarantee the safety of pesticides to the full extent, there is urgent need to establish testing procedures for newer aspects of toxicology of pesticides and to augment technical staffs and facilities.

In order to mitigate the effects of adverse impacts which are difficult to compensate, safety of pesticides to fish, shell-fish and other beneficial non-target organisms including wild life and those of aesthetic importance should be considered. However, methodology for testing safety of pesticides to these organisms should also be explored. Collection of data on the behavior and fate of pesticides and their metabolites in the environment as well as dependable analytical methods to detect minute amounts of pesticides and their metabolites and to trace their movement are required in order to find measures for spreading impacts in the environment.

Since development of pest control measures other than the use of pesticides is slow and insufficient except in a few cases, concerted research efforts of entomologists, plant pathologists, weed scientists, ecologists and agronomists are required to develop agronomical and biological control of pest species.

Banning or strict restriction on the use of pesticides which had serious impacts on human, industrial, economic and ecological aspects of the societal and environmental systems has been found effective both in removing such impacts and in encouraging the development of safer pesticides. Other institutional controls on the use of pesticides such as the enforcement of monitoring of pesticide residues in the environment are also effective to improve the use pattern of pesticides.

References

Mitre Corporation (1971): Technology Assessment Methodology, 285 pp.

Technology Assessment Study Group, Pesticide Subgroup (1972): Case Study of Technology Assessment. Pesticides. 285 pp. mimeographed. (in Japanese).

Ishikura, H. (1973): Case Study of Technology Assessment on the Use of Pesticides. Journ. Pl. Prot. 27 : 15–21 (in Japanese).

Biological Control and Environmental Quality *)

By **Robert van den Bosch**

Division of Biological Control, University of California, Berkeley

In simplest terms, biological control is the regulation of species numbers by their natural enemies. In other words, it is a natural phenomenon which plays a major role in the balance of nature (natural control). Biological control is of particular importance to man because of its key role in the suppression of our major competitors, the insects. Indeed, if it were not for biological control we might not be able to compete successfully with the insects.

The Prevailing Pest Control Strategy

It would seem then, that our insect control strategy should maximize the role of biological control, but in fact the very opposite is true. The prevailing chemical control strategy virtually ignores natural mortality factors, and its tools, the synthetic organic insecticides, as generally used, are highly disruptive to biological control. As a result, we seem to be losing ground in our battle with the insects while simultaneously creating serious ecological problems. The world over, there have been repeated episodes of pest control breakdown, environmental pollution and sociological disruption in the wake of pesticide usage, with insecticide disruption of biological control very much at the heart of the matter (Adkisson,

*) This paper was presented at the Third International Symposium on Chemical and Toxicological Aspects of Environmental Quality, Tokyo, Japan, November 19–22, 1973

1971; Georghiou, 1972; Huffaker, 1971; Huffaker et al., 1970; McMurtry et al., 1970; Reynolds, 1971; Smith and Reynolds, 1972; van den Bosch, 1971; van den Bosch and Messenger, 1973).

Pesticide Related Problems

Pesticide disruption of biological control has produced a trio of interrelated problems, (1) target pest resurgence, wherein the insecticides destroy the natural enemies of the target species, thus permitting the pest's rapid resurgence to damaging abundance, (2) induced secondary pest outbreaks, wherein previously innocuous species erupt to damaging levels when their natural enemies are destroyed by insecticides applied to control injurious forms, (3) pesticide resistance, which is aggravated by the repeated insecticidal treatments engendered by target pest resurgences and secondary pest outbreaks.

In combination, these three problems often create pesticide treadmills which cause severe economic, ecological and sociological problems. There is probably no better window into the shattering technological breakdown predicted in *The Limits of Growth* (Meadows et al., 1972), than the chaos that has attended chemical control of cotton pests worldwide. In northeastern Mexico, for example, the pesticide treadmill destroyed a $ 50 million cotton industry and its dependant society (Adkisson, 1971); in Central America it has caused massive human intoxication (Adams, 1972), and in the same area it has contributed to intensive and widespread insecticide resistance in the dangerous malaria vector, *Anopheles albimanus* (Georghiou, 1972). To a greater or lesser degree, this pattern has repeated itself in virtually all major cotton growing areas on earth (Smith and Reynolds, 1972; van den Bosch, 1971).

A close analysis of the pesticide resistance problem gives cause for deep concern, for although only about 225 of the world's ± 10,000 pest arthropod species have developed resistance (Brown, 1971; Georghiou, 1972), its incidence is disturbingly high among our most serious pests. In the United States, for example, more than 50 per cent of the major pests of cotton have developed resistance to one or more pesticides. A similar level of resistance has developed among the major pests (those causing losses of $ 1 million or more per year) of general agricultural crops in California. It is especially noteworthy that cotton is the most heavily treated crop in the U.S. and that California is the greatest pesticide user among the states.

The message is crystal clear — resistance is corrolary to heavy pesticide usage.

The gravity of the resurgence-secondary pest syndrome is perhaps best pointed up by the rise of spider mites to top rank among the world's pest arthropods. Spider mites, relatively minor pests at the time of the DDT breakthrough, have been boosted to their status of pre-eminence by the synthetic organic insecticides (Huffaker et al., 1970; McMurtry et al., 1970). In other words, man's most serious arthropod pest problem is a gift of his own technology.

Conclusion

Chemical control is a highly useful pest control tactic that has fallen into misfortune because of its mis-application as a strategy. Nothing so simplistic as chemical control, or for that matter autosterilization, cultural control, natural enemy colonization, use of pheromones or the development of pest resistant hosts will solve the insect problem. Instead, these elements and others combined with information can best be utilized in integrated control systems. Chemical insecticides and biotic mortality agents both have bright futures under integrated control; indeed, the best of each will be realized in pest management systems. But prevailing attitudes, policies and interests are hampering a rapid transition into the era of integrated control. Meanwhile, pest control and its impact on environmental quality continue to be a major public concern.

References

1 Adams, A. V. (1972): Summary of joint FAO/Industry seminar on the safe, effective, and efficient utilization of pesticides in agriculture and public health in Central America and the Caribean (TF: Lat/16). Food & Agric. Org. UN, Rome. AGPP: Misc/6 25 pp.
2 Adkisson, P. L. (1971): Objective uses of insecticides. In: Agricultural chemicals: — harmony or discord for food, people, enviroment, ed. J. E. Swift, pp. 43–51. Univ. Calif. Div. Agri. Sci. 151 pp.
3 Brown, A. W. A. (1971): Pest resistance to pesticides. Pesticides in the Environment, ed. R. White-Stevens, 1: 457–552. New York, Dekker. 629 pp.
4 Georghiou, G. P. (1972): The evolution of resistance to pesticides. Ann. Rev. Ecol and Systematics. 3: 133–168.
5 Huffaker, C. B. ed. (1972): Biological Control. New York Plenum Press. 511 pp.

6 Huffaker, C. B., M. van de Vrie, and J. A. McMurtry (1970): Ecology of tetranychild mites and their natural enemies: A review II. Tetranychid populations and their possible control by predators. Hilgardia 40: 391–458.

7 McMurtry, J. A., C. B. Huffaker, and M. van de Vrie (1970): Ecology of tetranychid mites and their natural enemies: A review. I. Tetranychid enemies — their biological characters and the impact of spray practices. Hilgardia 40: 331–390.

8 Meadows, D. H., D. L. Meadows, J. Randers, and W. W. Behrens III (1972): The limits to growth. Universe Books, New York. 205 pp.

9 Reynolds, H. T. (1971): A world review of the problem of insect population upsets and resurgenes caused by pesticide chemicals. In, Agricultural chemicals — harmony or discord for food, people, environment, ed., J. E. Swift, pp. 108–112. Univ. Calif. Div. Agric. Sci. 151 pp.

10 Smith, R. F., and H. T. Reynolds (1972): Effects of manipulation of cotton agroecosystems on insect populations. In The Careless Technology, Ecology and International Development, ed. M. T. Farvar and J. P. Milton, New York, Natural History Press. 1030 pp.

11 Van den Bosch, R. (1971): The melancholy addiction of ol' king cotton. Natural History (80, 10): 86–91.

12 Van den Bosch, R., and P. S. Messenger (1973): Naturally occurring biological control and integrated control. Chapt. 8, pp. 119–134. In: Biological Control, New York and London, Intext Educational Publishers. 180 pp.

Natural Enemies as a Control Agent of Pests and the Environmental Complexity from the Theoretical and Experimental Points of View

By **Fumiki Takahashi**

Entomological Laboratory, College of Agriculture, Kyoto University, Kyoto 606, Japan

Recently several ideas have been proposed in methods of insect pest control after the great hazard produced by the drastic effects of insecticides such as DDT and BHC. However, I have some doubts in regard to their underlying philosophy of the prospective features of the ecosystem. For example, the selective insecticides attempt to kill only the target pests without causing injurious effects on natural enemies and complexities of the ecosystem. Here, a possibility exists of repeating an error we have had in the use of insecticides. Formerly the studies on insecticides were directed to killing individual pests without thinking about their surrounding biological system. This process of thinking might be replaced by the idea that any insecticide will be useful so long as it preserves individuals of natural enemies. The target insects in this idea were considered on the individual level of the species and not on their whole life system. The species can not be maintained by an individual alone but is kept in the community of individuals, that is, as a population and as a component in the ecosystem. Even the selective elimination of pest insects will cause a big disturbance in natural enemy populations owing to the decrease of their food or host when they are mono- or oligophagous on a particular group of prey or host species.

The mathematical model of Volterra, which deals with the numerical interrelation between the prey and predator populations and insists on a balance produced by their interaction, states in the third law (Law of the disturbance of the averages) that the predator density decreases and the prey density increases when a certain proportion of their population is uniformly destroyed. The model of Nicholson and Bailey (1935) treats a similar relationship between the host and specific parasite populations under the following process:

*) This paper was presented at the Third International Symposium on Chemical and Toxicological Aspects of Environmental Quality, Tokyo, Japan, November 19–22, 1973

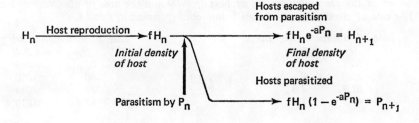

H_n and P_n are the population densities of host and parasite species, respectively, on the nth generation; f is the reproductive rate of the host; a is the host searching efficiency, "the area of discovery", of the parasite; and e is 2.718. The model proposes two fundamental characteristics, the maintenance of their population balance and the cyclic fluctuation of populations.

The effects of variable degree of selective mortality on the movement of average levels of host and parasite populations are analyzed by the model in this paper.

The average levels of host and parasite populations, $\tilde{H}$ and $\tilde{P}$, are obtained by the equations,

$$\tilde{H} = \frac{1}{f-1}\tilde{P} \quad \text{and} \quad \tilde{P} = \frac{1}{a}\ln f.$$

The elimination of hosts opereates equivalently in the change of the reproductive rate of host, f, and that of parasites in the change of the host searching efficiency of parasites. The result is shown in Fig. 1 with arrow marks. In the calculation an appropriate combination of the population levels of host and parasite at $f = 8$ and $a = 0.06$ is assumed as an original point of their balance. Arrow A shows the case when the mortality is given uniformly on both the host and parasite populations in proportion to their number. Host population increases depending on the strength of a given mortality and the parasite population rather maintains its level. Arrow B shows the case when the mortality operates only on the parasite population. While host population increases the parasite population also increases. Arrow C shows the case when the host population is selectively destroyed at various degrees as we expect from selective insecticide application.

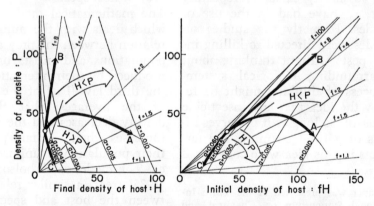

Fig. 1 *Movement of balancing point of host (H) and parasite (P) populations calculated by the model of Nicholson and Bailey; when the hosts were selectively destroyed (C), the parasites were selectively destroyed (B), and both the hosts and parasites were destroyed (A) with variable degrees in proportion to their population density. The original point of balance is fixed by the combination of the reproductive rate of host (f = 8) and the area of discovery of parasite (a = 0.06). The rate of elimination is shown below corresponding to f and a.*

Rate of elimination of hosts		Rate of elimination of parasites	
0%	f = 8	0%	a = 0.060
50	4	25	0.045
75	2	50	0.030
81.3	1.5	75	0.015
86.3	1.1	83.3	0.010

The initial host density decreases but the parasite population decreases more drastically even though the mortality does not operate on the parasite population. On the contrary, the final density of host increases in spite of the selective removal of host population. The selective mortality for hosts, pest insects, may be practically imperfect and the average level of host and parasite populations move in the area $H > P$; in this case the host population increases and the parasite population decreases. This result is contrary to our expectation from selective insecticide application.

There are some criticisms on the mathematical models of Volterra and of Nicholson and Bailey. They simplify to a great extent the mode of interaction between predator and prey which is only one component among the complex interrelations of food chain systems. It is a common fact that an insect population is governed by the effects of complex mechanisms of suppression and regulation of population and not merely by such a simple mechanism as illustrated by the models. However, the interaction between predator and prey is undoubtedly one of the fundamental mechanisms to control pest populations.

Criticisms will arise on these considerations from rather practical standpoints. They say that hitherto the insecticides were effective to suppress pest populations to low density and to produce plenty of harvest even though they killed pests and many other enemies. This is surely indicated in practical cases, but the control practices for insecticides exclude natural action of enemies and attempt pest control by insecticides alone. In this case the natural equilibrium level without the effective action of natural enemies will be much higher than that produced by the balance with natural enemies (Fig. 2). While the models illustrate the case where the natural balance of prey population is maintained mainly by the action of predator or parasite population, the selective insecticide does not maintain such natural balance and intends only to keep the natural enemy populations free from injury by insecticides such as is the case for biotic insecticide application.

The above-mentioned consideration gives emphasis to the numerical response of predator or parasite to prey or host density or the numerical relationships between these populations. In the model of Nicholson and Bailey the host searching efficiency of parasite (a) is assumed to be constant even when the host density is very high or very low. This unreal assumption is often criticized in the practical cases. The

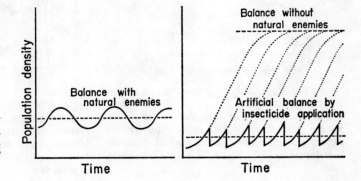

Fig. 2 Schematic representation of the insect population balance produced by the actions of natural enemies and/or insecticide application.

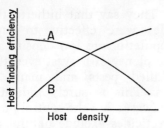

Fig. 3 Two types in the change of host finding efficiency of parasite with the change of host density.

significance of one type of functional response of parasite to host density is illustrated where the host finding efficiency of parasite decreases with the increase of host density (Fig. 3, A). On the other hand, another response is recently demonstrated in which the efficiency or activity of parasite decreases greatly with the decrease of host density (Fig. 3, B). This phenomenon is determined experimentally in some hymenopterous parasites. For example, in a small space an ichneumon parasite, *Nemeritis canescens*, attacks hosts, the almond moth *Cadra cautella*, frequently when the host density is high (Fig. 4). When the hosts of variable densities are distributed in a wide

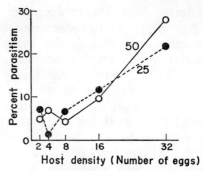

Fig. 5 The preference of Nemeritis canescens to the host density (free choice experiment). The preferences of 25 or 50 wasps are shown by the percentage of parasitism when they attack host larvae aged 14 days. (Takahashi, 1968).

space, the parasites prefer the higher density places (Fig. 5). This relation suggests that the action of effective natural enemies at high density of pest population will diminish its effectiveness with the decrease of population density of pests. Furthermore, the parasites of this type have poor ability to suppress the host population in their experimental system and they can keep their population only when the host population is at high density (Takahashi, 1973).

In the experimental population of the azuki bean weevil, *Callosobruchus*

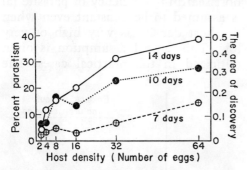

Fig. 4 The functional response of Nemeritis canescens to the host density (confined experiment). The response of a single wasp is shown by the percentage of parasitism and by the corresponding area of discovery when it attacks host larvae aged 7, 10, or 14 days. (Takahashi, 1968).

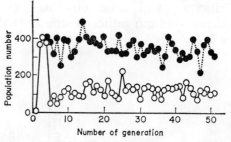

Fig. 6 The population fluctuation of host, Callosobruchus chinensis in the interacting system with its parasite, Anisopteromalus calandrae. ......●.....; the initial density of host, ————○————; the final density of host. (Utida, 1948).

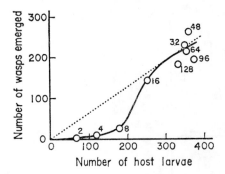

Fig. 7 The functional response of Anisoptero-malus calandrae *to* Callosobruchus chinensis *larval density. Numerals beside the points show the number of host parents (= final density of host) which oviposit on 10 gr of azuki bean and produce host larvae (= initial density of host) to be attacked by a single parasite. (from Utida, 1943).*

chinensis, and its pteromalid parasite, *Anisopteromalus calandrae,* supplying 10 grams of azuki bean in every generation, the host population maintains a fairly stable level (Fig. 6, Utida, 1948). The levels of the initial density (about 300–400) and the final density (about 80–150) of host are compared with other experimental results. The parasites attack efficiently at higher initial host densities over 300 (Fig. 7) which correspond to the initial host

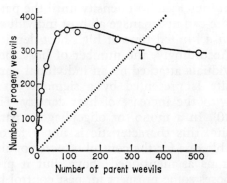

Fig. 8 The relationship between the parent density and its progeny density in Calloso-bruchus chinensis *on 10 gr of azuki bean. (from Utida, 1943).*

density shown in Fig. 6. In the reproduction curve, the relation of population densities between two successive generations of the host (Fig. 8), the equilibrium level of host population is expected at the point T. When the parasites are introduced in the system, the final host density decreases to about 80–150 as shown in Fig. 6, and, as a result, the maximum number of progeny (about 360) is produced. This level is nearly equal to the initial density of host in Fig. 6. When the host population is overcrowded, the parasites tend to reduce the host density to the optimum level for its reproduction. In other words, the parasite population seems to have a symbiotic life pattern with its host population. The results in the almond moth systems are shown in Fig. 9. In the single host system the host larvae perfectly consume food and the head width of host adults, which indicates the density effect of host larvae, becomes small (Fig. 9, C). When a "clever" parasite, N. canescens, is introduced into the system, the host population decreases and the host head width becomes larger (Fig. 9, B). But the initial host density is almost of the same level as that in the single host system. Food consumption by host larvae is a little less than or nearly equal to that in the single host system. In this case the inter-relation between the host and the parasite is similar to the case of C. chinensis and A. calandrae. Therefore, it might not be necessary to preserve such a symbiotic parasite from insecticide treatment unless the host density is very high. However, the characteristic of these parasitic species is not the representative one of overall parasitic species. There is another type of parasitic character which does not have a symbiotic relation between host and

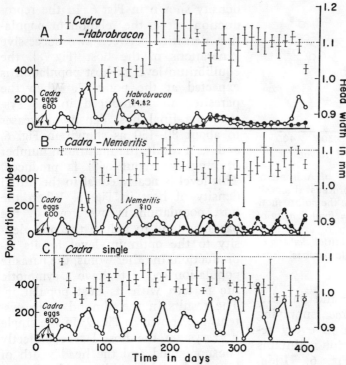

Fig. 9 Population fluctuations of host, Cadra cautella (—O—) and its parasites, (Nemeritis canescens (--●--) and Habrobracon hebetor .....●..... in their interacting system, and the changes of head width of host adults (vertical bar; the mean head width and its 95% fiducial limits). Twenty grams of rice bran were supplied every 10th day. (Takahashi, 1956, 1973).

parasite populations. When a "foolish" parasite, a good parasite for us, a braconid parasite, Habrobracon hebetor, is introduced into the system, the host density decreases greatly and plenty of food is left behind unused

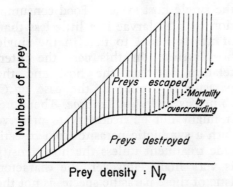

Fig. 10 Relationship between the density of preys and their outcome after the predation by a polyphagous predator which has a sigmoid curve (solid line) in its functional response to prey density.

(Fig. 9, A). A comparative study of the parasite ability to suppress the host density and the preference character for host density will provide information for finding the desirable nature of parasite populations.

In the symbiotic parasite the activity of individual parasites is more intensified at increased host density until the parasite can not manage all host individuals in a limited time at high host density. Accordingly, the number of host individuals attacked by an individual parasite is presented by a sigmoid curve with the increase of host density (Fig. 10). In a mono- or oligophagous parasite, this characteristic is not a favorable one for the control of pest populations as mentioned above. But it proposes some prospect for pest control by biological means when the natural enemies are polyphagous because they do not depend only on pest population for their food. Namely, they have an alternate food and their density is not so much affected by the decrease of pest population. This type of functional response is expected in many predator species, such as coccinelid beetles, spiders, birds, and mammals. The mechanism which accelerates the prey attacking activity of predator with increasing prey density is established by psychological and physiological characters of predators. But we have only a few studies on the predation especially at very low density of prey population in which the sigmoid functional response curve will be commonly observed. In this case the sigmoid curve of the functional response of predator to prey density (Fig. 10) gives rise to the sigmoid reproduction curve shown in Fig. 11 (Takahashi, 1964). In the schema the effect of overcrowding of

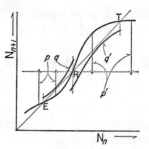

Fig. 12 *A representative sigmoid reproduction curve which has two equilibrium points (T and E). The movement of the equilibrium between two points through the population release point (R) is expected in two ways, p (p') and q (q'). (refined by Takahashi, 1964).*

pest population is included. This type of reproduction curve has been actually observed in some forest defoliaters such as *Choristoneura fumiferana* (Morris et al., 1963).

The reproduction curve of a in Fig. 11 has similar characteristic to the curve shown in Fig. 8, and the equilibrium point exists only at very high population density.*) With curve c the population inevitably decreases its density and disappears before long. Curve b has two equilibrium points. One (T) is at very high density of insect population at which it may greatly injure crop production. Another one (E) is at low density which may lie below the economic threshold. The mechanism to shift the population density between the two points can be achieved in two ways as shown diagramatically in Fig. 12. One is by a big change in the reproductive rate of the insect population so that the

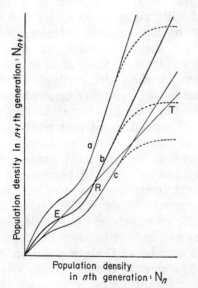

Fig. 11 *Three sigmoid reproduction curves of prey population obtained from the sigmoid functional response curve of a predator shown in Fig. 10. The reproductive rate of prey population is ×3 in a, ×2 in b, and ×1.5 in c on the number of prey escaped. (Takahashi, 1964).*

*) If an insect population has the reproduction curve shown in Fig. 8 or a of Fig. 11, the rate of population growth becomes higher at lower population density. Therefore the number of progeny produced can not be suppressed so much unless mortality intensities are very strong especially when the curve is greatly convexed as shown in Fig. 8.

equilibrium of population will move beyond the level R in a short period of one generation (p and p'). Ordinarily, insect populations have a very high reproductive potential which is suppressed strongly by environmental resistances. A little environmental change will produce a large change in their reproduction. Another mechanism is a little change in the reproductive rate but maintained for several generations. The reproduction curve shifts beyond the 45° line as shown by q and q' and moves the population density between two points. In practical cases, when a pest population is maintained at very high density as indicated by T, our strategy for control of pest insects should not be restricted to the drastic control method shown by p' but must search for even a weak but long lasting controlling effect as shown by q' via a modification of cultural management. This situation is supported only in the complex ecosystem which must accompany the regulatory mechanism by polyphagous natural enemies having the sigmoid functional response mechanism. The action of selective insecticides will provide considerable effects in the same manner. However, it is very important to determine which kinds of natural enemies are essential and are to be protected among many kinds of predators and parasites. The information for such purposes will be obtained from the ecological study of insect populations, but it is still very difficult to obtain such useful information without a laborious study of ecological systems. The difficulties of such a study originate from the fact that most of the insecticides fundamentally depend upon their toxic action. The control methods based upon the behavioral attraction, genetics, and sterile male technique are profitable in

this sense. The use of resistant crop varieties would be supported by this mechanism.

Though a certain natural enemy is characteristically polyphagous, it may not be effective when its alternate food is limited to a small number of species owing to the over-simplified ecosystem. In this case the situation is quite similar to the above-mentioned simple relation between single host and specific parasite species. Our strategy is now at the point of choosing either the complex ecosystem expecting natural stability or the simple ecosystem expecting artificial control of the balance. In this case, however, artificial manipulation of food for natural enemies should also be considered.

The studies on the life patterns or ecology of insect communities will provide basic information for the project of integrated control. These studies are now directed to a study of the life tables of an insect and analysis of the mortality factors which control their functions at different stages of insect development in the life cycles. Further, the analysis of variations in such mortality factors between generations are key factors in the control or regulation of pest populations. The effects of natural enemies are now recognized as important among mortality factors and will become an important consideration in insecticide treatments. These studies have been conducted mostly in a population of insect pests, especially when they exist in large numbers. Their population density is above the economic injury level. The mortality factors at such high population density may provide important information for an integrated control project. But these factors are not necessarily the indispensable and sufficient ones to maintain the insect

population at its very low density below the economic threshold. That is to
say, the insect becomes a pest and
maintains its high level of density because these mortality factors are inef-
fectual controlling agents. The analyses at high population density do not
necessarily propose important mortality factors. Important mortality factors
to suppress pest population density
would be found when the population is
maintained at low density below the
economic threshold.

We have to realize the difference in the
role of natural enemies at different
population levels of insect pests and the
significance of pest population analysis
at its low density. The analysis at low
population density has some difficulties. One originates from our social
demands and another from technical
means to estimate population density.
The analysis of key factors in insect
population control over long periods is
still more difficult. In this case, an
effort to analyze the environmental
resistance of insect pests during their
latent period in the natural field by
artificial inocculation of eggs and
larvae of the pests would give us useful
information, as has been done by Furuta (1968) in the pine caterpillar, Dendrolimus spectabilis. He found significant roles of polyphagous predators in
suppression of a pest population density rather than those of oligophagous
parasites. In this case, the material in-
occulated into the forest was the pest
itself; this had to be strictly checked to
avoid some risk of its unexpected outbreak. This technique is still undevel-
oped in comparison with the medical
treatment for tuberculosis control in

which the nontoxic antigen (tuberculin)
is used to test immunity development
and the nontoxic BCG is used to raise
immunity against tuberculosis. This
trial study will give initial information
on mortality factors to be preserved in
natural populations and on the sanitary
method to prevent a pest outbreak.

References

Furuta, K.: The relationship between population
density and mortality in the range of latency
of Dendrolimus spectabilis Butler. Jap. J. appl.
Entomol. Zool. 12 : 129–136 (1968).

Morris, R. F. et al.: The dynamics of epidemic
spruce budworm populations. Mem. Ent. Soc.
Canada No. 31, 332 pp. (1963).

Nicholson, A. J., and V. A. Bailey: The balance
of animal populations. Part 1. Proc. Zool.
Soc. Lond. 3 : 551–598 (1935).

Takahashi, F.: Reproduction curve with two
equilibrium points: A consideration on the
fluctuation of insect population. Res. Popul.
Ecol. 6 : 28–36 (1964).

Takahashi, F.: Functional response to host
density in a parasitic wasp, with reference
to population regulation. Res. Popul. Ecol.
10 : 54–68 (1968).

Takahashi, F.: An experimental study on the
suppression and regulation of the population
of Cadra cautella (Walker) (Lepidoptera;
Pyralidae) by the action of a parasitic wasp,
Nemeritis canescens Gravenhorst (Hymen-
optera; Ichneumonidae). Mem. Coll. Agric.
Kyoto Univ. 104 : 1–12 (1973).

Utida, S.: Host-parasite interaction in the experimental population of the azuki bean
weevil, Callosobruchus chinensis (L.). 3. The
effect of host density on the growth of host
and parasite populations. Ecol. Rev. (Sendai)
9 : 40–54 (1943).

Utida, S.: Host-parasite interaction in the experimental population of the azuki bean
weevil, Callosobruchus chinensis (L.). 5. Population fluctuations caused by host-parasite
interaction. Physiology and Ecology 2 : 1–11
(1948).

The Development of Agricultural Antibiotics in Japan *)

By **Tomomasa Misato**

The Institute of Physical and Chemical Research, Wako-shi, Saitama, Japan

Summary

Agricultural antibiotics are very hopeful biodegradable pesticides which are expected to be free from environmental pollution. The outline of the present status of agricultural antibiotics used or tested at present for practical application in Japan is presental in this paper.

Introduction

After the discovery of penicillin, many efforts to control diseases by the use of antibiotics have been made by plant pathologists all over the world.(1) In western countries, however, only a few of these antibiotics have been developed for practical use. These are streptomycin, tetracycline, cycloheximide and griseofulvin. Streptomycin, the first antibiotic introduced in agriculture, was first used in the United States for the control of pear fire blight. This antibiotic and a mixture of streptomycin and tetracycline have been used for the control of bacterial plant diseases, while cycloheximide and griseofulvin have been used for fungal plant disease control. Cycloheximide is a very powerful fungicide but, unfortunately, highly toxic to plants, which restricts its use against plant diseases. Griseofulvin is a much less phytotoxic systemic fungicide, but its use is also restricted because the relation of its cost to its performance under field conditions is not quite satisfactory.

In Japan (2), (3), (4), these four antibiotics had been used only on a very limited scale for practical control of plant diseases, until the curative effect of blasticidin S on rice blast was discovered by our group in 1958. The successful application of blasticidin S against rice blast has stimulated the development of agricultural antibiotics and led to the discovery of several excellent agricultural antibiotics such as kasugamycin, polyoxins, validamycin, etc. Nowadays, blasticidin S and kasugamycin have been in practical use for rice blast control instead of mercuric fungicides, and polyoxins and validamycin have been used for rice sheath blight instead of arsenic fungicides. The amount of antibiotics used in Japan is shown in Table 1.

The development of agricultural antibiotics have not been limited only to controlling plant disease, but have extended wider and more actively over various areas such as utilization of insecticides, herbicides and plant regulators. As shown in Table 2, many compounds of microbiological origin are already used as pesticides or show promise for practical application. Blasticidin S etc. as antifungal antibiotics, streptomycin etc. as antibacterial antibiotics, and gibberellines as plant growth regulators are practically used. Aabomycin as an antiviral antibiotic, tetranactin as a miticide, *Bacillus*

---

*) This paper was presented at the Third International Symposium on Chemical and Toxicological Aspects of Environmental Quality, Tokyo, Japan, November 19–22, 1973

Table 1   Agricultural Antibiotics used in Japan. (Apl. 25, 1973)

Registration	Antibiotics	Diseases	Amounts used in Japan (1973)	
	ANTIFUNGAL ANTIBIOTICS		(ton)	(10³ yen)
1959	Cycloheximide	Onion Downy Mildew	3	6,060
1959	Griseofulvin	Fusarium Wilt of Melon	1	2,300
1961	*Blasticidin S	Rice Blast	4,317	396,986
1965	*Kasugamycin	Rice Blast	39,710	3,835,978
1967	*Polyoxin	Rice Sheath Blight	2,387	1,015,863
1970	*Ezomycin	Stem Rot of Kidney Been	0	0
1972	*Validamycin	Rice Sheath Blight	2,567	399,302
	ANTIBACTERIAL ANTIBIOTICS			
1957	Streptomycin	Bacterial Disease of Fruits and Vegetables	332	602,377
1964	*Cellocidin	Rice Bacterial Leaf Blight	0	0
1964	Chloramphenicol	Rice Bacterial Leaf Blight	12	38,478
1968	Novobiocin	Bacterial Canker of Tomatoes	0	0

* Agricultural Antibiotics discovered in Japan.

thuringensis as a insectividal antibiotic and anisomycin derivatives as herbicides have been tested for practical use in the fields in our country.

It is the purpose of this paper to present the outline of the current status of agricultural antibiotics used at present or tested for practical application Japan.

Blasticidin S

Blasticidin S (5) was discovered in 1959 as an antibiotic produced by *Streptomyces griseochromogenes*. This antibiotic inhibits various species of bacteria and fungi at 5 to 100 ppm and gives excellent control of rice blast when sprayed at 10 to 20 ppm on rice plants.

Table 2   Pesticidal Compounds of Microbiological Origin

[Fungicide]	* antifungal antibiotic	: blasticidin S, etc.
	* antibacterial antibiotic	: streptomycin, etc.
	antiviral antibiotic	: aabomycin, etc.
[Insecticide]	miticidal antibiotic	: tetranactin
	bacterial toxin	: bacillus thuringensis
[Herbicide]	herbicidal antibiotic	: anisomycin
[Growth Regulater]	* Fungal product	: gibberellines

* practically used as pesticides.

Chemical Structure of Blasticidin S

The mode of action of blasticidin S on rice blast was found to be the inhibition of protein synthesis in *Piricularia oryzae*, a rice blast fungus. The toxicity to mammals and fishes is rather high. Carp is killed at a concentration of 8.7 μg/ml in water, and the LD_{50} for mice is 39 mg/kg. The antibiotic is sprayed at very low concentration from 10 to 20 ppm and rapidly broken down after application to the crop, so that there is little danger of food contamination.

The behavior and fate of blasticidin S in the environment were investigated using a radioactive compounds which was prepared biosynthetically from cytosine-2-^{14}C and L-Methionine-(methyl-^{14}C). (6, 7) The antibiotic was located on the surface of the rice plant and little was diffused or transported into the tissue. From the wound or infected part, however, the compound was incorporated and translocated mainly to the upper part. The compound located at the plant surface was efficiently decomposed by sunlight. A considerable quantity of sprayed blasticidin S fell to the ground and was adsorbed tightly on the soil surface. Microbes such as *Pseudomonas marginalis, Ps. ovalis* and *Fusarium oxysporum*, which are usually present in the paddy field, decreased the biological activity of blasticidin S.

Kasugamycin

Kasugamycin (8), a water-soluble basic antibiotic, was discovered from the culture broth of *Streptomyces kasugaensis* in 1965. This antibiotic controls rice blast when sprayed at 20 to 40 ppm. It can be safely used with no toxicity to man, livestock (LD_{50}, 2 g/

Table 3 Transition of effectiveness and successive application of kasugamycin on *pyricularia oryzae* at Shonai district in Yamagata.

Yearly Application	1967	1968	1969	1970	1971	1972 Head	Leaf	1973
Spraying times	2.7	3.5	4.6	4.2	5.0	1.0	2.6	3.2
Sprayed kasugamycin in total amount of anti-blast fungicides	36%	54%	94%	90%	87%	30%	23%	0%
Preventive value of kasugamycin	88	99	70	66	31	3		76

Caution: The preventive value of kasugamycin was assayed by bed-test in a branch of Yamagata Agriculture Experimental Station at Shonai. (S. Takahashi et al. 1974)

Chemical Structure of Kasugamycin

$$HOOC-\overset{NH}{\underset{H}{C}}-N-...$$

kg by oral administration to mice) and fish (TLM, 1 mg/ml to carp).

A weekness of kasugamycin may be that rice blast fungus easily acquires resistance to this antibiotic. The development of resistance in fungi to kasugamycin had been reported from laboratory experiments (9, 10), but not in the fields for some years after application of the antibiotic. However, since 1971, the development of a kasugamycin-resistant strain of rice blast fungus in the fields has become a serious prob-

lem. As shown in Table 3 (11), following the increase of kasugamycin application in paddy fields, the effectiveness of kasugamycin on rice blast at Shonai district in Yamagata was gradually decreased. In 1972, the population of the kasugamycin-resistant strain rapidly increased and formed approximately 97 per cent of the total rice blast fungus in the field. Therefore, kasugamycin use on the field was stopped for one year in 1973. Consequently, the kasugamycin-resistant strain was remarkably decreased. It was found that the kasugamycin-resistance depended on the change of affinity of ribosome to the antibiotic.(12)

Polyoxin

Polyoxins from A to M having closely related structures are formed by Streptomyces cacaoi var. asoensis. (13) Except for polyoxins C and I, these components show marked differences in their activities toward various fungi. Polyoxin D is most effective for rice sheath blight fungus, whereas polyoxins B and L are effective for pear black spot fungus and apple cork spot fungus at 50 to 100 ppm. As for its toxicity, oral administration at 15 g/kg and intravenous administration at 800 mg/kg to mice did not cause any adverse effect. Nor is it toxic to fish. Polyoxins are ideal low-toxic agricultural chemicals like kasugamycin.

Such excellent characteristics are perhaps due to the fact that polyoxin effectively inhibits the cell wall synthesis of sensitive fungi but does not greatly influence other enzymes.(14) Cell walls consist mainly of biopolymer, chitin in the case of fungi, peptideglycan in bacteria and cellulose in plants; and no cell walls exist in animal cells. This

Chemical Structure of Polyoxin

Polyoxin	R_1	R_2	R_3
A	CH_2OH	*	OH
B	CH_2OH	HO	OH
D	COOH	HO	OH
E	COOH	HO	H
F	COOH	*	OH
G	CH_2OH	HO	H
H	CH_3	*	OH
J	CH_3	HO	OH
K	H	*	OH
L	H	HO	OH
M	H	HO	H

	R
C	HO
I	COOH

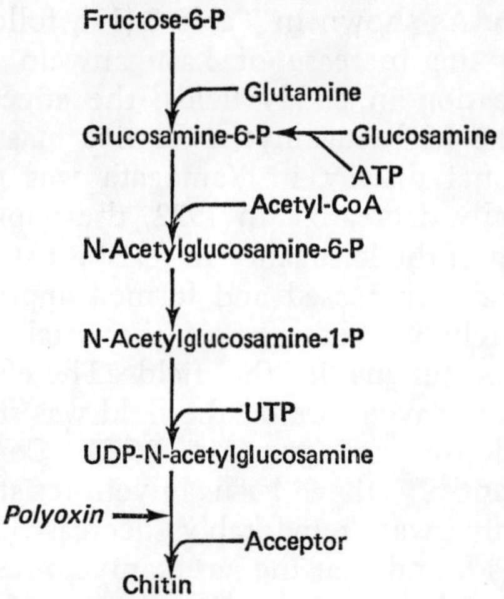

Fructose-6-P

— Glutamine

Glucosamine-6-P ◄— Glucosamine

ATP

— Acetyl-CoA

N-Acetylglucosamine-6-P

N-Acetylglucosamine-1-P

— UTP

UDP-N-acetylglucosamine

Polyoxin ⟶

— Acceptor

Chitin

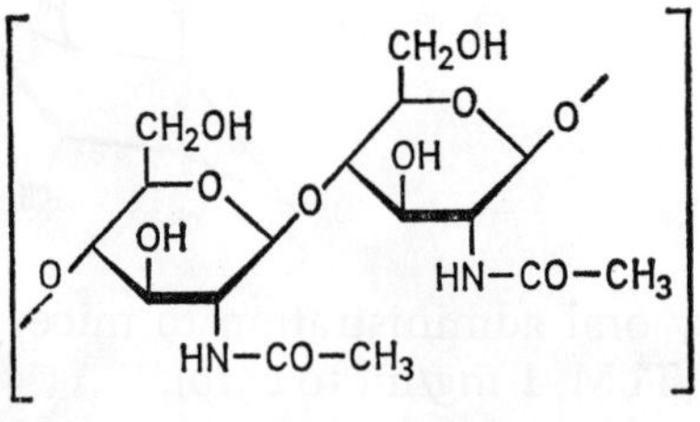

Fig. 1

suggested that chemicals which inhibit specifically the synthesis of cell wall chitin might be toxic to fungi but not to animals and plants. According to studies on the mode of action of polyoxin D in our laboratory, the site of action of this antibiotic is postulated as shown in Fig. 1. (15) The chemical structures of polyoxin D and UDP-N-acetylglucosamine closely resemble each other, giving a structural basis for the competitive activity of polyoxin D. Since polyoxins bind to the enzyme through their same atoms and atomic

groups, values of partial binding affinity, $\triangle g$, were calculated from the values of the inhibition constant, Ki, for polyoxins and their derivatives, which were competitive inhibitors of the enzyme. For example, the values of the partial binding affinity for polyoxin J are shown in Fig. 2. (16) Based on the results of these kinetic investigations on the structure-activity relationship of polyoxin, modifications of the chemical structures of the antibiotics are under way to obtain new ones.

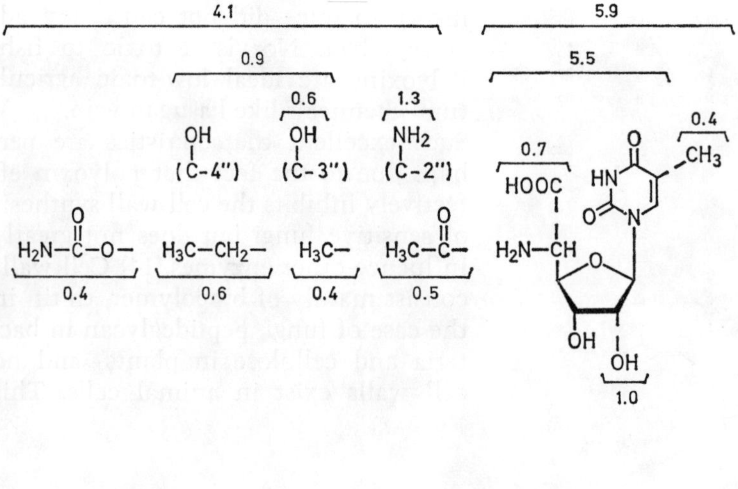

Fig. 2

Chemical Structure of Validamycin A

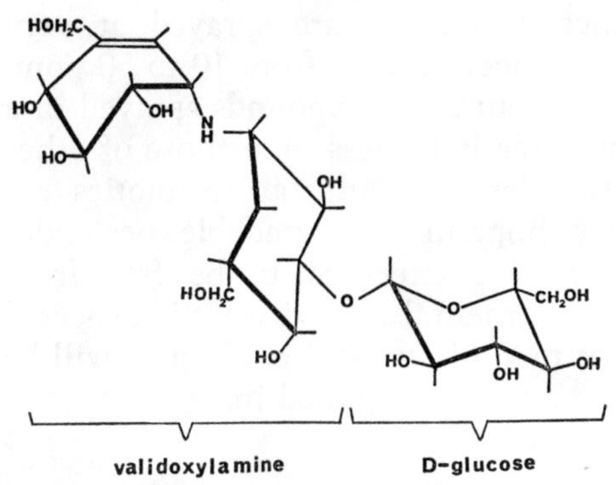

validoxylamine · D-glucose

$C_{20}H_{35}NO_{13} \cdot H_2O$

M.W. 515.5

Validamycin

Validamycin (17) is a new-antifungal antibiotic belonging to a group of the aminocyclitols isolated from *Streptomyces hygroscopius var. limoneus.* The antibiotic showed no antimicrobial activity on agar media against bacteria, yeasts, and fungi but caused an abnormal branching at the tips of the hyphae of *Pellicularia sasakii,* a pathogen of rice sheath blight. Although it is inactive *in vitro,* Validamycin is effective against the sheath blight of rice plants and damping off of cucumber seedlings at a concentration of 30 ppm. It can be safely used with no toxicity to plants and very low toxicity to man, livestock (LD$_{50}$, above 20 g/kg by oral administration to mice) and fish (TLM, above 40 μg/ml to carp).

Cellocidin

Cellocidin (18) is an antibacterial antibiotic produced from *Streptomyces chibaensis.* As its chemical structure is so simple, it is easy to produce chemically. Technical cellocidin is synthesized from fumaric acid or butyndiol. Cellocidin shows an excellent preventive effect against rice bacterial leaf blight at 100 to 200 ppm. However its use is restricted because it is toxic to rice plants.

Chemical Structure of Cellocidin

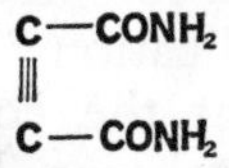

$$C-CONH_2$$
$$\| \|$$
$$C-CONH_2$$

Tetranactin

Tetranactin (19), a new miticidal antibiotic, was isolated as crystalline rhombic prisms from the filter cake of the fermented broth of *Streptomyces Aureus* strain S-3466. The antibiotic shows excellent insecticidal activity against the adults of carmine mite, since LD$_{50}$ for the mite is 9 ppm by the

Chemical Structure of Tetranactin

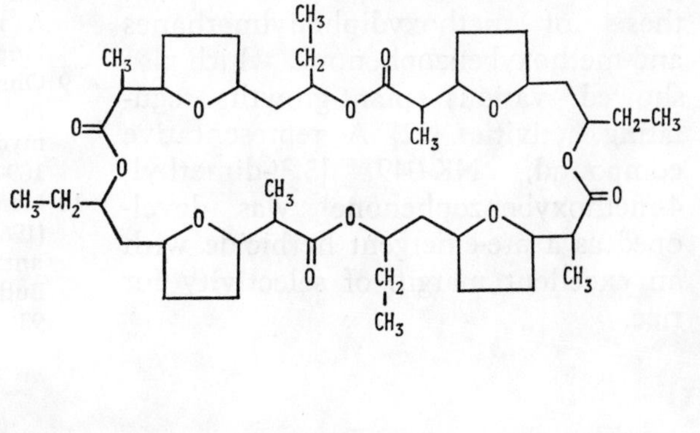

spray method. Residual activity was present 10 days after spraying because of the resistant property to weathering. (20) No systemic miticidal activity was observed in the system in which the mites were transferred on the host plants cultivated in an aqueous medium containing tetranactin. The antibiotic showed no phytotoxicity to apple, mandarin orange, and tea when sprayed at 1,000 ppm. Acute toxicity of tetranactin is low: mice tolerated an intraperitoneal administration of 300 mg/kg and an oral administration of 15 g/kg.

Others

Aabomycin A, a new antiviral antibiotic, showed a complete inhibition on TMV multiplication without any phytotoxicity at the concentration of 100 ppm by tobacco leaf-disc assay, while blasticidin S which was reported as a inhibitor of plant virus was less effective than aabomycin A at the concentration of non phytotoxic levels. (21) On tobacco plants, TMV multiplication and symptom development were inhibited by aabomycin A. The antiviral activity was the highest on applying the antibiotic 1 to 3 days before TMV infection.

Discovery that anisomycin, an antibiotic known to be effective against Trichomonas disease, exhibited plant-growth regulating activity led to synthesis of methoxydiphenylmethanes and methoxybenzophenones which also showed various plant-growth regulating activities. (22) A representative compound, NK-049 (3,3'-dimethyl-4-methoxybenzophenone) was developed as a pre-emergent herbicide with an excellent margin of selectivity for rice.

Conclusion

Since antibiotics are sprayed at very low concentrations from 10 to 50 ppm, the amount of compounds sprayed in a unit area is far less than those of other pesticides. Agricultural antibiotics are very hopeful biodegradable pesticides which are expected to be free from environmental pollution. It is expected that many improved antibiotics will be developed and applied in agriculture in the near future.

References

1 Dekker, J. (1969): Antibiotics. Fungicides, Volume II, Academic Press, 579–635.
2 Misato, T. (1969): The development of agricultural antibiotics in Japan., Japan Pesticide Information 1: 15–18.
3 Misato, T. (1969): Mode of action of agricultural antibiotics developed in Japan., Residue Reviews, Volume 25, Springer-Verlag, 93–106.
4 Huang, K. T., and T. Misato (1970): Agricultural Antibiotics., Rev. Plant Protec. (Tokyo), Res. 3, 12–23.
5 Misato, T. (1967): Blasticidin S, Antibiotics, Volume 1, Mechanism of Action, Springer-Verlag, 434–439.
6 Yamaguchi, I., K. Takagi, and T. Misato (1972): The site for degradation of blasticidin S, Agr. Biol. Chem. Japan, 36, 1719 to 1721.
7 Misato, T. (1972): A new approach in the development of biodegradable pesticides, Environmental toxicology of pesticides., Academic Press, 587–606.
8 Umezawa, H., Y. Okami, T. Hashimoto, Y. Suhara, M. Hamada, and T. Takeuchi (1965): A new antibiotic, kasugamycin, J. Antibiotics (Tokyo), Ser. A, 18, 101–103.
9 Ohmori, K. (1967): Studies on characters of Piricularia oryzae mode resistant to kasugamycin., J. Antibiotics (Tokyo), Ser. A, 20, 109–114.
10 Uesugi, Y., M. Katagiri, and K. Fukunaga (1969): Resistance in Piricularia oryzae to antibiotics and orgnaophosphorus fungicides. Bull. Nat. Inst. Agr. Sci. (Tokyo), C 23, 93–112.

11 Ito, H., H. Miura, and A. Takahashi (1974): Transition of the effectiveness of kasugamycin at Shonai district in Yamagata. The meeting of the Phytopathological Society of Japan.

12 Huang, K. T., Y. Hoshino, and T. Misato (1969): Studies on mechanisms of agricultural fungicide-resistance. 1, Kasugamycin-resistance., Ann. Phytopath. Soc. Japan., 25, 134.

14 Sasaki, S., N. Ohta, I. Yamaguchi, S. Kuroda, and T. Misato (1968): Studies on polyoxin action., Part I, Effect on respiration and synthesis of protein, nucleic acids and cell wall of fungi., J. Agr. Chem. Soc. Japan., 32, 633–638.

15 Endo, A., K. Kakiki, and T. Misato (1970): Mechanism of action of the antifungal agent, polyoxin D., J. Bacteriol., 104, 189–196.

16 Hori, M., K. Kakiki, and T. Misato (1974): Further Study on the relation of polyoxin structure to chitin synthetase inhibition., Agr. Biol. Chem. (Tokyo), 38, 691–698.

17 Iwasa, T., E. Higashida, H. Yamamoto, and M. Shibata (1971): Studies on Validamycins, new antibiotics. II, Production and biological properties of validamycins A and B., J. Antibiotics (Tokyo), Ser. A. XXIV, 107 to 113.

18 Suzuki, S., G. Nakamura, K. Okuma, and Y. Tomiyama (1958): Cellocidin, a new antibiotic., J. Antibiotics (Tokyo), Ser. A. XI, 81.

19 Ando, K., H. Oishi, S. Hirano, T. Okutomi, K. Suzuki, H. Okazaki, M. Sawada, and T. Sagawa (1971): Tetranactin, a new miticidal antibiotic. I. Isolation, characterization and properties of tetranactin., J. Antibiotics (Tokyo), Ser. A. XXIV, 347–352.

20 Hirano, S., T. Sagawa, H. Takahashi, N. Tanaka, H. Oishi, K. Ando, and K. Togashi (1973): Tetranactin, a new miticidal antibiotic. IV. Some Properties of tetranactin., J. Econ. Entomol., 66, 349–351.

21 Yamaguchi, I., R. Taguchi, K. T. Huang, and T. Misato (1969): Aabomycin A, a new antibiotic. II Biological Studies on aabomycin A., J. Antibiotics (Tokyo), Ser. A. XXII, 463–466.

22 Munakata, K., O. Yamada, S. Ishida, F. Futatsuya, K. Ito, and H. Yamamoto (1973): NK-049: From natural products to new herbicides, Proceedings of the Fourth Asian-Pacific Weed Science Society Conference, Rotorua, New Zealand, 215–219.

New Approaches

Pheromones and Related Chemicals in Some USDA Insect Control Programs*)

By **Morton Beroza**

U.S.D.A., Agricultural Environmental Quality Institute, Agricultural Research Service, Beltsville, Md. 20705

Abstract

Insect attractants, of both synthetic and natural origin, have been extremely valuable in detecting infestations of non-indigenous species and in guiding measures to control or prevent the spread of existing infestations. Much attention is now being focused on the use of attractants in the direct control of insects, especially as part of integrated control programs. These studies have been fostered by current attempts to find ways to reduce possible disruptive effects associated with pesticide use, by recent successes in identifying potent sex attractants for highly injurious species, and by some of the synthetic attractants becoming available in potentially large amount. The technology needed to utilize the lures to control insect behavior effectively is complicated by the fact that each species has its own idiosyncracies and must be dealt with individually. Efforts to use the sex attractant pheromone of the gypsy moth (called disparlure) to combat this insect are described to provide insight into the problems faced. Disparlure was tested in mass trapping experiments and by slow release into the atmosphere for disorientation of males which normally find mates by following the female sex attractant to its source. Since the efficiency of these techniques theoretically improves as the insect population diminishes, they are potentially valuable if eradication is contemplated. Use of other pheromones and other behavior-controlling chemicals is mentioned. Results show that progress in insect-attractant technology is being made and that behavior-controlling chemicals will play an important role in helping manage insects with a minimum of pesticide pollution.

With the world population at an all-time high and expanding rapidly, the need to control our insect competitors is now more crucial than ever in the past. Demands for food, fiber, and lumber are steadily increasing, and the public is insisting upon greater protection from the insect hordes that undermine our health and comfort. At the same time, safety considerations and efforts to upgrade our environment are imposing severe restrictions on the pest control practices that have proven so very effective in the past.

Insecticides have been and continue to be our major means of pest management. The saving of millions of lives from insect-borne diseases and the high uninterrupted production of food and fiber from our agricultural community attest to the tremendous value of these chemicals.

* This paper was presented at the Third International Symposium on Chemical and Toxicological Aspects of Environmental Quality, Tokyo, Japan, November 19–22, 1973

Unfortunately, difficulties relating to the use of insecticides have arisen. Insects with their genetic plasticity in many instances have become resistant to insecticides; beneficial insects and other nontarget organisms are sometimes attacked along with the targeted species; residues of some insecticides have found their way into our food, soil, water, and even into our bodies. If we must use insecticides — and responsible authorities agree that sufficient food to support the world's population cannot be produced without them — then their use should be made more effective and selective; they should not persist above negligible levels in the environment; and they should not accumulate in biological tissues.

In our efforts to develop noninsecticidal means of improving insect control, the prospects of utilizing innocuous chemicals such as insect attractants seemed particularly inviting. Many insect species depend heavily on their keen olfactory senses for survival; they follow odor trails to favored foods, to the opposite sex for mating, or to egg-laying sites selected to insure survival of their young. Social insects even coordinate the activity of their colonies with chemicals. Since behavior-controlling chemicals offer the possibility of manipulating important survival mechanisms to our own advantage, we (USDA) proceeded to seek such chemicals, attractants in particular, for economically important pests. As is often done in searching for physiologically active chemicals, we started by screening many chemicals for attraction and then synthesizing compounds related to those that had activity.

Our first challenge came in 1956 when an infestation of the Mediterranean fruit fly (or medfly), *Ceratitis capitata*

(Wiedemann), was discovered in Florida (Steiner *et al.* 1961). Fortunately, we were able to help by coming up with a potent attractant called siglure which was used in traps to help locate and delineate the infestation so control measures were applied only where the insect was found. Control measures became so efficient that eradication of the medfly from the more than 400,000 hectares found infested was achieved before the end ot the next year; but the cost was high — 11 million dollars. Subsequently, we synthesized a more potent medfly lure, called trimedlure (Beroza *et al.* 1961). USDA now has about 23,000 traps baited not only with trimedlure but with lures for the oriental fruit fly, *Dacus dorsalis* Hendel, (Steiner 1952) and the melon fly, *Dacus cucurbitae* Coquillet, (Beroza *et al.* 1960). The traps serve as an early warning system to detect an accidental entry of any of these highly injurious pests into the United States mainland. Several incipient infestations of the medfly were detected in this way and quickly eradicated. The saving in potential eradication costs has been estimated to be in the millions of dollars. An even greater benefit resides in the fact that we do not have to combat these insects at all and therefore do not have to contaminate our food or pollute the environment with the insecticides that normally would be needed for control, were any of these insects to become established.

Working with our entomologist co-operators, we have found additionally synthetic lures for the Japanese beetle, *Popillia japonica* Newman, (McGovern *et al.* 1970a); yellowjacket wasps, *Vespula* spp., (Davis et al. 1972); the coconut rhinoceros beetle, *Oryctes rhinoceros* (L.), (Barber *et al.* 1971); and the European chafer, *Amphimal-*

lon majalis (Razoumowsky) (McGovern *et al.* 1970b). Most of the synthetic lures, which are now used by USDA and others, are available from commercial sources. As with trimedlure, these attractants have been used to detect and monitor infestations so insecticides and other control measures may be applied only where needed and only as long as needed. In addition to preventing an accidental entry of a non-indigenous injurious pest, attractants have been used to time insecticide sprays and thereby make these applications more effective.

Some synthetic lures also hold promise for the direct control of insects. A recent report from Spain indicates that the medfly has been effectively suppressed with trimedlure-baited traps. Another report tells of the eradication of the oriental fruit fly from Rota, an 85-square-kilometer island in the Pacific Ocean (Steiner *et al.* 1965). In this pilot test, small adsorbent squares containing a combination of methyl eugenol, a powerful lure for the male fly, and a fast-acting insecticide were distributed over the island by airplane every few weeks. Males consuming the lure-insecticide combination died, and in a few months oriental fruit flies disappeared from the island. The obvious advantages of using the attractant are that the attack was concentrated on the injurious species, wildlife and other non-target species were not affected, and very little insecticide was required. The same method was also used successfully on the Mariana Islands (Steiner *et al.* 1970).

Unfortunately, the other known synthetic insect attractants are far less powerful than methyl eugenol and trimedlure, and prospects for using them to achieve eradication or control of insects are not considered promising.

There is, however, a growing number of chemicals identified in recent years which do have potential for use in the direct control of insects. These chemicals, known as pheromones, are produced by the insects themselves to induce a behavioral response in other individuals of their own species; e.g. a female will emit a sex pheromone to attract a male for mating and propagation. These pheromones are highly specific in that they affect only their own species and infinitesimal amounts often can induce responses from great distances.

Although the potential of attractant pheromones for insect control has long been recognized, until recently efforts to identify these chemicals were frustrated by the incredibly small amounts of pheromone normally present in insects, and accordingly the large number of insects required to secure enough material for identification. Recent successes have been due largely to the advances and developments in chemical instrumentation and associated methodology that have made possible the manipulation and analysis of microgram and submicrogram quantities of pheromone. In the past 6 or 7 years, sex attractant pheromones for at least 35 economically important species have become available (Table 1).

A crucial need now is to upgrade these important findings into practical pest control tools, and in general to find ways to incorporate the use of pheromones into integrated control programs.

If the current use of pheromones is limited and progress in using pheromones seems slow, it should be recognized that considerable technology has to be developed before each of these chemicals can be utilized effectively. Moreover the development of this tech-

Table 1   Economically Important Insect Pests for Which Effective Sex Attractants Are Known

Insect	Attractant	Reference
almond moth, *Cadra cautella* (Walker)	(*Z,E*)-9,12-tetradecadien-1-c1 acetate	Brady *et al.* 1971; Kuwahara *et al.* 1971
black carpet beetle, *Attagenus megatoma* (Fabricius)	(*E,Z*)-3,5-tetradecadienoic acid	Silverstein *et al.* 1967
boll weevil, *Anthonomus grandis* Boheman	(+)-2-(*cis*-isopropenyl-1-methylcyclobu-tyl)ethanol; (*E*)- and (*Z*)-2-(3,3-dimethyl-cyclohexylidene) acetaldehyde; (*Z*)-2-(3,3-dimethylcyclohexylidene)ethanol[1,2]	Tumlinson *et al.* 1969
cabbage looper, *Trichoplusia ni* (Hübner)	(*Z*)-7-dodecen-1-ol acetate	Berger 1966
California fivespined ips, *Ips confusus* (LeConte)	(—)-2-methyl-6-methylene-7-octen-4-ol; (+)-2-methyl-6-methylene-2,7-octadien-4-ol; (+)-*cis*-verbenol[1,2]	Silverstein *et al.* 1966
codling moth, *Laspeyresia pomonella* (L.)	(*E,E*)-8,10-dodecadien-1-ol	Roelofs *et al.* 1971
common grass grub beetle, *Costelytra zealandica* (White)	phenol	Henzell and Lowe 1970
Douglas fir beetle, *Dendroctonus pseudotsugae* Hopkins	1,5-dimethyl-6,8-dioxabicyclo-[3.2.1]octane (frontalin)[2,3]	Pitman and Vité 1970
eastern spruce budworm, *Choristoneura fumiferana* (Clemens)	(*E*)-11-tetradecenal	Weatherston *et al.* 1971
Egyptian cotton leafworm, *Spodoptera littoralis* (Boisduval)	(*Z,E*)-9,11-tetradecadien-1-ol acetate[4]	Nesbitt *et al.* 1973
European corn borer, *Ostrinia nubilalis* (Hübner)	(*Z*)-11-tetradecen-1-ol acetate[5]	Klun and Brindley 1970
European pine shoot moth, *Rhyacionia buoliana* (Schiffermüller)	(*E*)-9-dodecen-1-ol acetate	Smith *et al.* 1974
grape berry moth, *Paralobesia viteana* (Clemens)	(*Z*)-9-dodecen-1-ol acetate	Roelofs *et al.* 1971
greater wax moth, *Galleria mellonella* (L.)	undecanal[1]	Röller *et al.* 1968
gypsy moth, *Porthetria dispar* (L.)	*cis*-7,8-epoxy-2-methyloctadecane	Bierl *et al.* 1970
house fly, *Musca domestica* L.	(*Z*)-9-tricosene	Carlson *et al.* 1971
Indian meal moth, *Plodia interpunctella* (Hübner)	(*Z,E*)-9,12-tetradecadien-1-ol acetate	Brady *et al.* 1971; Kuwahara *et al.* 1971
larch bud moth, *Zeiraphera diniana* (Guenée)	(*E*)-11-tetradecen-1-ol acetate	Roelofs *et al.* 1971

Table 1   Continued

Insect	Attractant	Reference
Mediterranean fruit fly, *Ceratitis capitata* (Wiedemann)	(E)-6-nonen-1-ol and methyl (E)-6-nonenoate,[6]	Jacobson *et al.* 1973
mountain pine beetle, *Dendroctonus ponderosae* Hopkins	*trans*-verbenol[2,3]	Pitman *et al.* 1968
nun moth, *Lymantria monacha* L.	*cis*-7,8-epoxy-2-methyloctadecane[7]	Schönherr 1972
obliquebanded leafroller, *Choristoneura rosaceana* (Harris)	(Z)-11-tetradecen-1-ol acetate	Roelofs and Tette 1970
oriental fruit moth, *Grapholitha molesta* (Busck)	(Z)-8-dodecen-1-ol acetate[8]	Roelofs *et al.* 1969
palmerworm, *Dichomeris ligulella* (Hübner)	(E)-11-tetradecen-1-ol acetate[7]	Roelofs and Comeau 1971
pine tree emperor moth, *Nudaurelia cytherea cytherea* (Fabricius)	(Z)-5-decenyl 3-methylbutanoate	Henderson *et al.* 1973
pink bollworm, *Pectinophora gossypiella* (Saunders)	(Z,Z)- and (Z,E)-7,11-hexadecadien-1-ol acetate	Hummel *et al.* 1973
redbanded leafroller, *Argyrotaenia velutinana* (Walker)	(Z)-11-tetradecen-1-ol acetate	Roelofs and Arn 1968
red bollworm, *Diparopsis castanea* Hampson	(E)-9,11-dodecadien-1-ol acetate[4]	Nesbitt *et al.* 1973
smaller tea tortrix, *Adoxophyes fasciata* Walsingham	(Z)-9- and (Z)-11-tetradecen-1-ol acetate	Tamaki *et al.* 1971a
smartweed borer, *Ostrinia obumbratalis* (Lederer)	(E)- and (Z)-11-tetradecen-1-ol acetate	Klun and Robinson 1972
southern pine beetle, *Dendroctonus frontalis* Zimmerman	1,5-dimethyl-6,8-dioxabicyclo[3.2.1]-octane (frontalin)[2,3]	Kinzer *et al.* 1969
soybean looper, *Pseudoplusia includens* (Walker)	(Z)-7-dodecen-1-ol acetate	Berger and Canerday 1968; Tumlinson *et al.* 1972
spotted cutworm, *Amathes c-nigrum* (L.)	large species (Z)-7-tetradecen-1-ol acetate[7] small species (E)-7-tetradecen-1-ol acetate,	Roelofs and Comeau 1971
sugarbeet wireworm, *Limonius californicus* (Mannerheim)	valeric acid	Jacobson *et al.* 1968

Table 1 Continued

Insect	Attractant	Reference
summerfruit tortrix, *Adoxophyes orana* (Fischer von Röslerstamm)	(Z)-9- and (Z)-11-tetradecen-1-ol acetate	Meijer *et al.* 1972; Tamaki *et al.* 1971b
tufted apple budmoth, *Platynota idaeusalis* (Walker)	(E)-11-tetradecen-1-ol[7]	Roelofs and Comeau 1971
western pine beetle, *Dendroctonus brevicomis* LeConte	exo-7-ethyl-5-methyl-6,7-dioxabicyclo-[3.2.1]octane[2,3]	Silverstein *et al.* 1968
western spruce budworm, *Choristoneura occidentalis* Freeman	(E)-11-tetradecenal[7]	Weatherston *et al.* 1971
Autographa precationis (Guenée)	(Z)-7-dodecen-1-ol acetate	Roelofs and Comeau 1971
Grapholitha funebrana Treitschke	(Z)-8-dodecen-1-ol acetate	Granges and Baggiolini 1971
Lymantria obfuscata Walker	cis-7,8-epoxy-2-methyloctadecane[7]	Beroza *et al.* 1973b
Trogoderma inclusum LeConte	(Z)-(—)-14-methyl-8-hexadecen-1-ol; (—)-methyl (Z)-14-methyl-8-hexadecenoate[9]	Rodin *et al.* 1969

[1] Pheromone produced by males.

[2] Acts as an aggregating pheromone.

[3] Synergized by host volatiles.

[4] Three related compounds also found in pheromone complex.

[5] Small amount of E isomer needed for maximum activity (Klun *et al.* 1973); E isomer attracts New York strain (Roelofs and Comeau 1971).

[6] Both compounds and certain fatty acids required for field-cage attraction.

[7] Not rigorously shown to be a pheromone; active in field tests.

[8] Small amount of E isomer required for maximum activity (Beroza *et al.* 1973a).

[9] Two other compounds present in pheromone complex.

nology takes time and much patience. The major difficulty here is that each species has its own idiosyncrasies and has to be studied separately and treated differently. Even in using a sex attractant solely for monitoring or survey of populations, the investigator needs to consider a multitude of parameters in his efforts to achieve optimum response; e.g., trap design, trap height, trap placement, type of bait dispenser and its position in the trap, volatility and stability of the attractant, amount of lure, duration of its effectiveness, ratio of ingredients if the attractant is a multicomponent one, effect of aging on the lure, effect of host crop, relationship between trap catch and actual number of insects in the area, effective distance of attraction.

In considering the use of pheromones for direct control, the technology that must be developed is even more complex and again depends heavily on the peculiarities of the insect, the attractant, the insect's habitat or environment, and undoubtedly on such seasonal or transitory factors as the weather. Also, in suppressing one insect species, we may still have to contend with the remaining complex of insect pests.

Rather than attempt to offer generalities or to summarize all USDA's attempts to control insects with pheromones, I will give an in-depth description of efforts made with one insect pest — the gypsy moth, *Porthetria dispar* (L.) — along with brief mention of other projects under investigation. This approach should provide a better insight into the problems involved.

The gypsy moth is a serious pest of forest, shade, and orchard trees in Europe and northeastern United States. About early May, larvae emerge from eggs laid the previous year and begin to consume the leaves of their favored hosts; in heavy infestations, they defoliate and thereby kill many trees. Ultimately the larvae pupate and emerge as adult moths in mid-summer. While the male is a strong flier, the female is heavy with eggs and does not normally fly; instead she advertises her presence by releasing her sex pheromone which lures the male for mating. A mated female usually lays from 200–800 eggs.

In 1960 the use of DDT to combat the gypsy moth was discontinued; with no satisfactory substitute, the infestation has grown steadily and the moth now threatens to spread across the United States.

In 1970, following isolation, identification, and synthesis, the sex pheromone of the gypsy moth was reported by our laboratory to be *cis*-7,8-epoxy-2-methyloctadecane (Bierl *et al.* 1970). Availability of the synthetic pheromone, which is called disparlure, eliminated the costly process of securing the lure from field-collected insects and greatly increased the efficiency of

the USDA gypsy moth survey program. Last year, more than 100,000 disparlure-baited traps were set out to determine the location of moth infestations east of the Mississippi River.

Disparlure can not only exceed the attraction of virgin females (Beroza *et al.* 1971), but when mixed with non-volatile chemicals, it remains highly attractive for the entire mating season. As little as one nanogram suitably formulated, still attracted moths after 3 months in the field (Beroza *et al.* 1971 b). Findings of this kind led us to consider the possibility that disparlure might be useful in the direct control of the gypsy moth and possibly in halting further spread of the insect. Since the chemical is nontoxic and highly specific in its action, its use should not disturb wildlife and should be ecologically acceptable from all standpoints.

Two methods of using disparlure for control have been proposed (Beroza and Knipling 1972). In one, disparlure-baited, mass-produced, inexpensive small cylindrical traps with sticky insides are dropped from airplanes over very lightly infested areas, the object being to capture the males before they can find mates. Assuming each trap to be equal in attraction to a female and the moth population increase to be 10-fold per generation, at least 20 traps would be needed for each female in the area to reduce to a sufficiently low level the chances of males locating females before being trapped. This requirement for a large number of traps limits mass trapping to very light infestations, or to infestations that can be brought to low levels with insecticides or by other means.

Such mass trapping has an intriguing potential advantage over most other control measures. With an insecticide the degree of control that can be achieved remains constant; e.g. with a given treatment 97% of the insects may be killed regardless of whether the insect population is high or low. With traps dispensed at a given rate for each generation, control should theoretically become more efficient with each generation as the population is reduced, because the ratio of traps to females becomes greater, and ultimately it should be possible to eliminate a population completely.

That gypsy moth males can be intercepted before finding females by mass trapping has been demonstrated in a recent out-of-season pilot test (Beroza et al. 1973 c). A 16-hectare plot was left untreated while another such plot was treated with 400 of the small, disparlure-baited, tubular traps (2.5 cm diam. × 7.5 cm long) with sticky insides. Male moths were released weekly in both plots, and the numbers captured in 32 lure-baited monitor traps in each plot were determined. In 6 releases, 195 moths were captured in the monitor traps within the untreated area compared with only 12 moths taken in the monitor traps of the treated area — a 94% suppression of catch.

Even better results were obtained by the second method, called the confusion or air-permeation approach. With it, particles that slowly release disparlure vapor into the atmosphere are dispensed onto the leaves of trees from aircraft. Male moths surrounded by the odor of synthetic disparlure cannot find the natural scent trails that lead to the females, so mating is prevented. This approach is also limited to very light infestations. With denser populations, males may locate females by sight, or perhaps by chance in their searching.

Before tests were conducted in the field, hundreds of formulations of the sex attractant were subjected to weather-ing tests in the laboratory, and the resulting samples were analyzed for the attractant by gas chromatography and by bioassay to determine residual effectiveness. In our efforts to improve effectiveness, constituents of each formulation were varied, e.g., different carriers, volatility regulators (called keepers), thickeners, stickers, and microencapsulated materials were investigated. Adhesion of materials to foliage was also determined when this effect was desired. These laboratory tests were necessary because field trials are very expensive; also only few tests could be conducted each year because the number of lab-reared insects was limited. Thus, only the most promising formulations could be evaluated in the field.

Confusion tests with pheromones in small-size plots (e.g. 16 hectares) were not conducted during the mating season because the pheromone in the area would attract males from surrounding areas into the test site, while such an incursion of males would not occur in the untreated area. The number of males in the treated area would therefore greatly exceed the number in the untreated area, and captures by the monitor traps in the treated and untreated plots would not be comparable. Accordingly, in our first field tests, effectiveness of treatment was determined out of season, i.e., before or after the mating season when no wild insects would be present to interfere with our tests. In such tests, an equal number of laboratory-reared males was released weekly in each plot, and the number of males recaptured in female-baited monitor traps provided a measure of the effectiveness of each treatment.

It is recognized that these out-of-season tests might not give us a final indication of a treatment's effectiveness in a natural infestation. However, the side-by-side comparisons undoubtedly can provide meaningful data as to which of the treatments would be the most effective and the longest-lasting.

In our first out-of-season trial, in 1971, using disparlure on hydrophobic paper as the confusant, results were encouraging; captures by female-baited traps were suppressed effectively for 1 to 3 weeks (Stevens and Beroza, 1972).

In 1972, eleven confusant formulations were evaluated in out-of-season trials (Beroza *et al.* 1973 c). In one of the first tests, in Alabama, a coarse (6/12 mesh) cork coated with disparlure was applied at the rate of 7.5 to 10 grams lure on 0.63 kg cork/hectare; it confused male moths most successfully for 48 days. Monitor traps caught 195 males in the untreated plot vs. 1 in the treated plot. In a repetition of this same test in Massachusetts, the natural habitat of the moth, the lure-coated cork failed. In a concurrent test, a finer mesh (20/40) cork coated with disparlure and sprayed with a sticker did successfully confuse the released moths for about 6 weeks. These results illustrate the advantage of anchoring the lure-emitting particles to the foliage, thereby securing a vertical distribution of the lure through the forest canopy.

Of all the materials tested in 1972, the best results were obtained with micro-encapsulated disparlure formulations. These formulations consisted of water slurries of microcapsules, the capsules being tiny plastic spheres containing solutions of disparlure in xylene. Stickers were included in the water phase to hold the microcapsules on the leaves, so they could slowly release the lure to the atmosphere. Conventional spray equipment was used to dispense the microcapsules. A variety of gelatine-base and nylon capsules were tested in the laboratory, and the best four were field tested with the disparlure being applied at the rate of 2.5 to 5 grams per hectare. With all of the formulations, the suppression of catch (compared to catches in the untreated plots) was generally 90–100% for 6 to 8 weeks.

Results were promising enough to continue our trials, and in 1973 disparlure microcapsules were sprayed over a naturally infested 59-square-kilometer area in Massachusetts at the rate of 5 grams lure per hectare. A similar area in the same vicinity was left untreated and used as a control. These areas were deliberately large to minimize the influx of males into at least the central section of the treated area. (These tests were conducted by the Massachusetts Department of Natural Resources under the direction of C. S. Hood.) In both the treated and untreated areas 100 0.1-hectare plots were monitored as follows:

Early spring (before treatment with disparlure)

 Egg mass counts

June–July (before treatment)

 Counts of larvae and pupae under burlap bands around trees (insects tend to hide under these bands)

July–August (after treatment with disparlure)

 Captures in 10-μg disparlure-baited traps (2 per plot)

 Captures in female-baited traps (2 per plot, females replaced every third day);

 Mating of females placed untethered on trees (2 per plot, replaced every third day)

October–November

 Egg mass counts

At this point, all of the data have not been collected and analyzed. However, the 10-μg pheromone traps caught 2193 males in the control area and 63 in the treated area, for a 97 + % suppression of captures. With female-baited traps, 1136 males were captured in the control area and one in the treated area for nearly 100% suppression of catch. These data clearly show that the pheromone released by the microcapsules is effective in frustrating males in their attempts to locate females in the traps.

Results with the untethered females are somewhat less encouraging. For $2^1/_2$ weeks, mating of the exposed females was markedly suppressed in the treated area. After this time, percentage mating rose for $1^1/_2$ weeks and then dropped again to about half that in the control area for the last week of the test. Thus, it appears that gypsy moth mating in a natural infestation can be suppressed by the insect's pheromone, albeit for a $2^1/_2$-week period. The increased mating that occurred subsequently in the treated area may have resulted for a number of reasons. Too many males may have been present at the outset, outside males may have been drawn into the pheromone-treated area, output of lure may have decreased on aging, or some combination of these factors may have existed. If anything, the data reinforce our stated expectations that the pheromone will only suppress matings effectively in very light infestations where males require the pheromone to find females. Indirectly, the data also highlight the need for a safe insecticide or other means to suppress gypsy moth populations to a low level to allow use of the disparlure confusion approach against high-level infestations.

Efforts to improve the confusant formulation are underway. Ideally the formulation should last the entire mating season with sufficient lead time for application. Although our present formulation seems adequate for very light infestations, we do not have the data to decide on the optimum amount of lure to suppress the moths in any given area, and we recognize that our means of estimating moth populations are inadequate. We can improve performance by increasing the amount of lure applied per hectare and by applying the lure more than once per season, but we hope to avoid the extra expense these alternatives would necessitate by devising better formulations. We also anticipate that a truly integrated approach will be utilized in combating the gypsy moth (insecticides, parasites, viruses, etc., in addition to pheromone). Cost of the pheromone treatment is much less than that of an insecticide treatment. On a very large scale, for example, the pheromone treatment is expected to cost about $ 6.75 per hectare if 15 grams lure per hectare is used; insecticide treatment would cost about three times as much. Despite the lesser cost, use of pheromones to stop the spread of the gypsy moth from the presently infested northeastern United States will be very expensive because very large areas will have to be treated. At the same time, the longer treatment is delayed, the more the insect will spread; in a few years, the infestation may be too large to prevent spread of the moth. Unfortunately, no means other than use of the pheromone has been advanced to stop the spread of the moth, and unchecked, the moth will undoubtedly spread across the country. The approach of permeating the atmosphere with pheromone presupposes the treatment of large extensive areas. I believe that such attempts to achieve area-wide control should and will be given serious consideration in the future as the most economical means of coping with mobile insect pests. Co-existence of an infestation near a treated area should not be tolerated.

In discussing the gypsy moth problem, I have — because of lack of time — been able to describe only some of the work underway. Other work (including basic studies on the insect) is being conducted in cooperation with Pennsylvania State University and Canada and by some States and Universities.

In a related endeavor, we have tried to reduce mating with (Z)-2-methyl-7-octadecene, which inhibits response of gypsy moth males to traps baited with either disparlure or virgin gypsy moth females (Cardé et al. 1973). The chemical was dispensed in a slow-release microcapsular formulation similar to that used with disparlure, but mating was not suppressed sufficiently to consider this approach promising for the gypsy moth. Thus far, the inhibitor-type chemicals when broadcast have failed to disrupt communication between the sexes, not only for the gypsy moth but for several other species [codling moth, Laspeyresia pomonella (L.); pink bollworm, Pectinophora gossypiella (Saunders); cabbage looper, Trichoplusia ni (Hübner)].

Concurrent with the foregoing trials, similar experiments have been conducted with other species. Microencapsulated pheromones of the codling moth and of the oriental fruit moth, Grapholitha molesta (Busck), were found to suppress trap captures of the respective species to a low level when compared to captures in untreated areas (H. R. Moffitt and C. R. Gentry; unpublished).

Notable among the field trials and a model of the application of integrated control techniques is the recently completed cooperative pilot boll weevil (Anthonomus grandis Boheman) eradication experiment covering an area of 40-kilometer radius around Columbia, Mississippi. The experiment relied upon the cumulative impact over a 2-year period of several suppressive procedures: late-season insecticide treatment, cultural practices, pheromone traps and trap crop planting, and the release of sexually sterile male boll weevils. Ways to utilize these procedures more effectively were developed during the course of the experiment, and much information was accumulated. Based on the results obtained, a Technical Guidance Committee concluded that elimination of the boll weevil as an economic pest in the United States by use of the aforementioned techniques was technically and operationally feasible and that economic and environmental benefits of achieving this goal will far exceed the costs involved. Since about a third of the insecticide used in the United Staes is used in control measures against the boll weevil, this program, if adopted and successful, could significantly reduce the pesticide load entering the environment. Needless to state, the synthetic pheromone, grandlure (Tumlinson et al. 1969) played an important role in this undertaking both as a suppressant of the insect and as a means of monitoring progress in this suppression.

Many significant advances in the finding and use of pheromones are also being made outside USDA, in the research facilities of the various states and universities, and in quite a few countries. The industrial community is also beginning to participate by making available different attractants, traps, and related services.

In exploring our future options, it would be a serious misjudgment to presume that the pheromone and related chemicals are now ready to replace insecticides and other control measures. I fully expect insect attractants to be an important tool in helping guide the use of insecticides and other control measures more effectively. They should fit nicely into integrated control programs, and they should upgrade our control efforts; at the same time they can be expected to cut down on the amount of pesticide needed for pest management. In isolated instances, they may even help eliminate an injurious species from a limited area.

In conclusion, then, progress in insect-attractant technology is being made, and this technology is setting the stage for some new and safer approaches to the control of some important insect pests. However, as valuable as they may be, insect attractants and related chemicals should not be regarded as a panacea. They have an important role to play in helping us attain our long-range goal of managing insects with a minimum of pesticide pollution; but in most instances, attractants or pheromones will have to be adopted as part of integrated control programs, very likely over large areas, to achieve their full potential.

References

Barber, I. A., T. P. McGovern, M. Beroza, C. P. Hoyt, and A. Walker: *J. Econ. Entomol.* 64, 1041–4 (1971).

Berger, R. S.: *Ann. Entomol. Soc. Amer.* 59, 767–71 (1966).

Berger, R. S., and T. D. Canerday: *J. Econ. Entomol.* 61, 452–4 (1968).

Beroza, M., B. H. Alexander, L. F. Steiner, W. C. Mitchell, and D. H. Miyashita: *Science* (Washington, D.C.) 131, 1044–5 (1960).

Beroza, M., B. A. Bierl, E. F. Knipling, and J. G. R. Tardif: *J. Econ. Entomol.* 64, 1527–9 (1971 a).

Beroza, M., B. A. Bierl, J. G. R. Tardif, D. A. Cook, and E. C. Paszek: *J. Econ. Entomol.* 64, 1499–1508 (1971 b).

Beroza, M., N. Green, S. I. Gertler, L. F. Steiner, and D. H. Miyashita: *J. Agr. Food Chem.* 9, 361–5 (1961).

Beroza, M., and E. F. Knipling: *Science* (Washington, D. C.) 177, 19–27 (1972).

Beroza, M., G. M. Muschik, and C. R. Gentry: *Nature New Biol.* 244, 149–50 (1973 a).

Beroza, M., A. A. Punjabi, and B. A. Bierl: *J. Econ. Entomol.* 66, 1215–6 (1973 b).

Beroza, M., L. J. Stevens, B. A. Bierl, F. M. Philips, and J. G. R. Tardif: *Environ. Entomol.* 2, 1051–7 (1973 c).

Bierl, B. A., M. Beroza, and C. W. Collier: *Science* (Washington, D. C.) 170, 87–9 (1970).

Brady, U. E., J. H. Tumlinson, R. G. Brownlee, and R. M. Silverstein: *Science* (Washington, D. C.) 171, 802–4 (1971).

Cardé, R. T., W. L. Roelofs, and C. C. Doane: *Nature* (London) 241, 474–5 (1973).

Carlson, D. A., M. S. Mayer, D. L. Silhacek, J. D. James, M. Beroza, and B. A. Bierl: *Science* (Washington, D. C.) 174, 76–8 (1971).

Davis, H. G., R. J. Peterson, W. M. Rogoff, T. P. McGovern, and M. Beroza: *Environ. Entomol.* 1, 673–4 (1972).

Granges, J., and M. Baggiolini: *Rev. Suisse Vitic. Arboric.* 3, 93–4 (1971).

Henderson, H. E., F. L. Warren, O. P. Augustyn, B. V. Burger, D. F. Schneider, P. R. Boshoff, H. S. Spies, and H. Geertsema: *J. Insect Physiol.* 19, 1257–64 (1973).

Henzell, R. F., and M. D. Lowe: *Science* (Washington, D. C.) 168, 1005–6 (1970).

Hummel, H. E., L. K. Gaston, H. H. Shorey, R. S. Kaae, K. J. Byrne, and R. M. Silverstein: *Science* (Washington, D. C.) 181, 873–5 (1973).

Jacobson, M., C. E. Lilly, and C. Harding: *Science* (Washington, D. C.) 159, 208–10 (1968).

Jacobson, M., K. Ohinata, D. L. Chambers, W. A. Jones, and M. S. Fujimoto: *J. Med. Chem.* 16, 248–51 (1973).

Kinzer, G. W., A. F. Fentiman, Jr., T. F. Page, Jr., R. L. Foltz, J. P. Vité, and G. B. Pitman: *Nature* (London) 221, 447–8 (1969).

Klun, J. A., and T. A. Brindley: *J. Econ. Entomol.* 63, 779–80 (1970).

Klun, J. A., O. L. Chapman, K. C. Mattes, P. W. Wojtkowski, M. Beroza, and P. E. Sonnet: *Science* (Washington, D. C.) 181, 661–3 (1973).

Klun, J. A., and J. F. Robinson: *Ann. Entomol. Soc. Amer.* 65, 1337–40 (1972).

Kuwahara, Y., C. Kitamura, S. Takahashi, H.

Hara, S. Ishii, and H. Fukami: *Science* (Washington, D. C.) *171*, 801–2 (1971).

McGovern, T. P., M. Beroza, T. L. Ladd, Jr., J. C. Ingangi, and J. P. Jurimas: *J. Econ. Entomol. 63*, 1727–9 (1970 a).

McGovern, T. P., B. Fiori, M. Beroza, and J. C. Ingangi: *J. Econ. Entomol. 63*, 168–71 (1970 b).

Meijer, G. M., F. J. Ritter, C. J. Persoons, A. K. Minks, and S. Voerman: *Science* (Washington, D. C.) *175*, 1469–70 (1972).

Nesbitt, B. F., P. S. Beevor, R. A. Cole, R. Lester, and R. G. Poppi: *Nature, New Biol. 244*, 208–9 (1973).

Pitman, G. B., and J. P. Vité: *Ann. Entomol. Soc. Amer. 63*, 661–4(1970).

Pitman, G. B., J. P. Vité, G. W. Kinzer, and A. F. Fentiman, Jr.: *Nature* (London) *218*, 168–9 (1968).

Rodin, J. O., R. M. Silverstein, W. E. Burkholder, and J. E. Gorman: *Science* (Washington, D. C.) *165*, 904–6 (1969).

Roelofs, W. L., and H. Arn: *Nature* (London) *219*, 513 (1968).

Roelofs, W. L., R. Cardé, G. Benz, and G. von Salis: *Experientia 27*, 1438–9 (1971).

Roelofs, W. L., and A. Comeau: In "Chemical Releasers in Insects; Proc. IUPAC 2nd Internatl. Congr Pest. Chem.", A. S. Tahori, ed., pp. 91–112, Gordon and Breach Science Publishers, London, 1971.

Roelofs, W., A. Comeau, A. Hill, and G. Milicevic: *Science* (Washington, D. C.) *174*, 297–9 (1971).

Roelofs, W. L., A. Comeau, and R. Selle: *Nature* (London) *224*, 723 (1969).

Roelofs, W. L., and J. P. Tette: *Nature* (London) *226*, 1172 (1970).

Roelofs, W. L., J. P. Tette, E. F. Taschenberg, and A. Comeau: *J. Insect Physiol. 17*, 2235–43 (1971).

Röller, H., K. Biemann, J. S. Bjerke, D. W. Norgard, and W. H. McShan: *Acta Entomol. Bohemoslov. 65*, 208–11 (1968).

Schönherr J.: *Z. Angew. Entomol. 71*, 260–3 (1972).

Silverstein, R. M., R. G. Brownlee, T. E. Bellas, D. L. Wood, and L. E. Browne: *Science* (Washington, D. C.) *159*, 889–91 (1968).

Silverstein, R. M., J. O. Rodin, W. E. Burkholder, and J. E. Gorman: *Science* (Washington, D. C.) *157*, 85–7 (1967).

Silverstein, R. M., J. O. Rodin, and D. L. Wood: *Science* (Washington, D. C.) *154*, 509–10 (1966).

Smith, R. G., G. E. Daterman, G. D. Daves, Jr., K. D. McMurtrey, and W. L. Roelofs, *J. Insect Physiol. 20*, 661-8 (1974).

Steiner, L. F.: *J. Econ. Entomol. 45*, 241–8 (1952).

Steiner, L. F., W. G. Hart, E. J. Harris, R. T. Cunningham, K. Ohinata, and D. C. Kamakahi: *J. Econ. Entomol. 63*, 131–5 (1970).

Steiner L. F., W. C. Mitchell, E. J. Harris, T. T. Kozuma, and M. S. Fujimoto: *J. Econ. Entomol. 58*, 961–4 (1965).

Steiner, L. F., G. G. Rohwer, E. L. Ayers, and L. D. Christenson: *J. Econ. Entomol. 54*, 30–5 (1961).

Stevens, L. J., and M. Beroza: *J. Econ. Entomol. 65*, 1090–5 (1972).

Tamaki, Y., H. Noguchi, T. Yushima, and C. Hirano: *Appl. Entomol. Zool. 6*, 139–41 (1971 a).

Tamaki, Y., H. Noguchi, T. Yushima, C. Hirano, K. Honma, and H. Sugawara: *Kontyu 39*, 338–40 (1971 b).

Tumlinson, J. H., D. D. Hardee, R. C. Gueldner, A. C. Thompson, P. A. Hedin, and J. P. Minyard: *Science* (Washington, D. C.) *166*, 1010–2 (1969).

Tumlinson, J. H., E. R. Mitchell, S. M. Browner, and D. A. Lindquist: *Environ. Entomol. 1*, 466–8 (1972).

Weatherston, J., W. Roelofs, A. Comeau, and C. J. Sanders: *Can. Entomol. 103*, 1741–7 (1971).

Use and Field Application of Pheromones in Orchard Pest Management Programs *)

By **Kenneth Trammel**

New York State Agricultural Experiment Station, Cornell University,
Geneva, New York, U.S.A.

Concern over the extensive use of chemical pesticides in controlling orchard pests has stimulated increased interest in developing alternative methods. Near total reliance on chemicals has given way in recent years to the concept of "integrated control". As the term implies, two or more control methods are combined in a pest management program. In practice, integrated programs are usually developed through modification of the chemcal insecticide program to favor natural predators of phytophagous mites. Most orchard insect pests, especially fruit-infesting species, are not adequately controlled by natural enemies, thus other alternatives must be sought. Identification of the sex pheromones or sex attractants of several major orchard insect pests has provided a new tool that shows promise in supplementing chemicals in orchard pest management. These species, all tortricid moths, include codling moth *(Laspeyresia pomonella)*, oriental fruit moth *(Grapholitha molesta)*, lesser appleworm *(Grapholitha prunivora)*, redbanded leafroller *(Argyrotaenia velutinana)*, obliquebanded leafroller *(Choristoneura rosaceana)*, 3-lined leafroller *(Pandemis limitata)*, tufted apple bud moth *(Platynota idaeusalis)*, fruit

tree leafroller *(Archips argyrospilus)*, grape berry moth *(Paralobesia viteana)*, summer fruit tortrix *(Adoxophes orana)*, and *Clepsis spectrana*.

The potential of sex pheromones in orchard pest management can be divided into two major roles. The first and, perhaps, most eminent role is that of pest population monitoring and assessment. The second role is the use of pheromones for direct control of pest populations through disruption of the mating process. Both applications are under evaluation at the New York State Agricultural Experiment Station.

Population Monitoring

Precicion in orchard pest management programs is relative to the accuracy in assessing pest population pressure. Efficiency in pesticide usage is greatly increased if application is based on actual need rather than on a strictly preventive schedule. To determine the need for pesticides, a reliable system for monitoring population levels is required. Since pheromones are one of the most potent and specific attractants available for insect detection, our current research program is attempting to develop a pheromone-based pest monitoring system that will provide a sound basis for applying chemical or other control measures as needed.

In developing a pheromone monitoring system, several factors must be con-

*) This paper was presented at the Third International Symposium on Chemical and Toxicological Aspects of Environmental Quality, Tokyo, Japan, November 19–22, 1973

sidered for each pest species. These include pheromone purity, synergism, pheromone dispenser, release rate, distance of pheromone attractiveness under varying climatic conditions, trap design and trap placement. Furthermore, an economic threshold must be determined for each pest species and, finally, trap catch must be related to this economic threshold as a basis for sound pest management decisions.

Experience has shown that pest population levels vary from orchard to orchard and pest problems can be either indigenous or exogenous depending on orchard conditions and surroundings. Orchards with a history of good pest control practices are likely to be relatively pest-free and the influx of pests from outside the orchard is a primary concern. Orchards under marginal or poor pest control can be plagued by both internal and external pest pressure. The monitoring and detection system should indicate the source of pest problems.

Bionomics of individual pests must be considered. Redbanded leafroller has a very broad host range (Chapman and Lienk, 1971) and capture of male moths in pheromone traps is usually quite high. Both internal and peripheral trapping are necessary in the assessment of the problem. Codling moth, on the other hand, has a very narrow host range and relatively low trap catches are common in our area. Proximity to abandoned orchards and wild apple trees influence codling moth pressure on the monitored orchards and monitoring traps should be placed accordingly. Knowledge of the host range of pest species and proximity of alternate hosts to the orchard is essential to successful application of pheromones in monitoring and control programs.

Our research program over the last 3 years has yielded a system of peripheral and internal orchard monitoring that is experimentally feasible, based on the unit area effectively covered by the pheromone trap. After 2 years research with higher trap densities, the most efficient and practical unit area appears to be 10 acres (ca 4.0 ha) and monitoring stations are placed accordingly, depending on the size and shape of the orchard, farm or area being monitored. However, a minimum of 6 monitoring stations (4 peripheral and 2 internal) are used at any given monitoring site so that direction of pest movement can be detected. This system is used in 9 different orchards and insecticide output has been reduced ca 25—40% from previously applied programs.

We now have a large area monitoring system set up over ca 3000 acres in conjunction with the New York State Apple Pest Management Program. Pheromone monitoring stations are set up on a 10-acre grid system over the entire area. Areas of intense pest activity have been detected and sources of infestation identified. This information will be used as a basis for ridding the area of wild and abandoned apple trees or other host plants or in localized chemical abatement programs. It is expected that total chemical output can be significantly reduced from current levels applied in routine preventive programs, without sacrificing fruit quality.

Research in pheromone monitoring of insects is far from complete but the potential use of sex attractants in assessment of certain pest problems is very promising. The tortricid moths are an important part of the total orchard pest spectrum and pheromones in conjunction with detection devices for other pests can provide a reliable monitoring system that will indicate actual orchard conditions to the pest management specialist.

Direct Control

The role of sex pheromones in perpetuation of a given species provides a vulnerable point in its life cycle. In the tortricid moths, the pheromone is emitted by the female to attract the male for mating. This communication system can be altered to prevent sexual encounter. One way is to lure the male moths to an artificial source of pheromone and remove them from the population prior to mating (mass-trapping). Another approach would be to saturate the environment with a sufficient level of pheromone to prevent male orientation on any specific pheromone source, i.e. virgin female (confusion). A further possibility is based on the fact that after males react to a pheromone stimulus, further response is suppressed for some time afterwards (habituation). An artificial source of pheromone could accomplish just this, either as a constant exposure to a high level of pheromone or by periodic release into the environment. Confusion and habituation are considered as one under the heading of mating inhibition because of the difficulty in distinguishing one from the other in field tests. These techniques are currently under investigation in our orchard pest management research.

Mass-trapping. — Redbanded leafroller (RBLR) is an important pest of apples in the eastern U.S. and was the intital orchard pest species on which pheromone research was conducted at our laboratory. The pheromone was identified (Roelofs and Arn, 1968) as cis-11-tetradecenyl acetate. The synthetically prepared chemical, "RiBLuRe"

(Roelofs and Comeau, 1968), was used in the initial mass-trapping test conducted in the summer of 1968 (Glass et al., 1970). A heavily-infested 20-acre orchard was blanketed with 1700 pheromone traps. Although the pheromone traps captured ca 4000 males, larvae developed abundantly and fruit injury was extensive. Traps baited with live female moths captured males also, indicating that male-female encounter was occurring. Concurrently, synergism was discovered as an important factor in increasing the attractancy of RiBLuRe (Roelofs and Comeau, 1968, 1971). A 60 : 40 solution of dodecyl acetate: cis-11-tetradecenyl acetate was far more attractive than RiBLuRe alone and the synergized attractant has been used in all subsequent trapping work.

Mass-trapping experiments were continued in 1969 in 2 tests under both high and low population pressures. Failure was encountered under the high pest pressure but successful control was obtained under low pressure and it was concluded that mass-trapping for control would be practical only under relatively low population levels. This principle has been adhered to in subsequent mass-trapping control experiments and for the past 3 years RBLR has been commercially controlled in 2 orchards totalling ca 65 acres. Currently, we are evaluating a multiple-species mass-trapping program for control of codling moth, oriental fruit moth, lesser appleworm, RBLR, obliquebanded leafroller and 3-lined leafroller.

Mating Inhibition. — This technique of pheromone control is currently under evaluation against two pest species, RBLR and codling moth. The RBLR test is applied to ca 40 acres of apple orchard. Pheromone dispensers are deployed 1/tree, releasing > 100 μg/hr throughout the moth flight period. In the area of pheromone release male capture by pheromone traps was reduced $> 95\%$. Traps baited with live females failed to capture male moths. Fruit and foliage inspection reveal only a trace of injury, indicating excellent control. The codling moth test is currently under evaluation and results are not yet available but reduction in capture of male moths is apparent.

Summary

Pheromones are recognized as potentially valuable tools in orchard pest management. Their most immediate

practical application is in pest monitoring to indicate the need for chemical control. In this role, they can have considerable impact in reduction of pesticide output.

References

Chapman, P. J., and S. E. Lienk (1971): Special Publication of the New York State Agric. Expt. Sta. 142 pp.

Glass, E. H., W. L. Roelofs, H. Arn, and A. Comeau (1970): J. Econ. Entomol. 63: 370 to 373.

Roelofs, W. L., and H. Arn (1968): Nature 219: 518.

Roelofs, W. L., and A. Comeau (1968): Nature 220: 600–601.

Roelofs, W. L., and A. Comeau (1971): J. Insect Physiol. 17: 435–448.

Approaches to Insert Control Based on Chemical Ecology — Case Studies

Izuru Yamamoto

Department of Agricultural Chemistry, Tokyo University of Agriculture, Setagaya, Tokyo, Japan

Summary

In this article two attempted approaches to control a stored product insect, azuki bean weevil, are discussed based on host selection study and oviposition ecology. The latter study indicated the presence of an oviposition marker, a new kind of pheromone, which was derived from the weevil and was lipid in nature, and suggested its possible use as an oviposition inhibitor.

Introduction

Control of stored product insects has been carried out by fumigation with chlorpicrin, methyl bromide, and phostoxin, or by the use of pyrethrins with piperonyl butoxide, γ-BHC, malathion, and DDVP. The use of such toxicants causes problems of contamination of food, deterioration of food quality, and local air pollution when the fumigants are released into the air from the warehouse; whereas the need for pest control increases after recent introduction of new storage and packing devices. Protection from stored product insects without the use of toxicants will be an important area of study. Pest control is basically an ecological problem of great complexity which requires a multidisciplinary approach. The following discussion is related to attempted case studies to control the azuki bean weevil, *Callosobruchus chinensis* L., based on a blending of chemistry and population ecology, an area of chemical ecology.

Host Selection and Growth Inhibitory Factor

The azuki bean weevils lay eggs on azuki beans, *Phaseolus angularis*, and cause a marked injury when the insects develop by feeding inside of the beans. They also lay eggs on various other kinds of beans which have a smooth surface with the appropriate curvatures, even though the bean may not be suitable for further development of the hatched larvae.[1] A typical example is the ingen bean or kidney bean, *Phaseolus vulgaris*, which belongs to the same genus as the azuki bean but does not provide a suitable media for feeding inside of the bean. The reason for the unsuitability of the kidney bean for larval development was studied by Ishii [1] and traced to the presence of a water-soluble growth inhibitory factor in the bean. Though unidentified, the factor must be edible by man because we have eaten the kidney bean for a long time. Accordingly, incorporation of such a factor into the component of the azuki bean was attempted by Umeya

*) This paper was presented at the Third International Symposium on Chemical and Toxicological Aspects of Environmental Quality, Tokyo, Japan, November 19–22, 1973

and Imai [2] by grafting the azuki plant to the kidney bean plant. Against the expectation, the bean seeds obtained from the azuki part of the grafted plant were edible by the weevil. Breeding has been considered as a next attempt. Someday this interesting approach will be realized.

Population Ecology and Oviposition Marker

Another approach is based on the population ecology of oviposition of the azuki bean weevil. In 1943 Utida [3] published a paper where he arranged about 120 azuki beans in a petri dish and released the males and females of the azuki bean weevil. With a higher density of the weevils, 32 pairs, they laid their eggs statistically randomly on the beans, and the distribution followed approximately poisson distribution. However, in the case of one pair, the female laid one egg on each bean. Nakamura [4] also found that the weevils, after finishing oviposition on about 90% of the beans (one egg on each bean), started further oviposition on the same beans. Utida [3] conceived that "a female weevil has a habit of depositing an egg on a bean and distributing the eggs as evenly as possible. Whether the number of eggs deposited on a bean is controlled by the discriminating ability of sight, or by that of smell, or by that of touch is not known yet." Then Yoshida [5] found that the weevils conditioned the beans with a substance or substances during creeping and egg-laying behavior and preferred fresh or less-conditioned beans for oviposition. In collaboration with Oshima and Honda, we found the same phenomenon and mated females were allowed to lay their eggs among (a) fresh beans (ether-washed); (b) conditioned beans (egg scraped off); (c) conditioned, then ether-washed beans; and (d) beans treated with the ether-washing. They selected (a) and (c) almost exclusively for oviposition. The result indicated the presence of an ether-extractable conditioning substance on the surface of the beans. The substance is called "oviposition marker" in the following discussion.

Isolation and Identification of Oviposition Marker

First we developed a sensitive bioassay method which required 1 to 10 μg substance on one bean. Ten mated females of the azuki bean weevil were provided with the choice of laying eggs on 10 ether-washed control beans or 10 sample-coated beans during one hour. The result is expressed by n_s/n_c, where the total egg number on the sample beans is n_s and that on the control beans is n_c. We used the term "active" when there was a significant difference between n_s and n_c at the 0.05 level.

Next, production of large amounts of starting material was obtained by keeping the weevils with glass balls in a large petri dish. The marking activity was obtained from the beans infested by the weevils, but it was more convenient and advantageous to use glass balls because the active extract from the glass balls could not contain substance directly derived from the beans. A scanning electron microscope picture [7] of the surface of the glass balls (3000 x) showed a pile of waxy substance. Washing the glass balls with ether gave an active extract, which was further fractionated into the acidic,

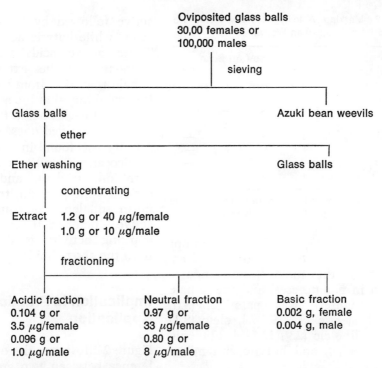

Fig. 1 Fractionation of Oviposition Marker.

neutral, and basic fractions as shown in Figure 1.

The extract derived from the male and the female gave about the same activity at the same dose level (10 μg/bean), but the female released about 4 times more activity than the male. The activity was distributed about equally between the acidic and neutral fractions for the female, but was mostly in the neutral fraction for the male. This was due to the difference in the chemical composition of each fraction. Further fractionation and identification by column chromatography, thin-layer chromatography and gas chromatography indicated the lipid nature of the active components. Table 1 summarizes the composition of the extract from the male and the female. The main activity was found in the hydrocarbon, triglyceride, and fatty acid fractions for the female but only in the hydrocarbon fraction for the male.

The hydrocarbon fraction gave the same gas chromatographic pattern for the male and the female but was not yet characterized. The fatty acid composition of the triglyceride fraction obtained from the female was myristic (1.0%), palmitic (23.8%), stearic (16.3%), oleic (47.5%), linoleic (8.6%),

Table 1 Composition of Oviposition Marker Released from Azuki Bean Weevil.

Component	Composition %	
	Female	Male
Hydrocarbons	32.3	80.2
Triglycerides	44.1	
Diglycerides	9.2	5.1
Monoglycerides	1.5	1.1
Fatty acids	12.7	13.2
Basic substance	0.2	0.4

Table 2 Marking Activity of Fatty Acids Released from Azuki Bean Weevil.

Acids	Dose (μg)/Bean			
	100	10	1	0.1
Palmitic	+	—	—	—
Stearic	+	—	—	—
Oleic	+	+	—	—
Linoleic	+	+	—	—
Linolenic	+	+	+	—
Butyric	—	—	—	—
Mixed*)	+	+	—	—

*) Fatty acids mixed in the ratio found in the acidic fraction from the female weevil.

linolenic (1.9%), and other minor (0.9%) acids. The free fatty acid fraction from the male (former figures shown in parentheses) and the female (latter figures) was palmitic (3.0; 11.4%), stearic (3.7; 9.3%), oleic (8.1; 28.8%), linoleic (2.9; 12.5%), linolenic (none; 5.0%), and butyric and minor (66.4; 33.0%) acids.

Table 2 gave the activity of individual fatty acids; linolenic acid was the most active followed by linoleic and oleic acids, while butyric acid was inactive. These active acids were far more abundant in the extract from the female than that from the male. There is a possibility that the active substance concealed itself behind these lipids and not the identified ones. However, such activity was found in many lipids such as eicosane, caprylic acid, capric acid, corn oil, triolein, and l-caprylogly-ceride, while glycerin, triacetin, tripal-mitin, squalene, and squalane were inactive. Therefore, it is conceivable that marking activity is derived directly from the above lipids.

Implication to Practical Application

Figure 2 illustrates the oviposition preference between two groups. Without any conditioning, the weevils lay eggs uniformly among beans or glass balls, but prefer beans to glass balls when both are provided. When all the beans provided are conditioned by the natural marker, the weevils reluctantly lay eggs on the conditioned beans. However, almost one-sided preference is given for the unconditioned ones between the following pairs: unconditioned bean vs. conditioned bean; unconditioned glass ball vs. conditioned glass ball; unconditioned glass ball vs. conditioned bean. The observed fact that the weevils laid eggs on glass balls which were mixed among the marked beans may constitute the basis for protecting the beans from infestation by the weevils, because the larvae hatched on the glass balls will die. These lipids are known to be non-toxic and edible by man. Furthermore, capric acid was so active that it prevented oviposition on the beans even without any glass balls. In

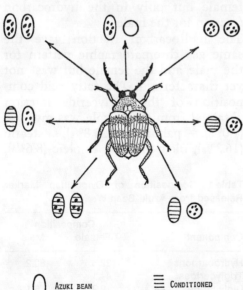

○ AZUKI BEAN
○ GLASS BALL
≡ CONDITIONED
⋰ OVIPOSITED

Fig. 2

another experiment, 20 pairs of the male and the female were released among 250 azuki beans coated with the natural marker (about 300 μg/bean). No development of the adult weevils was observed (from av. 4 eggs/bean), whereas 500 adults emerged from the uncoated beans (from av. 20 eggs/bean) under the same conditions.

It will require more investigations for such a protective approach to be practically utilized. Encouragingly, there is a successful case for another chemical, chlorphenamidine, which protects the rice plant in the field from the injury of the rice stem borer primarily by its repellent action.

Conclusion

Although insecticides will remain as a major means to control pest insects for years to come, the input of chemistry into the understanding of insect behavior and the broadening of our scope from "insecticidal" to "protective" may develop desirable ways to deal with specific problems.

References

1 Ishii, S.: Studies on the host preference of the cowpea weevil (Callosobruchus chinensis L.). Bull. Nat. Inst. Agric. Sci. Ser. C, No. 1: 185–256 (1952).
2 Umeya, K., E. Imai: Growth of the azuki bean weevil (Callosobruchus chinensis L.) and the Mexican bean weevil (Zabrotes subfasciatus Boh.) on beans of grafted Phaseolus plant. Jap. J. appl. Ent. Zoo. 9: 238–246, 1965.
3 Utida, S.: Studies on experimental population of the azuki bean weevil, Callosobruchus chinensis (L.). VIII. Statistical analysis of the frequency distribution of the emerging weevils on beans. Mem. Coll. Agric. Kyoto Imp. Univ. No. 54: 1–22 (1943).
4 Nakamura, H.: A comparative study on the ovipositional behavior of two species of Callosobruchus (Coleoptera: Bruchidae). Jap. J. Ecol. 18: 192–197, 1968.
5 Yoshida, T.: Oviposition behaviors of two species of bean weevils and interspecific competition between them. Mem. Fac. Lib. Arts Educ. (Nat. Sci.) Miyazaki Univ. 11: 41–65, 1961.
6 Oshima, K., H. Honda, I. Yamamoto: Isolation of an oviposition marker from azuki bean weevil, Callosobruchus chinensis (L.). Agr. Biol. Chem. 37: 2679–2680 (1973).
7 Taken by courtesy of JEOL Co. Ltd. on a JEOL scanning electron microscope.

Embryological Aspects of Environmental Toxicology

Relation between Environmental Protection and the Successful Conservation of Species in the View of Experimental Embryology

By **Jan W. Dobrowolski**

The experimental datum concerning, the influence of various physical and chemical factors; on the development of vertebrates, invertebrates and higher plants — indicate, that the early ontogenesis is characterized by a much greater sensitiveness, than the later periods of development.(1, 2, 3, 4, 5, 6) Let me stress one fact, that usually the most sensitive to environmental influence, there are early, but not earliest stages of embryogenesis. It is comprehensible on the background of developmental genetic and theory of information.(7)

But at all during embryogenesis is the highest sensibility. Although it is rather general regularity, it is not taken into consideration, while settling standards of e.g. permissible concentrations of chemical pollutants in water. The *experimental embryology* can provide *the most sensitive biological criterion* for standardization of many kinds of the pollutants of fresh and salt water, as well as some others environmental pollutions, on international scale.

The laboratory experiments indicate also that the simultaneous action of various chemical pollutants, changes of temperature and many others environmental agents — caused lethal and teratogenic effects quite different, than the same factors — at similar intensities e.g. concentrations of pollutants —, but when there were influenced separately. (1, 3)

Meanwhile, the standards are based on considerably less sensitive tests e.g. microbiological, and using the adult individual of some indicatory species. There do not take into consideration, either phenomenon of the synergistic effects of many chemical pollutants as well as coincident physical factors in natural environment.

Very interesting problem may be analize on exampl of the gradual extinction of some species e.g. such birds as marine eagles on the Alands Islands. Despite the excellent protection of natural reserve this process is occuring, as a result of the lack of environmental protection — e.g. sea water — against chemical pollutions. The adult individuals are still able to survive, but all the embryos die.(4)

It is very dangerous that concentration of some pesticides and heavy metals, just in eggs, is very high.(8, 9) Some of pesticides are metabolized by developing embryos of birds.(10) Pesticides can also make eggshell to thin for hutching out them.(11)

As it is known from very interesting Korte's publications, including wide review of literature, there are not datum about the influence of some commonly used pesticides on embryonic development, and same time also about wide life aspect.(12, 13) Thus there arises an urgent necessity for a considerable intensification of the research in this field of experimental embryology, particularly dealing with the species used as indicators of water quality.

It seems to be one of important factors supporting successful preservation of nature, by protection of proper environmental quality. So this studies should include the influence of different levels of chemical pollutants, both actually existing, as well as these expecting in the neare future.

A good exemples of investigation in this field, there are e.g. curred out by: Kinne and Rosenthal (14), about the influence of sulfuric water pollutants on fertilization, embryonic development and larvae of the herring, and by Mileikovsky (15), about the influence of some chemical pollutions on pelagic larvae of bottom invertebrates in sea.

The risks of teratogenicity of some embryotoxic herbicides in human civilian use, which discussed Båge et all. (16), should be stimulating factor for future similar studies.

Investigation about molecular mechanisms of stage-dependent difference in sensitibility, are very important, but it seems to be also necessary to curry out research on organismal level.(17) It may be useful for explanation of some changes observed in populations and biocenoses.

The complex studies, including embryological data are necessary for protection of social health as well as the proper reproduction of the resources of living nature.

Hagström and Lönning propose the utilization of sea urchin egg for testing of the teratogenic effects exerted by drugs — on example of 4 substances —.

In general it is possible to stress, that embryological study in environmental toxicology, are not sufficiently development, neither they are adequate exposed in the actual international programmes on the environment. It is condition sine qua non, for successful protection of species; to obtain relevant datum from laboratory as well as from different areas, and used both them as criteria for verification and improvement of the technological methods and other protective activity, in practice. It appears that a great deal of chemical compounds, which pollute environment more and more frequently just in concentrations indicated, in laboratories; teratogenic effects similar to those caused by commonly known pathogenic factors, such as e.g. ionizing radiation. Experimental results implicate some conclusions; about the international standardization of the levels of pollutants — i.e. denote permissible concentration on the background of the most sensitive biological criterions — and the next about the global monitoring system, alternative technology, as well as for recommendation and supporting permanent improvement of scientific and practical cooperation in this field. Several hundreds of new chemical compounds are added each year in the environment, but no one is testing in all biological and medical aspects, like teratogenic, mutagenic, long-term lethal and others effects.(19) The experimental data are not collected in one center and quickly distributed. It is an occasion for many dramatical even situations. So it seems to be necessary to realize in the neare future, the SCOPE Registry Commission's proposition about the creation by UN a global computerized system for "International Registry of Data about Chemicals in the Environment" / January 1972 /.

In my opinion it would be also very useful to publish every year, international bulletin including information about actual and planned for the neare future, studies in this field, as well as addresses of scientists interested in cooperation on similar problems. It may be a first step for organizatory facilitation of increasing the international exchange of information and scientific cooperation on regional and global scale. Otherwise, the innate abnormalities quite similar to those caused by "Thalidomide", may became widespread and the future generations will evaluate the behavior and the degree to which we made use with the actual prophylactic possibilities: by integrated, interdisciplinary studies, as well as by popularisation of results among public and by practical implication. During II session of the International Coordinatorial Council Programme "Man and the Biosphera" it was postulated i.e. elaborating methods of early diagnosis of embryonic malformations but the most important it is successful preventive activity. For this reason it seems to be necessary to increase investigation about the influence of environmental factors on the most labile period of development, during embryogenesis. These datum, together with the results of other biological and medical studies, should be utilise for complex: educational, technological, planistic, jurisdical etc. activities in practice. Implications from interdisciplinary research should be wider discussed and taking into acount by decision-making man — including degree of utilization of better knowledge of permissible changes and new technological and others possibilities.

More efficient solving of essential for developed countries, problem of pollution — implicate modernisation of education system, and preparing of students for wide scientific and practical cooperation.(20)

References

1 Dobrowolski, J. (1971): Wplyw zasolenia i p_H na rozwój plazów Kosmos, s. A, z. 5.
2 Dobrowolski, J., (1971): Wplyw promieni jonizujacych na rozwój plazów, Kosmos, s. A., z. 6.
3 Dobrowolski, J. (1972): Wplyw temperatury i cisnienia na rozwój plazów, Kosmos, s. A, z. 1.
4 Dobrowolski, J. (1972): Zarodki sa najwrazliwsae, Problemy, nr. 8.
5 Dobrowolski, J. (1973): The most sensitive and the critical stages in early development of Xenopus laevis, in relation to some chemical factors (NaCl, KCl, LiCl, CaCl$_2$, MgCl$_2$, some of detergents, phenol and DDT) — in preparation.
6 Dobrowolski, J., (1973): The most sensitive and the critical stages in embryonic development of Xenopus laevis in relation to some physical factors (ultraviolet radiation, X-rays, ultrasounds, changes in temperature, p_H, centrifugation) in preparation.
7 Dobrowolski, J., W. Byrski, K. Cetnarowicz, R. Tadeusiewicz (1973): One proposal of the cybernetic interpretation of the most sensitive and the critical stages of early development in relation to environmental deteriorating factors — in preparation.
8 Undertal, B. (1969): Mercury determination in some foods, Nord. Hyg. Tidskr., 50.
9 Westöö, G. (1967): Determination of methylmercury salts in various kinds of biological material, Acta Chem. Scand., 22.
10 Nisbet, I. C. T. (1973): DDE in Eggs and Embryos of Brown Pelicans, Nature, 242.
11 Lakhani, K. H. (1973): Use of "Whole Egg Residues" in Pesticide Eggshell Studies, Nature, 242.
12 Korte, F. (1970): Technische Umweltchemikalien, Vorkommen, Abbau und Konsequenzen, Naturwissenschaft. Rundschau, bd. 23, h. 11.
13 Korte, F. (1972): Are Pesticides Suitable Model Substances for the Evaluation of Industrial Chemicals in the Environment? OEPP/EPPO Bull., 4.
14 Kinne, O., H. Rosenthal (1967): Effect of sulfuric water pollutants on fertilization, embryonic development and larvae of the herring, Clupea harengus, Marine Biol., 1/1.
15 Mieikovsky, S. A. (1971): The influence of pollution on pelagic larvae of bottom invertebrates in marine nearshore and asturine waters, Inst. Oceanol. Acad. Sci. USSR.

16 Båge, G., E. Cekanova, K. S. Larsson (1973): Teratogenic and Embryotoxic Effects of the Herbicides Di- and Trichloropheno, xyacetic Acids (2,4-D and 2,4,5-T), Acta Pharmacol. et. Toxicol, 32.

17 Dobrowolski, J. (1973): The utilization of embryological criterion for more successful protection of species against environmental-caused birth malformations – in preparation.

18 Hagström, B. E., S. Lönning (1973): The sea urchin egg as a testing object in Toxicology, Acta Pharmacol. et Toxicol., 32.

19 Global Environmental Monitoring (1971): SCOPE/ICSU, Stockholm.

20 Dobrowolski, J. (1973): Modern Science for future environment and social university youth activities – in press.

Modern Science for Future Environment and Social University Youth Activities

By **Jan W. Dobrowolski**

Actually it is of fundamental importance if natural resources' managing is really rational, and whether technical, financial and other possibilities of protection of proper human environment; are utilized fully.

Analysis of environment deteriorating factors show one essential reason of all them — it is lack of regard of the whole consequence of caused by man changes in our environment. More complex scientific research, seems to be necessary both: for proper decissions concerning all kinds of new sources of influences on the environmental quality, as well as for adequate behaviour each of us. Results of such, interdisciplinary studies strongly support new ethic formulated by UNEP Director General Dr. M. Strong. It is known that the social activity in this field depends on knowledge of environmental problems as well as methods, and on individual understanding of the importance of increase of personal involvement in wide cooperation, as a condition of improvement of quality of the life. This activity depends also on the perception of quality of natural and cultural environment. So it is a very complex problem, including working environment, free time and proper condition for recreation and many others. Very important seems to be coordination of environmental efforts at all educational levels and out-of-school youth activity, in connection with different practical needs in this field and permanent education of adults. *Interdisciplinary scientific analysis of methods and results of all kinds of environmental activity, seems to be condition sine qua non for higher efficiency of efforts for permanent improvement of our indivisible on global scale, natural environment.* Unfortunately none of existing international programmes include enough wide scientific investigation in connection with needs of really integrated practical activity at all planes. However it is optimistic prognostic, that such programmes become more complex.

Also educational system is still not adequate, especially concerning preparation for environmental cooperation on wide, international scale. But hopeful symptom, it is also modernisation and development in this direction all of educational activities. Now these efforts are very important and urgent.

It seems to be very useful to increase social activity in this field, as a implication of understanding that each man, and especially scientists, is responsible for environmental quality. So concerning opinion expressed by Dr. G. Budowski, "survival and ultimately quality of life, will be achieved only if decisions are based on scientific facts".(1) But opposite to this truth, only a small percentage of scientists, also from young generation, is active participating in environmental cooperation. It is a result both of not adequate educational programmes and methods and not proper perception of environmental quality.

It may be a danger symptoms of situation rather similar to denoted as circulum vitiosis...

So a quite agree with conclusion of the UN's Declaration on the Human Environment, that: "Education in environmental matters, especially for younger generations, is essential", as well as with remark included in the former UN's General-Secretary U Thant's report on the Man and Environment that: "The quality of the human environment, were not efficient because scientific studies were made in a traditional way: they were fragmentary and did not results in integration".

In many countries exist similar problems of a great importance for environmental quality, and as we know, some of them may be solving only on international scale. So it seems to be very needed to increase exchanging of methodological experience concerning ways of getting young people really involved in wide cooperation for more effective utilization of actual possibilities for better human environment.

More integrated interdisciplinary study, seems to be necessary both for better knowledge of relation between man and the environment, and for methodological verification and incresing of the efficiency off all kinds professional and social environmental activities. Therefore we may say, that social youth activity on modern, scientific basis is especially important. So Dr. J. Čerovsky in recommendation of IUCN I European Working Conference on Environmental Education 1971, confirmed a great significance and urgency of development out-of-school youth environmental education, including international and interdisciplinary students' scientific courses. (2) He rightly stressed very actual

problem of proper it means adequate to practical needs, methodology of this activity. Also Expert Panel on Educational Activities under the Man and Biosphere Programme recommended interdisciplinary research seminars and post-graduate courses, as a implementation on MaB. This panel support seminars preparing young scientists from different disciplines — e.g. from natural, social sciences and technics — for integrated investigation as well as for cooperation with decision-making man and politics at all levels — national, sub-regional and international. Very important seems to be connection between theoretical discussions during interdisciplinary seminars and studies with complex, practical activity e.g. during environmental field camps.

Polish university youth realised such ideas by means of multidisciplinary, long-term and all-country environmental action, on basis of students' organization. So we hope, that information about our activity, in connection with the United Nations and cooperating international organizations proposals and recommendation in this field, may be useful for carry on similar activities by interested university youth in other countries, as well as on international scale. As we were informed-by participants of IYNGO's Meeting on Youth Involvement in Environmental Issues in Palais des Nations, Geneva in 1971 and during another youth meetings, e.g. ISMUN's Seminar on European Security and Cooperation, Poland in 1972 as well as during IYF General Assemblies, Sweden in 1972 and France, 1973 etc. — wide scientific and education action of Polish students id one this kind. The similar opinion expressed professors and students from 11 countries, who visited our multidisciplinary environmental camps — e.g. Prof. Dr. R. G. Miller, Director of Foresta Institute, in USA, confirmed that the total results of such camps will be significant both for Poland and the globe.

Let me stress this fact, that we carry on action of interdisciplinary environmental investigations and complex students' activity since 1967. It is in connection with concept of modern protection of natural environment, which was elaborated by some of cofoundators of IUCN: Prof. Dr. W. Goetel and Prof. Dr. W. Szafer and others well-known Polish scientists.(3) So since 1969 in our multidisciplinary camps have also participated representatives of social sciences, industrial economy and technics. So our studies dealt in practice with the problems of more efficient cooperation, discussed now in the view of UNESCO and SCOPE programme Man and the Environment. In our scientific environmental camps participated thousands of students and many young scientists. During action of such interdisciplinary camps, we took into account the analysis of the quality of the natural environment and its biological-medical implications. On this background we carry on complex studies in aspects of the social, economic sciences and engineering for preparing not only complex reports, but also real conclusions, proposals and projects for improvement of environmental quality on investigated areas. Interested local authorities are always informed about the results of students'; multidisciplinary camps. University youth try to help authoritie directly in as quick as it is only possible realisation of proper programme, and students' volunteers labour office is especially interested in environmental, practical activity.(4) In above-mentioned scientific and educa-

tion action participated representatives of over 30 subjects of study from all kinds higher schools: / biologists and economists, physicians and architects, linguists and foresters, lawyers and psychologists, etc. /. So every interested student can take part in interdisciplinary cooperation for environmental protection.(5)

By this way, we try not only participate at changing the fragmentary outlook upon environment, but also university youth teams involve future scientists and decision-making man in complex action for betterment of the environmental conditions and human welfare. For such wide cooperation, it is very important to perfect effective managering, concerning e.g. suggestions by Dr. P. Drucker.(6) Very useful seems to be utilize now possibilities of systems analysis and modelling approaches in environmental problems, as there were presented not long age during IFAC/UNESCO Workshop in Poland. But essential for more rational management in environment, would be better preparation of students for integrated cooperation at all steps of needed activities. We have tried to work out rather universal model of interdisciplinary investigation of main environmental problems. It is working model, because we know, that it is very important to secure permanent improvement of methods of more and more complex activity on basis of feedback between concepts of cooperation and its scientific and practical results. So each of multidisciplinary students' camps is a new testing ground for perfection of methodology in this field.

In accordance with recommendations of MaB's meeting of expert on education (7), we include in our research and training interdisciplinary projects, which contain both theoretical and practical aspects. During students' camps were organized seminars on the main questions in connection with complex efforts for solving these general environmental problems in practice.

Such seminars at the I Multidisciplinary International Students' Camp "Human Environment 73" concerned e.g.: limits of permissible changes caused by man in environment / as levels of different but co-active pollutants, or the number and localization of works, or resting houses in relation to specificity and absorptiveness of respective area, the needs of new, more sensitive biological and medical criterions and relation between modern protection of natural environment and successful protection of species, including embryological aspects, the importance of cooperation among medical and non-medical sciences for environmental prophylaxis of public health, methods of interdisciplinary research and complex environmental activity, including problem of environmental prognosis, optimalisation of cooperation and range of cooperation deneted by practical needs of successful activity / e.g. protection of the Baltic Sea, border parks, reduction of air pollution, new problems of the human environment protection in connection with cosmic space exploration, etc. /, the profitability of the environmental pretection in the view of the social costs caused by its lack and its relation to economic and others analysis, legal activity and the realisation of legal rules in practice / e.g. by tourists on area of national parks /, reasons for the existance of "psychological barriers" and the inadequate preparate and relatively small involvement of the scientists in the interdisciplinary cooperation for complex studies as well as for increasing the practical effi-

ciency of environmental activity, and some others problems. This I Multidisciplinary Intercamp was held in the southern Poland, both on areas of national parks / e.g. the older in Europe border park in Pieniny Mts. /, and in the bigest in our country industrial region Upper Silesia. The guests participated in research and discussions concerning very actual issues like: biological and medical implications of different condition of natural environment, possibilities of reduction of chemical pollution /, industrial, agrotechnic, communal and motorized tourism /, more successful protection of national parks and proper conditions for recreational areas; including natural as well as cultural values, and activities in spatial planning, utilization of alternative technology, increasing and modernisation of educational activity — in relation to perception of environmental quality (8), economic development accommodatingly with specificity of these areas, etc. It was rather a good background for exchanging the methodological experiences about environmental problems of international interest, and for discussion of ideas for making university youth more involved in modern scientific and education activity needed in this field.

Participants of this Intercamp were also informed about the actual UN's activities and recommendation on the human environment. A very good source of mentioned information were very interesting meetings kind organized by Mr. J. Milwertz, Head CESI and Mrs. F. Sicilia-Bolomey, Coordinator Special Projects Section Information Service, in Palais des Nations, Geneva, at 6 July 1973. At this meetings took part also Mr. W. Kines, Director Communications UNEP, Mr.

Curnow and Mr. Ducret from ECE, Mr. P. B. Stone, Editor-in chief "Development Forum", Mr. M. B. Gosovic from UNEP and others UN's officers. From Polish students' side participated 30 representatives of university youth from all academic towns. Such meeting seems to be very helpful for increase students' participation in UN's environmental efforts.

Also very interesting and useful were information about activities and environmental projects which were presented during meetings at headquarters of other international organizations with Dr. B. M. Dieterich, Director, Division of Environmental Health WHO, Dr. G. Budowski, IUCN Director-General and Mr. A. Hoffmann, IUCN Planning Executive Officer, and Mr. R. Wiederkehr, Secretary-General, WWF Swiss. We were suggested to utilize Polish experience concerning multidisciplinary students' camps as a help for preparation specialists in modern environmental activity, especially for developing countries. We will be glad to help interested students' from these countries. Recently stressed e.g. by Mr. S. G. Madge importance of youth environmental education in Africa is really clear.(9) Polish students had organized Multidisciplinary Environmental Intercamp also during summer vacation 1974, as well as helt some conferences and seminars concerning complex analysis of actual environmental questions, especially modernisation of education as well as chemical pollution of environment. University youth is very interested in wide international cooperation and exchanging of experiences in this field.

Let me inform that in Poland all kinds of students' environmental activities, including international cooperation, are supported and coordinated by SUPS Committee for the Formation and Protection of the Human Environment. There are already proper consultation between this Committee and interested Ministries, Committees Polish Academy of Sciences and other organizations like: League for Protection of Nature, Polish "Know you country" Association, Nature Guards, Polish Anglers Association, PAS Team "Space Exploration and the human environment" and with some others.

Our Committee participate in wide information action on environmental problems and students' activities, by press, TV, documental short-films, on radio, as well as by public meetings and discussions in students' clubs, in particular. Every year thousands of Polish students are participating in touristic action not only for rest, but also for better knowledge of natural environment and methods for its protection. We are also very interested in permanent stimulation of increasing of youth environmental activities. Our Committee is also involved in modernisation of environmental education in connection with social out-of-schools activity, especially at university level. Complex scientific analysis of effects of all kind environmental activities seems to be necessary for permanent improvement of methods and intergation of efforts. So we try to do it on all-country scale — on basis of elaborated by us rather a general model of interdisciplinary environmental cooperation. In my understanding, UNs Secretary General Dr. Kurt Waldheim appeal for effective mobilization of world public opinion (10) include i.e. increasing of complex scientific and practical cooperation, especially this one of young generation against the deterioration of undivisible environment. It is known, that such cooperation co-work cannot be found in traditional ways, but it required also new and permanently improvement of organizatory framework.

For rational man's management in natural environment, creation of international regional centers as well as Youth World Center for Environmental Information and Cooperation, seems to be necessary. On basis of such centers, it would be possible to arrange exchanging of methodological experiences, concerning especially effeciency, of new activity for solving similar or cammon environmental problems. It would be also useful step for wider exchanging of participants among environmental seminars, conferences, scientific camps, for scholarships, training periods, etc., as well as for carry on long-term, international, more and more complex actions, especially these by university youth for future better environment. It would be optimistic fact, if is possible to discuss about just active youth international environmental centers — at the next World Development Information Day / as the exchanging of information would be the first function of these centers /. Such modern centers may be very useful for progressive preparation and selection in practice, generalists needed for cooperation on international and global scale on the human environment. It seems to be suitable to realize, under the support of UNEP, UNESCO, UN's University/SCOPE and other international organisations; international and interdisciplinary teams for wider participation in actual projects. Modern scientific basis for social and professional activity is helpful for solving environmental problems of international significance, like e.g.: increasing of mass touristic, network of national parks and recreational areas, transport and retention of pollutants, recycling, energetic sources and use the most limited natural resources, etc. Suggested centers and teams, may assist ☐ in protection of international river basins and sea. Hopefully Youth Scientific Center for Protection of the Baltic Sea and its tributaries, would be involved also in preparation United Nations Water Conference in 1977. Increasing of scientific cooperation in this field, would be a continuation i.e. of two international youth scientific conferences on the Baltic protection held in Poland in 1972 and one of implications from "Gdansk's Convention" 1973. I believe, that would be also useful, to hold every year at the World Environmental Day — International Scientific Youth Conference for make balance of effects of scientific and education cooperation on the human environment and for making agreement about main topics and methods of wider and wider integration of efforts. In my opinion, it would be also very important for increasing of international cooperation, to create UNs University, dealing particularly with the education of young scientists' staff for supporting effeciency of environmental programmes, especially those international and for developing countries.

I believe that every students and scientists, out of the ethical consideration should take part in the cooperation for a better acquaintance with the relation between man and changing environment, in all aspects. Such involving in research, training and practical activities pro bone publico; it is the indication for proper understanding of the place and the part of man in nature. It is particularly now, in the year of Copernicus jubilee, which reminds us about the real place and role of the unprivileged earth in the universe and the next — mankind in all nature. Thus we remember about the biological relation dependence of man upon the quality of natural environment as well as on the role of modern science in optimal utilization of possibilities for better future environment.

References

1 Budowski, G.: The current state of world conservation, Nature and Resources, Vol. IX, No. 1: 19–22, 1973.
2 Čeřovský, J.: I European Working Conference on Environmental Education, IUCN, 1972.
3 Goetel W., ed: Sozology and Sozotechnics, Scientific Bulletins of the S. Staszic Academy of Mining and Metalurgy, No. 293, Special Series – Bulletin 21, Kraków (1971).
4 Dobrowolski, J.: For the Future (in 6 languages), Poland, No. 2 (1972).
5 Dobrowolski, J.: The importance of developing the multidisciplinary international cooperation of the young scientists to overcome the world's environmental crisis, SUPS Committee for the Shaping and Protection of the Human Environment: Information for Foreign Countries, Poland, Warszawa (1973).
6 Drucker, P.: The Effective Executive, W. Heinemann Ltd., London (1967).
7 Expert Panel on Educational Activities under the Man and the Biosphere Programme, Unesco, Paris, 5–8 December 1972, MAB report series No. 7.
8 Expert Panel on Project 13: Perception of Environmental Quality Programme on Man and the Biosphere, Unesco, Paris, 26–29 March (1973), MAB report series No. 9.
9 Madge, S. G.: Letter to Editor, Development Forum, Vol. 1, No. 7 (1973).
10 Waldheim, K.: UNs Secretary-General, Development Forum, Vol. 1, No. 1 (1973).

The Evaluation of the Teratogenic Effects of 2, 4, 5-Trichlorophenoxyacetic Acid in the Rhesus Monkey

By **W. J. Dougherty, F. Coulston, L. Golberg**

International Center of Environmental Safety, Holloman Air Force Base, New Mexico

Abstract

Observation and examination of the offspring of female rhesus monkeys (*Macaca mulatta*) administered the herbicide 2,4,5-trichloro-phenoxyacetic acid (2,4,5-T) revealed no evidence of teratogenesis. The 2,4,5-T preparation used in the study contained 0.05 ppm of the contaminant 2,3,7,8-tetrachlorodibenzo-p-dioxin. Dose levels of 0.05, 1.0, and 10.0 mg/kg were administered orally in a No. 5 gelatin capsule from day 22 through day 38 of gestation to 10 monkeys per group; the controls received the vehicle.

The herbicide, 2,4,5-trichlorophenoxyacetic acid, is used primarily for the control of broad-leaved woody species. In the late 1960's this compound was used extensively as a defoliant in Vietnam. Press releases attempted to link a reported higher incidence of malformations in some areas of Vietnam to the defoliating agent.[1] Although these reports were unsubstantiated, a study by Courtney et al. [2] revealed that high doses of 2,4,5-T administered to rodents were teratogenic. Reports of this type led the U.S. Governement to limit the use of 2,4,5-T as of January 1970.[1] However, it should be noted that during the normal manufacturing process of 2,4,5-T, the contaminant, 2,3,7,8-tetrachlorodibenzo-p-dioxin (dioxin) is produced [3], and the sample used by Courtney et al. [2] contained 27 ppm of the dioxin contaminant.

Courtney et al. [2] observed significant increases in the percentages of abnormal fetuses and fetal mortality in two strains of pregnant mice and in one strain of pregnant rats treated with their sample. There was an increase in fetal mortality, cleft palate, and cystic kidney in C57B46 mice treated from day 6 to 14 of gestation with 113 mg of 2,4,5-T/kg administered orally or subcutaneously and with 46.4 mg/kg administration orally. A subcutaneous dose of 21.5 mg/kg did not affect the viability or development of the fetuses. The subcutaneous administration of 113 mg/kg from day 9 to 17 produced a significant increase in the percentage of abnormal fetuses with cleft palate and cystic kidney. The oral administration of 113 mg/kg to AKR strain of mice from day 6 to 15 produced a significant increase in fetal mortality and cleft palate but no increase in cystic kidney. The oral administration of 4, 6, 10 and 46.4 mg/kg to Holtzman rats from day 10 to 15 produced a significant increase in fetal mortality and abnormal fetuses having enlarged renal pelvis and cystic kidney. Emerson et al.[4] reported that no adverse effects were seen in fetuses of Sprague-Dawley rats receiving a high dose of 24 mg/kg of 2,4,5-T orally from day 6 through day 15 of gestation. The sample of 2,4,5-T used in this investigation contained less than 1.0 ppm of the contaminant 2,3,7,8-tetrachloro-dibenzo-p-dioxin. In a companion study, Sparaschu et al. (5), treated pregnant rats with 0.125, 0.5, 2.0, and 8.0 μg/kg of 2,3,7,8-tetrachlorodibenzo-p-dioxin (dioxin), administered orally from day 6 to 15 of gestation. The purpose of this study was to test the possibility that the dioxin contaminant was responsible for the effects observed by Courtney et al. The highest dose used by Courtney, 113 mg, would have contained approximately 3.4 μg of the dioxin. There was an increase in the incidence of intestinal hemorrhage and subcutaneous edema in the fetuses of rats receiving 0.125 μg of the dioxin. At the 0.5 and 2.0 μg/kg level there was a dose related increase in the number of fetal deaths and resorptions as well as in the incidence of intestinal hemorrhage and subcutaneous edema. Pregnant females receiving 8.0 μg/kg had no viable fetuses when examined on day 20 of gestation. The authors conclude from these results that the high level of dioxin (27 ppm) in the 2,4,5-T used by Courtney et al. could have accounted for the embryotoxicity observed in that study.

Khera et al. (6) administered 2,4,5-T containing less than 0.5 ppm of the dioxin to pregnant Wistar rats from the 6th to the 15th day of gestation. At a high dose of 100 mg/kg they reported no adverse effects detectable after 12 weeks in naturally littered progeny. In a second study, the pregnant rats were killed at term and the embryos examined for visceral and skeletal abnormalities. At a level of 100 mg/kg there was a slight depression in average litter weight and an increased incidence of spontaneously occurring skeletal abnormalities.

Neubert and Dillmann (7) using the purest available sample of 2,4,5-T ($<$ 0.02 pmm of dioxin) produced a high incidence of cleft palate in NMRI mice at doses higher than 20 mg/kg when administered orally from day 6 to 15 of gestation.

Collins (8) demonstrated that 2,4,5-T is feticidal and teratogenic in the Syrian hamster and that the incidence of effects increased with the content of the dioxin contaminant. However, Hart et al. (9) reported that pregnant mice treated subcutaneously with 100 mg/kg of 2,4,5-T greater than 99% pure produced offspring with abnormalities so scattered as to preclude any causal relationship to the treatment.

The only published study to date using an experimental animal other than rodents was by Wilson (10) who treated 17 pregnant rhesus monkeys with 2,4,5-T at dose levels between 5 and 40 mg/kg, three times per week, between day 20 and 48 of gestation. The fetuses were removed by hysterotomy on day 100 of gestation. He reports that no malformations were evident at these dose levels.

The primary purpose of this investigation was to evaluate the teratogenic effect of 2,4,5-T containing 0.05 ppm of dioxin in the offspring of the non-human primate. The rhesus monkey (Macaca mulatta) was selected for this study because of the similarities of its reproductive system and embryologic development to that of man (11, 12, 13, 14, 15, 16, and 17) thus making possible a reasonable extrapolation of the results to man. The test protocol was developed to investigate primarily the teratogenic effects of the oral administration of 2,4,5-T while at the same time enabling us to determine any gross toxicity to the pregnant female and to observe the development of the offspring for up to 1 year of age.

The 40 pregnant rhesus monkeys used in this study were selected from the International Center of Environmental Safety's breeding colony at Holloman AFB, New Mexico. All females were bred for 48 hours beginning on day 11 of the menstrual cycle.(18) Pregnancy was diagnosed by the Tullner-Hertz modification of the Ascheim-Zondek pregnancy test.(19) Blood was drawn from the brachial vein of the female on day 19 after mating; the first day of mating was counted as day 1. Immature mice, 21 days old, were injected with 0.5 ml of monkey serum for 3 consecutive days. On day 4, the mice were killed, the uteri removed and weighed. A 100% increase in mouse uterine weight as compared to control mice was indicative of a pregnancy in the monkey. Using this procedure pregnancy was diagnosed by day 22 of gestation. All pregnant animals were housed individually and allowed free access to food and water.

The 2,4,5-T (20) used for this study contained 0.05 ppm of the dioxin contaminant. It was administered orally in No. 5 gelatin capsules using a stomach tube. The 40 pregnant monkeys were divided into four groups of 10 monkeys each and received either the empty gelatin capsule or the gelatin capsule containing either 0.05, 1.0, or 10.0 mg/kg doses of 2,4,5-T (Table 1). All animals wer medicated once daily from day 22 through 38 of gestation. All pregnancies are allowed to go to term.

Within 24 hours after birth, the mother and offspring are separated. The infant is given a physical examination, weighed and returned to the mother. Babies are weaned at 4 months of age. In addition to the daily observation of the offspring, monthly body weights are obtained and two offspring from each group are scheduled for necropsy shortly after birth, 6 months, and 1 year. A careful dissection of major organ systems is done on each infant necropsied. In addition, a skeletal X-ray examination, clinical chemistry, and hematology is done on each infant killed.

During this study none of the pregnant females receiving 2,4,5-T exhibited any evidence of toxicity. All the babies born to date show no evidence of teratogenesis. Nine of the 10 control pregnant monkeys delivered live infants. The one remaining monkey aborted at day 68 of gestation. For our colony we define an abortion as the loss of the conceptus prios to 110 days of gestation. A stillborn fetus is defined as one that is found dead on first observation following delivery and by autopsy it is confirmed that no air entered the lungs. One of the infants born at day 143 of gestation was considered premature according to the criterion of van Wagenen (21), i.e. a pregnancy terminating beyond ± 2 SD of the mean gestation length which for our colony is 163.7 ± 6.3 days. This premature infant died at 21 days of age.

Table 1   Experimental Design 2,4,5-T

Group	No. of Pregnant Monkeys	Dose level	Dosing Schedule	Duration (Gestational age)
Control	10	# 5 gelatin capsule	daily	22—38
1	10	.05 mg/kg	daily	22—38
2	10	1.0 mg/kg	daily	22—38
3	10	10.0 mg/kg	daily	22—38

Table 2   2,4,5-Teratology Study   Outcome of Pregnancies

Group	No. of Monkeys	No. of Livebirths	No. of Stillbirths	No. of Abortions	Gestation Length (days) Mean ± SD t	Initial Body Weight of Infant (gms) Mean ± SD t
Control	10	9	—	1	163.0 ± 10.3	410.3 ± 66.2
0.05	10	10	—	—	162.9 ± 7.8 .02*	455.7 ± 44.0 1.67*
1.0	10	8	1	1	161.6 ± 16.3 .28*	413.1 ± 63.7 .08*
10.0	10	8	—	2	164.3 ± 8.2 ,26*	447.2 ± 21.9 1.49*

+ One monkey is still pregnant
++ Two monkeys are still pregnant
* Not statistically different at 0.1 level

The mean gestation length for the controls was 163 ± 10.3 days.

All 10 pregnant monkeys that received the 0.05 mg/kg doses of 2,4,5-T delivered live babies. One died at 37 days of age. Eight of the 10 monkeys assigned to the 1.0 mg/kg group have delivered live infants. One monkey aborted, and one delivered a stillborn infant of 119 days gestational age. Of the 10 pregnant females receiving 10.0 mg/kg, two aborted, and eight have delivered live infants. To date the only additional mortality was an infant born in the 1.0 mg/kg dose group. This infant was raised in isolation because its mother refused to care for it. In nursery conditions the infant showed satisfactory weight gains and development; however, when reintroduced to the colony the infant was unable to adapt and died within two months.

The results of this investigation indicate that 2,4,5-T administration to the rhesus monkey at these dose levels produces no evidence of gross teratogenecity in the offspring. Necropsy of all infants that died, were stillborn, or killed revealed no evidence of toxicity or teratogenesis. The abortion and stillbirth rate for the study is 13.3% which is not significant when compared to the normal abortion rate, 12% (22), in our colony. The mean gestation lengths and the mean-initial infant body weights for each group are shown in Table 2. These means were statistically analyzed for significance of difference by the Student "t" test (23). Examination of these data revealed that in all instances the mean differences between the medicated groups and the control group were *not* statistically significant. In order to guard against a Type II error, i.e., finding no

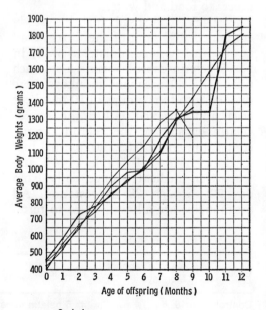

Fig. 1

Controls
0.05 mg / kg
1.0 mg / kg
10.0 mg / kg

statistical differences when actual differences between the means exist, the level of statistical confidence was set at 0.1, as opposed to the conventional 0.05 or 0.01 levels. At present, all four groups of infants exhibit no evidence of teratogenesis, are developing normally, and show satisfactory weight gains (Figure 1).

A final report on this study will be forthcoming when the 1-year observation period for the last born infant has been completed and all the data including histology, skeletal X-rays and clinical chemistry are analyzed.

References

1 Nelson B.: Science 166; 977 (1969).

2 Courtney, K. D., D. W. Gaylor, M. D. Hogan, H. L. Falk: Science 168; 864 (1970).

3 Kimmig J., and K. H. Schulz: Naturwissenschaften 44; 337 (1957).

4 Emerson, J. L., D. J. Thompson, C. G. Gerbig, and V. B. Robinson: Toxicol. Appl. Pharmacol. 17; 317 (1970) Abstract.

5 Sparschu, G. L., F. L. Dunn, and V. R. Rowe: Toxicol. Appl. Pharmacol. 17; 317 (1970) Abstract.

6 Khera, K. S., B. L. Huston, and W. P. McKinley: 10th Ann. Meet. Soc. Toxicol. Washington, D. C., March 7–11 (1971) Abstract. (Academic Press, Inc., Great Britain, 1971).

7 Neubert, D., and I. Dillmann: Naunyn – Schmiedeberg's Arch. Pharmacol. 272; 243 (1972).

8 Collins, T. F. X., and G. C. Gray: Bull. Envir. Contam. and Toxicol. 6; 559 (1971).

9 Hart, E. R., and M. G. Valerio: 11th Ann. Meet. Soc. Toxicol. Williamsburg, Va. March 5–9 (1972) Abstract.

10 Wilson, J. G.: Symposium on the Use of Non-human Primates for Research on Problems of Human Reproduction, Sukhumi, USSR, December 13–17 (1971).

11 Heuser, C. H., and G. L. Streeter: Carnegie Inst. Contrib. Embryol. 29; 15 (1941).

12 Heuser, C. H., and G. W. Corner: Carnegie Inst. Contrib. Embryol. 26; 29 (1957).

13 Streeter, G. L.: Carnegie Inst. Contrib. Embryol. 30; 211 (1942).

14 Streeter, G. L.: Carnegie Inst. Contrib. Embryol. 31; 29 (1945).

15 Streeter, G. L.: Carnegie Inst. Contrib. Embryol. 32; 123 (1948).

16 Streeter, G. L.: Carnegie Inst. Contrib. Embryol. 34; 164 (1951).

17 Wilson, J. G.: Fed. Proc. 30; 104 (1971).

18 Wagenen, G. van, J. Yale: Biol. and Med. 17; 745 (1945).

19 Tullner, T. W., and R. Hertz: Endoc. 78; 204 (1966).

20 The 2,4,5-T was supplied by the C. H. Boehringer Sohn Co. Ingelheim am Rhein, Germany.

21 Wagenen, G. van: Proc. Conference on Nonhuman Primate Toxicology, June 12 to 14 (1966). C. O. Miller, Ed. U.S. Department of Health, Education, and Welfare (U.S. Government Printing Office, Washington, D.C. 1966).

22 Dougherty, W. J., L. Goldberg, and F. Coulston: 10th Ann. Meet. Soc. Toxicol., Washington, D.C., March 7–11 (1971) Abstract (Academic Press, Inc., Great Britain, 1971).

23 Downie, N. M., and R. W. Heath: Basic Statistical Methods, Harper Bros. (1959).

24 The authors wish to acknowledge Miss Joanna Lowry for her help in preparing the manuscript and also Mr. Wayne McDonald, Mr. Jerry Gardner and Mrs. Flo Jackson for their technical assistance.

Table 1   2,4,5-T Teratology Study   Control Pregnant Monkeys

Monkey Number	Date Conceived	Date Due to Deliver	Date Delivered	Length of Gestation (days)	Outcome of Pregnancy	Sex of Infant	Body Weight of Infant (gms)
1169	4/26/71	10/08/71	10/08/71	165	Live Birth	F	482
294	5/05/71	10/15/71	9/25/71	143	Live Birth	M	336
1170	5/07/71	10/19/71	10/22/71	168	Live Birth	M	476
580	5/12/71	10/24/71	10/21/71	162	Live Birth	F	336
820	12/31/71	6/13/72	6/17/72	169	Live Birth	F	420
730	9/28/71	3/10/72	2/23/72	147	Live Birth	M	340
630	9/28/71	3/10/72	2/17/72	163	Live Birth	M	482
475	11/21/71	5/03/72			Aborted		
1216	12/13/71	5/26/72	6/01/72	171	Live Birth	M	410
306	6/21/72	12/04/72	11/20/72	153	Live Birth	F	355

Table 2   2,4,5-T Teratology Study   Pregnant Monkeys Receiving 0.05 mg/kg

Monkey Number	Date Conceived	Date Due to Deliver	Date Delivered	Length of Gestation (days)	Outcome of Pregnancy	Sex of Infant	Body Weight of Infant (gms)
1173	5/12/71	11/01/71	11/08/71	172	Live Birth	M	481
1174	7/26/71	1/06/72	12/26/71	154	Live Birth	F	410
1146	8/16/71	1/28/72	1/29/72	166	Live Birth	M	481
1144	8/21/71	2/02/72	1/19/72	152	Live Birth	F	340
1139	8/22/71	2/03/72	2/07/72	169	Live Birth	M	482
1161	8/22/71	2/03/72	2/05/72	167	Live Birth	M	538
1155	9/03/71	2/16/72	1/31/72	152	Live Birth	F	368
1172	9/03/71	2/16/72	2/17/72	166	Live Birth	F	453
1154	10/29/71	4/11/72	4/17/72	171	Live Birth	F	484
1180	11/24/71	5/07/72	5/02/72	160	Live Birth	F	520

Table 3   2,4,5-T Teratology Study   Pregnant Monkeys Receiving 1.0 mg/kg

Monkey Number	Date Conceived	Date Due to Deliver	Date Delivered	Length of Gestation (days)	Outcome of Pregnancy	Sex of Infant	Body Weight of Infant (gms)
1137	11/06/71	4/18/72	3/03/72	119	Stillborn	F	181
1136	11/09/71	4/21/72	4/23/72	167	Live Birth	F	482
613	11/11/71	4/23/72	5/01/72	173	Live Birth	M	443
817	11/11/71	4/23/72	4/13/72	155	Live Birth	F	—
824	11/14/71	4/27/72	12/28/71	40	Aborted		
1215	11/25/71	5/08/72	5/15/72	172	Live Birth	F	382
1179	11/26/71	5/09/72	4/19/72	145	Live Birth	F	299
1195	11/27/71	5/10/72	5/07/72	162	Live Birth	M	410
1198	11/29/71	5/12/72	4/26/72	147	Live Birth	M	396
1260	7/17/72	12/29/72	1/05/73	172	Live Birth	M	480

Table 4   2,4,5-T Teratology Study   Pregnant Monkeys Receiving 10.0 mg/kg

Monkey Number	Date Conceived	Date Due to Deliver	Date Delivered	Length of Gestation (days)	Outcome of Pregnancy	Sex of Infant	Body Weight of Infant (gms)
944	11/28/71	5/21/72	5/07/72	151	Live Birth	M	426
711	12/04/71	5/27/72	1/04/72	30	Aborted	—	—
1220	12/08/71	6/02/72	5/21/72	153	Live Birth	F	457
1200	12/24/71	6/06/72	2/17/72	54	Aborted	—	—
1218	12/30/71	6/13/72	6/19/72	171	Live Birth	F	414
827	1/08/72	6/21/72	6/23/72	167	Live Birth	M	431
1223	1/29/72	7/11/72	7/17/72	171	Live Birth	M	450
1188	2/12/72	7/26/72	8/03/72	173	Live Birth	M	455
1265	5/13/72	10/20/72	10/19/72	164	Live Birth	F	480
933	7/18/72	12/30/72	12/30/72	165	Live Birth	M	465

Table 4   2,4,5-T Teratology Study Infant Weights   a. Control Group Births (weights in grams)

Mother Number	Baby Number	Birth Weights	1	2	3	4	5	6	7	8	9	10	11	12
294	1207	336	died											
1169	1209	482	606	750	909	1160	1106	1259	1470	1580	1670	1800	1950	2110
1170	1211*	476	682	765	—	907	876	886	990	1120	1350	1700	1852	1950
580	1210*	336	454	625	709	879	770	909	974	1110	1240	1350	1400	1540
730	1344	340	574	639	655	720	840	930	1185	1300	1490	1570	1600	1680
630	1346	482	494	594	650	719	875	950	1095	1370	1410	1500	1640	1770
820	1478**	420	480	550	685	728	800	850	980					
1216	1477**	410	606	770	885	950	1220	1350						
306	1532*1	355												

*1   Taken by cesarean at day 153 of gestation. Infant vital signs excellent; however, infant killed and necropsied.

*   Sacrificed

**   Scheduled for sacrifice

Table 4   2,4,5-T Teratology Study Infant Weights   b. 0.05 mg/kg Group (weights in grams)

Mother Number	Baby Number	Birth Weights	1	2	3	4	5	6	7	8	9	10	11	12
1174	1306	410	652	—	693	724	753	805	died					
1173	1224**	481	520	—	708	793	946	1140	1394	1380	1500	1740	1800	1850
1146	1338**	481	625	765	835	910	917	1060	1145	1245	1486	1240		
1144	1307	340	452	died										
1139	1342*	482	652	840	919	1064	1230							
1161	1343*	538	595	677	751	812	895							
1155	1340	368	—	609	702	785	855	—	1035	1075	1044	985		
1172	1345	453	631	764	833	875	930	995	1080	1110	1140			
1154	1441*	484	574	725										
1180	1476*	520	530	740										

*   Sacrificed

**   Scheduled for sacrifice

Table 4   2,4,5-T Teratology Study Infant Weights   c. 1.0 mg/kg Group (weights in grams)

Mother Number	Baby Number	Birth Weights	Months												
			1	2	3	4	5	6	7	8	9	10	11	12
1136	1444*	482	615	740	830	990	1082	1145						
817	1434**		620	650	670	750	925	950	975	1220	1370			
1198	1445*	396	456	630	755	950	985							
613	1446**	443	443	602	850	915	1018	1082	1220	1370				
1195	1448*	410	575	710										
1260	1715	480												
1215	1449*	382	450	580										
1179	1442¹	299												

*   Sacrificed

**   Scheduled for sacrificed

¹   The mother would not care for this infant. The infant was raised in a nusery environment and when reentered the colony could not adjust and died.

Table 4   2,4,5-T Teratology Study Infant Weights   d. 10.0 mg/kg Group (weights in grams)

Mother Number	Baby Number	Birth Weights	Months												
			1	2	3	4	5	6	7	8	9	10	11	12
944	1447**	426	571	735	885	998	1170	1260	1330	1620				
1220	1450**	457	556	762	890	1010	—	1050	1275	1470				
1218	1479	414	560	735	886	910	—	800	1040	995	1190			
827	1480	431	520	610	860	915	985	1340	1460					
1223	1481*	450	500	610	750	890	1115	1360						
1188	1482*	455	—	510	670	890	950	1270						
933	1713*	465	575											
1265	1485	480	540	610	760									

*   Sacrificed

**   Scheduled for sacrifice

Experiments on the Effect of Carbon Monoxide on Aminolevulinic Acid Dehydrase (ALAD)

By **Alex Azar, M. D., N. W. Henry III, B.A., F. D. Griffith, Ph. D., John W. Sarver** and **Ronald D. Snee, Ph.D.**[*]

Haskell Laboratory for Toxicology and Industrial Medicine
E. I. du Pont de Nemours and Company Newark, Delaware 19711

Abstract

The inverse relationship between blood lead concentration and aminolevulinic acid dehydrase (ALAD) is well known. Recently, it has been suggested that a similar relationship exists between carboxyhemoglobin (COHb) and ALAD activity. This study was undertaken to examine more closely the possible effect of carbon monoxide on ALAD.

Blood from 19 human volunteers was analyzed for both carboxyhemoglobin and ALAD activity. Smokers had significantly lower concentrations of ALAD than nonsmokers and a rise in carboxyhemoglobin concentration was associated with a fall in ALAD activity. The in vitro bubbling of carbon monoxide into human blood did not significantly effect ALAD activity.

Four groups of rats (10 per group) were exposed to carbon monoxide or dietary lead acetate according to the following design: (I) Control — no Pb or CO; (II) 500 ppm Pb acetate in diet; (III) 250 ppm CO four hours/day $\times$ 5 days/week $\times$ 4 weeks; (IV) Both Pb and CO. Analysis of the rat data showed a significant depression of ALAD by lead. The activity of ALAD in the rats exposed to CO was significantly increased suggesting the possibility of an adaptive phenomenon.

Introduction:

The inhibition of the enzyme ALAD by lead has been known for several years (1, 2). However, a recent epidemiologic study by our laboratory (3, 4) showed that smokers had significantly lower levels of ALAD activity than nonsmokers despite no significant difference in the blood lead concentration of smokers versus nonsmokers. Further analysis of the data demonstrated a negative correlation between increasing concentration of carboxyhemoglobin and a fall in ALAD activity. There was no significant interaction between the concentration of blood lead and carboxyhemoglobin and each accounted for approximately 30 per cent of the variance in ALAD activity. In vitro experiments by Rausa et al. (5) have shown that CO and lead together result in a greater depression of ALAD activity than that resulting from lead alone.

In view of these findings, the following in vitro and in vivo experiments were carried out to obtain additional information on the possible influence of CO exposure on ALAD activity.

Methods

Smokers versus Nonsmokers

Blood from 10 non-smoking and 9 smoking human volunteers was obtain-

---

[*] Engineering Department Louviers Building. E. I. du Pont de Nemours and Company Newark, Delaware 1971

ed by venipuncture. Carboxyhemoglobin was determined by the spectrophotometric method by Amenta (6) and aminolevulinic acid dehydrase (ALAD) activity was measured using the procedure of Bonsignore et al.(7)

In Vitro Bubbling of CO

The blood samples from the nonsmokers were divided into two equal aliquots. One was saturated with air and the other with carbon monoxide. The air and carbon monoxide (99.5% CO, C.P. grade, Matheson Gas Products®) were slowly bubbled through the blood in glass impingers for approximately 30 minutes. The saturated aliquots were immediately analyzed for carboxyhemoglobin and ALAD using the methods described above. Additional measurements of ALAD and carboxyhemoglobin were made one and two hours post-saturation. Microscopic slides of both saturated aliquots were stained for carboxyhemoglobin using a modification of the technique described by Bitke and Kleinhauer.(8)

Rat Experiments

Four groups of 10 young adult ChR-CD male rats each were exposed to carbon monoxide or fed diets containing lead acetate. Group I served as a control group and was exposed to houseline air and fed the basal laboratory diet of Purina Laboratory Chow® plus 5% corn oil. Group II was also exposed to houseline air but had 500 ppm lead acetate added to the diet. The third group of animals received the basal laboratory diet and were exposed to 250 ppm carbon monoxide, four hours a day, five days a week (Monday thru Friday) for four weeks. Group IV was fed the lead containing diet in addition to being exposed to carbon monoxide.

The dietary level of lead acetate was chosen on the basis of earlier experiments which showed that it inhibited ALAD activity.(9) The level of carbon monoxide was approximately one-fifth the median lethal dose.

The exposures were carried out in 50-liter dynamic exposure chambers. Carbon monoxide was metered from a cylinder into a one-liter round-bottom flask where it was mixed with houseline air to achieve the desired concentration. Chamber atmospheric CO was analyzed by gas chromatography. Houseline air was delivered in a similar manner. Between exposures the animals were housed two per cage in suspended stainless steel cages. Tap water and the experimental diets were available to the animals between exposures.

Samples of tail blood were taken prior to the start of the experiment and immediately following the fourth exposure (Thursday) each week. It was analyzed for carboxyhemoglobin and ALAD using the methods described previously. Hemoglobin was determined using a hemoglobinometer and a micro hematocrit was also done on the blood sample. An overnight urine sample was collected every Thursday to Friday for delta-aminolevulinic acid (DALA) and osmolality determinations. DALA was measured with disposable ion exchange chromatography columns prepared by Bio-Rad Laboratories, Richmond, California. Osmolality was determined by a Fiske osmometer.

The data from the rat experiments were subjected to analysis of variance appropriate for a $2 \times 2 \times 10 \times 4$ partially nested and crossed factorial design (2 Pb levels, 2 CO levels, 10 rats/group, 4 weeks). In the case of ALAD, the data were transformed to logarithms prior to analysis to insure

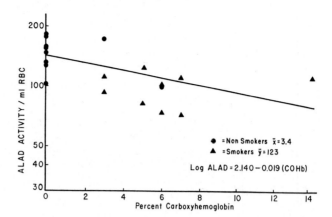

Fig. 1 *The relationship between % carboxyhemoglobin and log ALAD for 19 human smokers (●) and nonsmokers (▲) is shown. The equation for the line is log ALAD = 2.140–0.019 Carboxyhemoglobin.*

that the variability would be independent of the level of the response. Pretreatment data were analyzed separately to determine significant differences among the groups. When the analysis of variance showed a significant effect, the technique of least significant difference (LSD) was used to determine which means were producing the effect.

Results

Smokers versus Nonsmokers

The mean carboxyhemoglobin and ALAD activity of the smokers and nonsmokers is shown in Table 1. The smokers had a significantly ($p < .05$) lower level of ALAD activity. Figure 1 is plot of the relationship between carboxyhemoglobin and log ALAD. Regression analysis showed a significant ($p < .001$) negative correlation between the carboxyhemoglobin and log ALAD. Approximately 36% ($R^2 = .36$) of the variance in ALAD is explained by carboxyhemoglobin. These findings are consistent with those reported earlier.[3]

In Vitro Bubbling of CO

Table 2 shows the average ALAD activity obtained when carbon monoxide

is bubbled through the blood of nonsmokers. Also shown are the air controls. There was no significant difference in the ALAD activity of the blood treated with carbon monoxide or air. Regression analysis showed no correlation between carboxyhemoglobin and ALAD.

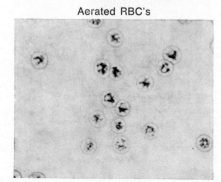

Fig. 2 *Photomicrographs of aerated red blood cells and those saturated with carbon monoxide in vitro.*

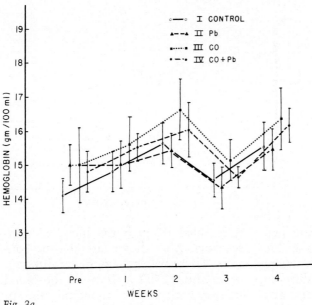

Fig. 3a

Fig. 3 The weekly average ± 1 S. D. of hemoglobin (3a), hematocrit 3b), carboxyhemoglobin (3c), urine osmolality (3d) and urine DALA (3e) is shown. Group I rats received control diet; Group II had 500 ppm Pb acetate added to the diet; Group III were exposed to 250 ppm CO, 4 hours per day, 5 days a week x 4 weeks; and Group IV received both the lead diet and CO.

Photomicrographs of the aerated and carboxylated blood cells provide proof that indeed the blood was carboxylated (Figure 2).

Rat Experiments — Pretreatment Data

Table 3 shows the average ALAD, DALA, osmolality, hematocrit, hemo-

globin, and carboxyhemoglobin of the rats prior to exposure. With the exception of hemoglobin and carboxyhemoglobin, there were no significant differences between the groups. The average hemoglobin of the control group was significantly ($p < .05$) lower than the treatment groups and average

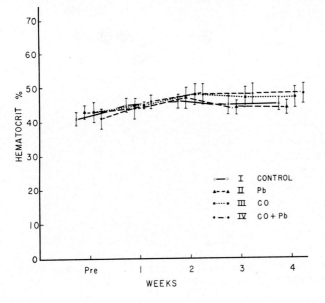

Fig. 3b

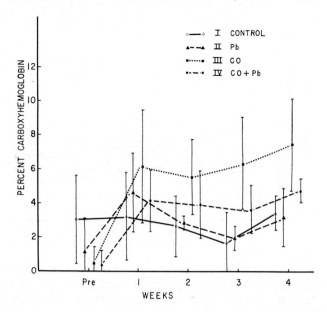

Fig. 3c

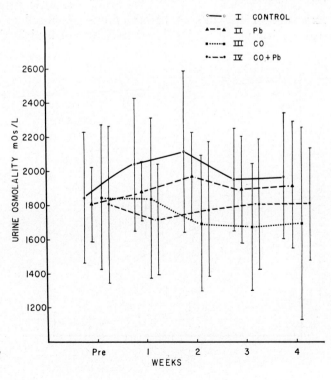

Fig. 3d

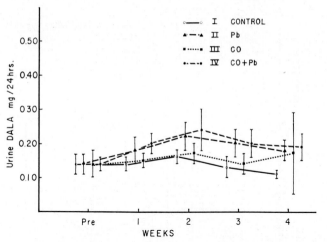

Fig. 3e

carboxyhemoglobin of the control group was significantly higher than the treated groups. The reason for these differences is not known.

There was no significant effect of lead or carbon monoxide on the weight gain of the animals.

Rat Experiments — Treatment Data

Exposure of the rats to carbon monoxide (Groups III and IV) resulted in a significant (p < .05) rise in hemoglobin, hematocrit and carboxyhemoglobin (Fig. 3a, b, c). The rise in carboxyhemoglobin concentration in the rats exposed to both lead and carbon monoxide was significantly less than that obtained in the rats exposed to carbon monoxide alone. Urine osmolality was significantly decreased in those animals exposed to CO (Fig. 3d). The presence of lead in the diet resulted in a signifi-

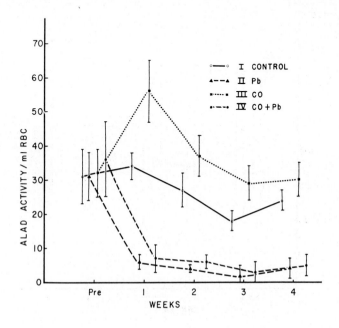

Fig. 4 The weekly response ± 1 S. D. of ALAD activity is shown. Group I rats received control diet; Group II had 500 ppm Pb acetate added to the diet; Group III were exposed to 250 ppm CO, 4 hours per day, 5 days a week x 4 weeks; and Group IV received both the lead diet and CO.

cant rise in the excretion of delta-aminolevulinic acid in the urine (Fig. 3e).

The aminolevulinic acid dehydrase (ALAD) activity of the rats fed lead acetate in their diet was significantly ($p < .05$) less than that of the control rats. Figure 4 shows that ALAD activity fell rapidly during the first week and more slowly thereafter. This depression of ALAD by lead was not affected by the presence of carbon monoxide (Fig. 4). The rats exposed to carbon monoxide alone had a significantly ($p < .05$) higher concentration of ALAD activity than control rats exposed to air (Fig. 4).

Discussion

Aminolevulinic acid dehydrase (ALAD) is widely distributed in nature being found in plants and animals. It is involved in heme synthesis and catalyzes the conversion delta-aminolevulinic acid (DALA) into porphobilinogen.[1, 2] Inhibition of ALAD results in an increased excretion of its substrate, DALA, in the urine. In the present experiment, the rats fed lead in their diet demonstrated a fall in ALAD activity and a rise in the excretion of DALA in the urine.

Carbon monoxide has been shown to effect porphyrin metabolism and also to accentuate the depression of ALAD by lead in vitro.[11, 5] It is also well known that the carboxyhemoglobin levels of smokers is substantially higher than nonsmokers.[12] Therefore, it was not unreasonable to examine the ALAD activity of smokers versus nonsmokers. In addition to lead, however, alcohol, copper, mercury, silver, manganese and cobalt have also been shown to effect ALAD activity.[1, 10] The present experiment did not examine the possibility that ALAD was being affected by some substance in cigarette smoke other than carbon monoxide or lead.

It could be argued that smokers have higher blood lead concentrations than nonsmokers and that the higher blood lead level of smokers is responsible for their lower ALAD activity. The literature comparing the blood lead concentration of smokers versus nonsmokers is contradictory. Some investigators [13, 14] report an increased blood lead concentration in smokers; whereas, others [15, 16] have found no significant difference. Blood lead concentrations were available on only 11 of the 19 subjects in this study. Because, 11 subjects were too few and the range of blood lead and carboxyhemoglobin values too narrow, the data from these 11 subjects were combined with that available on 30 subjects from an epidemiological study. The results of this study have been reported elsewhere [3]; however, it was shown in that study that there was no significant difference between the blood lead concentration of smokers versus nonsmokers and blood lead and carboxyhemoglobin each accounted for approximately 30 percent of the variance in ALAD activity.

The failure to demonstrate a significant fall in ALAD activity by bubbling carbon monoxide into a test tube of human red blood cells does not rule out the possibility of an in vivo effect. It was not possible to determine in the rat experiments whether carbon monoxide potentiated the depression of ALAD by lead because the dose of lead alone resulted in barely detectable levels of ALAD (Fig. 4).

The finding that the ALAD activity of the rats exposed to carbon monoxide increased, raises some interesting possibilities. Since the determination of ALAD includes a correction factor for hematocrit, it is not believed that the rise in ALAD activity observed in the rats exposed to carbon monoxide is related to the rise in hemoglobin and hematocrit seen in these animals. It is interesting to conjecture that this rise in ALAD represents a compensatory or adaptive response and had the exposure to carbon monoxide been more prolonged, a depression would have occurred. Recently, Hapke [17] has suggested an adaptive phenomenon of ALAD in sheep exposed to lead. It is important to emphasize that the suggestion that the increase in ALAD seen in rats exposed to carbon monoxide represents an adaptive phenomenon is entirely speculative at this time.

Other possible explanations for the failure to demonstrate a fall in ALAD activity of rats exposed to carbon monoxide is that it may be a species dependent phenomenon or that more prolonged exposures are necessary. Therefore, additional studies using several species of animals and more prolonged exposures to carbon monoxide are indicated.

In summary, the ALAD activity of smokers was found to be significantly lower than nonsmokers and there was a fall in ALAD activity as carboxyhemoglobin concentration rose. In vitro bubbling of carbon monoxide into tubes of human blood failed to result in a depression of ALAD. Rats exposed to carbon monoxide showed an increase, rather than a decrease, in ALAD activity suggesting the possibility of a compensatory or adaptive phenomenon.

Acknowledgements

The authors wish to acknowledge the assistance of Miss Mary E. Finan in carrying out the statistical analyses.

References

1 Haeger-Aronsen, B., M. Abdulla, and B. I. Fristedt: Effect of Lead on δ-Aminolevulinic Acid Dehydrase Activity in Red Blood Cells. Arch. Environ. Health 23: 440–445 (1971).

2 Hernberg, S., G. Nikkanen, G. Mellin, and H. Lilius: δ-Aminolevulinic Acid Dehydrase as a Measure of Lead Exposure. Arch. Environ. Health 21: 140–145 (1970).

3 Azar, A., R. D. Snee, and K. Habibi: Relationship of Community Levels of Air Lead and Indices of Lead Absorption. To be published in Proceedings of International Symposium on Environmental Health Aspects of Lead held at International Con-

Table 1   ALAD of Smokers VS. Nonsmokers (Average ± 1 S.D.)

	Non-smokers	Smokers
Number of Subjects	10	9
Average Age	35	41
Hematocrit	41 ± 4	43 ± 3
Hemoglobin	15.8 ± 1.2	15.8 ± 1.1
Carboxyhemoglobin (%)	0.9 ± 2.0	6.0 ± 3.0
ALAD (units/ml rbc)	147 ± 31	97 ± 18

Table 2   Effect of Bubbling Air or Carbon Monoxide Into Human Blood for 30 Minutes on ALAD Activity (Average ± 1 S.D.)

	Air	Carbon Monoxide
Number of Samples	10	10
Carboxyhemoglobin (%)		
Immediately post bubbling	0.9 ± 2.0	90.0 ± 9.0
ALAD		
Immediately post bubbling	147 ± 32	135 ± 30
One hour post bubbling	147 ± 31	140 ± 33
Two hours post bubbling	149 ± 30	139 ± 28

Table 3   Average Pretreatment Responses by Group

Response	Group I Control	Group II Lead	Group III CO	Group IV Lead/CO	LSD*
1. ALA Dehydrase	30.80	31.30	31.60	35.70	3.74
2. DALA	0.14	0.14	0.14	0.14	0.03
3. Hematocrit	41.10	42.50	42.60	41.00	2.40
4. Osmolality	1 845.30	1 806.10	1 846.60	1 805.10	348.60
5. Hemoglobin	14.07	15.04	14.96	14.82	0.33
6. Carboxyhemoglobin	3.00	1.10	0.40	0.30	0.79

* The difference between a pair of means is significant at the .05 level if it exceeds the given least significant difference.

grescentrum RAI — Amsterdam-Europaplein, October 1972.

4 Azar, A., R. D. Snee, and K. Habibi: An Epidemiologic Approach to Community Air Lead Exposure Using Personal Air Samplers. Submitted to Environmental Quality and Safety.

5 Rausa, G., L. Dana, and G. Perin: Action in Vitro of Carbon Monoxide and Lead on the Delta-Aminolevulinic Acid Dehydrase Activity in Human Erythrocytes. Revista Italiana d'Igiene 28: 319 (1968).

6 Amenta, J. S.: The Spectrophotometric Determination of Carbon Monoxide in Blood. Standard Methods of Clinical Chemistry 4: 31–38 (1963).

7 Bonsignore, D., P. Callessano, and C. Cartasegna: A Simple Method for the Determination of Blood Delta-Aminolevulinic Dehydrase. Med. d. Lavoro 56: 199–205 (1965) and Bulletin of Hygiene 40: 1352 (1965).

8 Bitke, K., and E. Kleinhauer: Distribution of HbCO in Human Erythrocytes Following Inhalation of CO. Nature 227: 386 (1970).

9 Azar, A., H. J. Trochimowicz, and M. E. Maxfield: Review of Lead Studies in Animals Carried Out at Haskell Laboratory — Two Year Lead Feeding Study and Response to Hemorrhage Study. International Symposium on the Environmental Health Aspects of Lead, Amsterdam, October 2–6 (1972). Abstract of Papers. Submitted to Proceedings.

10 Moore, M. R., A. D. Beattie, G. G. Thompson, and A. Goldberg: Depression of δ-

Aminolaevulic Acid Dehydrase Activity by Ethanol in Man and Rat. Clinical Science 40: 81–88 (1971).

11 Pecora, L., S. Fati, and C. Vecchione: The Behavior of the Erythrocytic Protoporphyrins and Urinary Coproporphyrins in Carbon Monoxide Intoxication. (Il comportamento delle protoporfirine eritrocitarie e delle coproporfirine urinarie nell intossicazione da ossido di carbonio.) Boll. Soc. Ital. Biol. Sper. 32: 655–656 (1956).

12 The Health Consequences of Smoking. A Public Health Service Review: (1967) Public Health Service Publication No. 1696. Revised January 1968.

13 Hofreuter, D. H., E. J. Catcott, R. G. Kennan, and C. Xintaras: The public health significance of atmospheric lead. Arch. Environ. Health 3: 568–574 (1961).

14 Survey of lead in the atmosphere of three urban communities. Public Health Service Publication No. 999 AP–12, January 1965.

15 Lehnert, G., K. H. Schaller, A. Kuehner, and D. Szadkowski: Effects of cigarette smoking on the blood lead level. Inter. Arch. Ind. Path. & Hyg. 23: 358–363 (1967).

16 McLaughlin, M., and G. J. Stopps: Smoking and Lead Presented at the 1972 American Academy of Occupational Medicine Meeting. Accepted for publication in Arch. Environ. Health.

17 Hapke, H. J.: Subclinical Lead Poisoning in Sheep. Abstract of Paper Presented at International Symposium on the Environmental Health Aspects of Lead. Amsterdam, October 2–6 (1972).

Recent Chemical and Biological Studies on Natural Products at the Central Drug Research Institute, Lucknow.

By **M. L. Dhar**

The vital role of plants in the mechanics of operation of the biosphere is of prime relevance to scientific effort in the field of eco-environment. Undoubtedly, plants constitute a most important balancing link in stabilizing the economic functioning of the biosphere and are also the most generous benefactors of the animal kingdom. A large variety of plants act as, or provide, food and many have been used, since antiquity, for relieving distress of physical pain and/or pathology in both humans and animals. However, a fair number are also strongly poisonous and indeed most of the drugs derived from the plant kingdom owe their role to their inherent toxic character. Only, these toxic properties are rather selective and whereas these may destroy a pathogen or relieve a metabolic disorder in man and animal selectively, they are by and large less toxic to the host and therefore beneficial to him. So, besides cleansing the air from the noxious carbon dioxide and providing food, plants also help man meet the challenges of pathogenicity such as may be caused by intemperate environment. In the underdeveloped or even in the developing countries in the tropics, inappropriate hygienic conditions lead to wide spread pollution of air, water and food and promote such disabling or killing diseases as malaria, amoebiasis, filariasis, cholera, other gastrointestinal infections etc. Drugs of plant origin such as quinine, emetine and conessine have been of consider-able help in the treatment and indeed in near elimination of some of these diseases and this has been reinforced by the control of insect vectors by pyrethrins provided by a beautiful flowering plant *Chrysanthemum cinerari folium.*

The above few examples have been given only to emphasize the role of plants in the regulation and control of environmental hazards arising out of underdevelopment which a large section of human race must face for a long time yet.

Search of curatives and palliatives from plant sources is universal and the Central Drug Research Institute at Lucknow is the largest centre in India for such search.

Ten years ago we decided to make a material departure from the classical methods employed in the investigation of medicinal plants at the Central Drug Research Institute. We adopted a system of pre-screening of total extracts of plant materials, over a wide spectrum of biological tests, before launching on detailed studies on products with significant biological activity. A combination of preliminary extraction and biological test procedures permit the isolation of active chemical entities and their subsequent identification. The inactive or other biologically unimportant molecules get eliminated in the process.

Our experience over the years has demonstrated that this procedure is perhaps more economical and is in any case simpler and rather less fortuitous than the classical method of investigation of natural products.

At the Central Drug Research Institute, we have collected over 1700 plant species over the last eight years and the extracts of over 1600 plant materials have been put through a comprehensive biological screen. This screen has been modified from time to time and comprises, at the moment, 91 standardised test system each plant extract is put through as many of these tests as possible.(1)

Once an activity is detected in a crude extract in this primary screen, the extract is fractionated by a standard procedure and the fractions retested for the activity observed in the first screen. In most of our biological test systems, we obtain an efficiency of better than 85% at this confirmatory stage. Thus, of the plant extracts found active in the primary screen, 85% or more confirm at the first fractionation stage and are then followed up.

We have collected these plants from nearly all parts of India. The choice of collection of individual plants is guided by, among other factors, the availability of a given plant in a particular area, in sufficient quantity to permit the follow up studies. Care is also taken throughout this study, to eliminate, as far as possible, ecology-dependent interfering factors, by ensuring that repeat collections of plant materials are restricted to the areas from which the first collections were made and at the same time of the year. The overall picture of the distribution of the collected plants within higher taxa is as follows:

Mycophyta (Fungi & Lichens)	3
Bryophyta (Mosses & Liverworts)	1
Pteridophyta (Lycopods & Ferns)	35
Gymnospermae	20
Angiospermae	ca. 1650

These plants cover about 1000 genera and the higher plants, namely the angiosperm and the gymnosperms, belong to about 175 families in terms of Engler and Prantle classification. These 175 families comprise almost all the families that are recorded in India except for a few families of aquatic plants which we have not collected so far.

In our hands, 73 plant materials have confirmed for antitumour activity, ten have shown good antibacterial and antifungal properties and about 70 plants have shown promising pharmacological properties, such as hypotensive, spasmolytic, diuretic, anti-inflammatory, anticonvulsant and cardiotonic. Sixteen plants have been found to possess significant anti-viral activity and one plant extract has moderate hypoglycaemic property. These 170 plants belong to 63 families.

Some of the more recent chemical and biological work carried out in this area at the Central Drug Research Institute is outlined in this article.

Arnebia nobilis Reich. of the family Boraginaceae has furnished eight naphthaquinones from the root extract. These have been designated as arnebins 1, 2, 3, 4, 5, 6, 7 and 8 and the first six have been identified as alkannin β,β-dimethylacrylate, dihydro-hydroxyalkannin-β, β-dimethylacrylate, alkannin monoacetate, alkannin, dihydro-hydroxyalkannin and dihydro-hydroxyalkannin acetate respectively.(2)

Although all the arnebins possess antibacterial and antifungal activities, the most potent is arnebin-2 which inhibits *Staphylococcus aureus, Bacillus subtilis, Candida albicans* and *Cryptococcus*

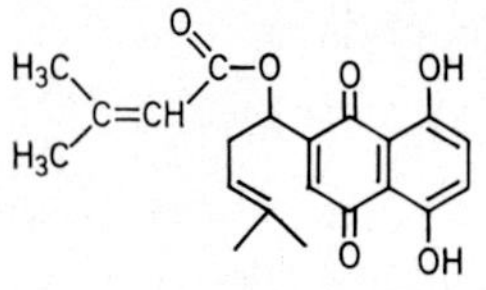

Arnebin-1

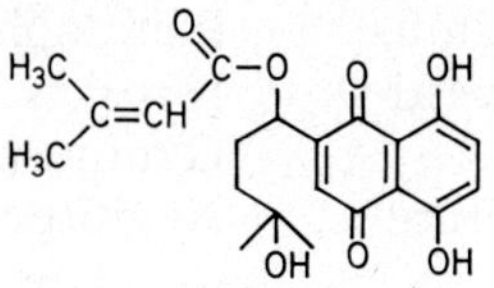

Arnebin-2

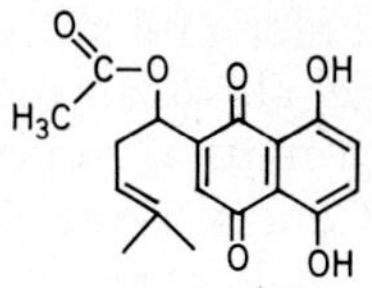

Arnebin-3

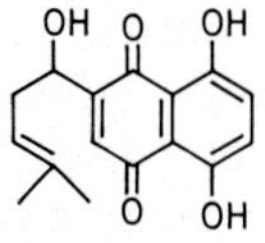

Arnebin-4

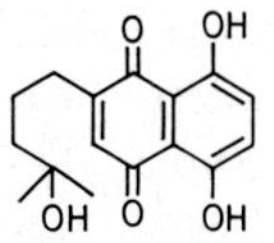

Arnebin-5

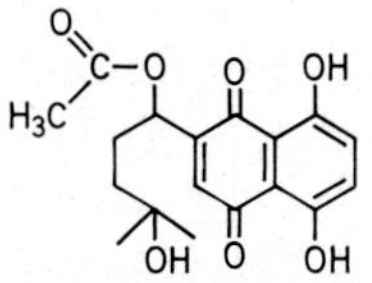

Arnebin-6

neoformans at a concentration of 6.25 μg/ml.

The major constituent, arnebin-1, has potent antitumour activity in WM system. Its LD_{50} in mice is 12.5 mg/kg i.p. and detailed work on its pharmacodynamics has been carried out. Toxicity studies with a view to its development as an anticancer drug, are now in progress.

Mappia foetida Miers. of the family Olacaceae has given two potent anticancer alkaloids, camptothecin and 9-methoxy camptothecin (3) in a total yield of 0.25%.

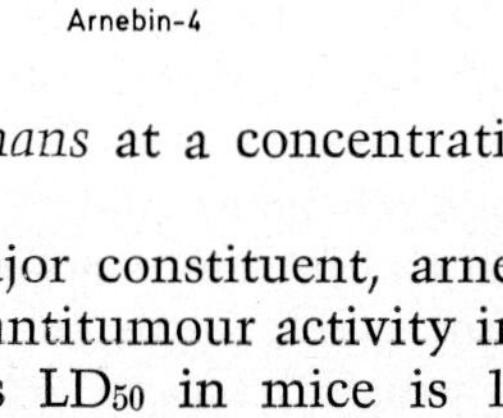

camptothecin R = H
9-Methoxy camptothecin R = OCH$_3$

Until this study was undertaken, camptothecin was found only in a rare Chinese species Camptotheca acuminata and in a low yield of 0.005%. Camptothecin is active in PS, KB and LE systems. In the PS system particularly, this base has high order of activity at a dose level of 1 mg/kg in mice. Some clinical trials have been conducted with camptothecin by Na-

tional Institutes of Health, Bethesda, Maryland, U.S.A. Recent reports, however, indicate that the earlier enthusiasm generated by this compound, for its wide spectrum anticancer activity, is now on the decline. Even so, however, it is expected that camptothecin may prove to be of value for some forms of malignancies particularly those of the rectum and the intestine.

Nicotiana plumbaginifolia Viv. of the Solanaceae family gave an antitumour compound which we have named Solaplumbin. Solaplumbin has been shown to be β-D-glucosyl-β-L-rhamnosyl solasodine.(4)

In WM system, Solaplumbin is active at a dose of 10 mg/kg in rats. Its LD_{50} has been found to be 75 mg/kg i.v. Detailed pharmacological studies including gross effects, CNS, CVS, anticonvulsant, analgesic effects etc. do not show any untoward properties which would preclude its use in therapeutics. Detailed study of Solaplumbin in the

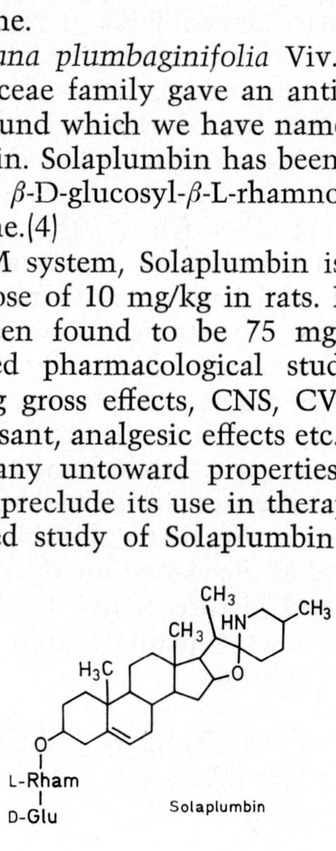

Solaplumbin

Celsioside - I: R=glu-Fuc-rham
Celsioside - II: R=glu-glu-Fuc -rham

$$\begin{array}{ccccc} & \text{D-Glu} & \text{C}_5\text{H}_{11} & & \text{OH} \\ & | & | & & | \\ \text{L-Rham} & -\text{O} - \text{CH} - (\text{CH}_2)_7 - \text{CH} - \text{CH}_2 - \text{COOH} \\ | & & & \\ \text{D-Fuc} - \text{L-Rham} \end{array}$$

Ipolearoside

PS system is now in progress and we expect to arrive at an evaluation of this material at an early date.

The crude extract of *Celsia coromandeliana* Vahl of the family Scrophulariaceae, showed significant antitumour activity in WM system and it gave two saponins, celsiosides I and II. These have been identified as triglycosides (glucose, fucose and rhamnose 1:1:1; 2:1:1 respectively) of olean-11,13(18)-dien-3β, 22β, 28-tetrol.[5]

The common sapogenin and indeed both these saponins have been reported for the first time. The exact configuration and the precise position of attachment of the carbohydrate moiety have, however, not been worked out so far.

Celsioside I is active at 6.25 mg/kg in WM system both by oral and i.p. routes. It is active in the PS system. The pharmacodynamic data and follow up work for its evaluation, as a potential anticancer agent, are now in progress.

Ipomoea leari Paxt. of Convolvulaceae family, gave an antitumour glycoside named ipolearoside, which was isolated in a yield of 0.05%. It showed activity at 2 mg/kg in WM system in rats. This compound has been shown to be tetra-

glycoside (glucose, rhamnose, fucose 1:2:1) of 3,11-dihydroxy-hexadecanoic acid.[6]

Tithonia tagitiflora Desf.

Fam: Compositae

Crude extract of this whole plant was found to be active in PS system and it gave 6 new sesquiterpene lactones (germacranolides A—F). Structures of all these six germacranolides, as also their interactions, have been established.

In all these germacranolides, the evidence for the common structural features namely the ester grouping, the lactone function and the double bonds was obtained from spectral data. All these six compounds furnish, on catalytic hydrogenation, products characterised by the presence of a newly generated carbonyl and a tertiary hydroxyl group establishing the presence of a hemiketal function in all these molecules. The relative position of the functional groups was determined by NMR and mass spectral data and the relevant chemical studies.[7]

Substance A has been found to possess pronounced antitumour activity. Interest in sesquiterpene lactones has been renewed recently because of the discovery of anticancer activity, especially in sesquiterpene lactones with multiple oxygen functions. Prof. Kupchan, Herz and their coworkers have shown that all the plants elaborating active lactones belong to the Compositae family. However, as happens all too often, anticancer activity detected in a given material does not always prove meaningful in terms of potential clinical desirability. In the present case it does seem interesting, however, that tumour inhibiting activity is associated with a high degree of unsaturation in this class of compounds; the less unsaturated molecules are inactive.

① $-O-CO-CH(CH_3)_2$ — IR, MS, PMR

②   [structure: H_3C, CH_3 lactone with =CH_2]   — IR, UV, PMR

③   Hemiketal group   — IR, UV, CD, PMR, PMR with TAI

[chemical structures]

Common structure $\xrightarrow{H_2}$ Hydro derivate

Substance B
$C_{19}H_{26}O_6$

Substance A
$C_{19}H_{24}O_6$
(Active in PS system)

Substance C
$C_{19}H_{28}O_6$

Substance D
$C_{19}H_{24}O_6$

Substance E
$C_{19}H_{26}O_7$

Substance F
$C_{19}H_{28}O_7$

Cocculus pendulus (Forsk) Diels.
Fam: Menispermaceae
The anticancer and hypotensive activities of the crude extract of the whole plant were found to reside in the alkaloidal fraction. Three new alkaloids namely pendulin, cocsulin and cocsu-

linin, belonging to the bisbenzyl isoquinoline group, were isolated and their structures elucidated.(8)
Pendulin is hypotensive but has the undesirable side-effect of causing histamine release. Cocsulinin, on the other hand, possesses marked antitumour activity in the KB system.

[chemical structures]

Pendulin

Cocsulin

Cocsulinin

This is an interesting example of the variation of biological activity arising out of only slight change in the gross structural features of the constituents belonging to a same chemical group.

Annona squamosa Linn.

Fam: Annonaceae

Seven aporphine alkaloids, anonaine, roemerine, norcorydine, corydine, nor-isocorydine, iso-corydine and glaucine have been isolated from the leaves of this plant.[9] The yields of these bases are rather poor, ranging from 0.001 to 0.002 per cent.

Corydine has been found to possess antitumour avtivity in KB system (cell culture). Further *in vivo* experiments are under way.

Coleus forskohlii Brig. (= *C. barbatus*)
Fam: Labiatae

Benzene-soluble fraction of the alcoholic extract of the roots of this plant was found to be hypotensive. The active constituent, named coleonol, was found to be a terpenoid, $C_{22}H_{34}O_7$, m.p. 220°. Elucidation of its structure is now in progress.[10]

Coleonol has an LD_{50} of 75 mg/kg in mice i.p., and in the anaesthetised cat, administration of the compound by i.v. route produces a hypotensive effect in low doses. Thus 50 μg/kg of coleonol produced a mean fall of 53 mm lasting for 30 minutes. In addition, pressor and depressor responses to noradrenaline, acetylcholine, histamine and isoprenaline were inhibited to the extent of 25—30%. There was no significant effect on auricular and ventricular contractions at this dose level.

On the rabbit ileum, concentration of 1–10 μg/ml gave a marked relaxant effect and potentiated the relaxant effect of isoprenaline and adrenaline. 1 μg/ml concentration of coleonol also produced 68, 84 and 88 per cent antagonism of acetylcholine, histamine and serotonin induced contractions respectively of guinea pig ileum. In addition 5 μg/ml produced 92% antagonism of nicotine contractions also.

It would thus seem that both the hypotensive and the spasmolytic activities of

Anonaine (R=H)

Roemerine (R=CH₃)

Glaucine

Cocculidine

Iso-cocculidine
$C_{18}H_{23}NO_2$

Corydine (R=CH₃)
Norcorydine (R=H)

Isocorydine (R=CH₃)
Norisocorydine (R=H)

Base C

Base A

coleonol are due to direct relaxant effect on smooth muscles. Further detailed work on this interesting terpenoid is now in progress.

Cocculus laurifolius DC.

Fam: Menispermaceae

Crude extract of the leaves of this plant showed hypotensive and neuromuscular blocking activities and these properties were located in the alkaloidal fraction. The ether — insoluble fraction showed hypotensive action whereas the ether — soluble fraction was found to possess both hypotensive and neuromuscular blocking activities. Four new alkaloids and the known base cocculidine have been isolated from the latter fraction.[11]

The major new base isococculidine has hypotensive and neuromuscular blocking properties. A dose of 10 mg/kg i. v. caused a hypotension of 60—80 mm for 90—120 minutes which was accompanied by respiratory failure, blockade of the nictitating membrane and failure of neuromuscular transmission. Isococculidine exerts a direct depressant action on the skeletal muscle. Direct and indirect stimulation of the skeletal muscles was blocked equally and there was no recovery up to 120 minutes. This interesting compound is now under study with a view to modify some of its structural features with a view to separate the two identified activities and obtain a specific and useful hypotensive or neuromuscular blocking agent.

Vittadinia australis A. Rick.

Fam: Compositae

Butanol fraction of the total extract of the whole plant was found to possess diuretic activity and the active constituent was identified as α-spinasterol ($\Delta^{7,22}$ — stigmasterol) glucoside.[12] Its diuretic activity is of the order of 85% of chlorthiazide at equivalent dose levels (125 mg/kg orally). Its ultimate emergence as a therapeutically effective diuretic will naturally depend on the results that may be obtained when its pharmacodynamic and other properties are compared with those of the commonly used diuretics.

Ancistrocladine

Ancistrocladinine

Ancistrocladisine

Cedricladine

HO CH₃ — Himachalol $C_{15}H_{26}O$

H OH CH₃ — Allohimachalol $C_{15}H_{26}O$

HO CH₃ — Substance B $C_{15}H_{26}O_2$

Substance A $C_{15}H_{26}O$

Substance C $C_{15}H_{26}O_2$

Ancistrocladus heyneanus Wall.
Fam: Ancistrocladaceae
An alkaloidal fraction obtained from the benzene and chloroform — soluble portions of the crude alcoholic extract of the whole plant, showed spasmolytic activity. Four alkaloids, ancistrocladine $C_{25}H_{29}O_4N$, ancistrocladinine $C_{25}H_{27}O_4N$, ancistrocladisine $C_{26}H_{29}O_4N$, and a new alkaoid, cedricladine $C_{25}H_{27}O_4N$ have now been isolated from this fraction. The structures of the first three alkaloids have recently been elucidated by Prof. Govindachari and his colleagues.[13] The structure of the fourth base cedricladine has now been elucidated and it has been found to be related to the other three alkaloids.[14] The *in vitro* spasmolytic activity of the major alkaloids, ancistrocladine and cedricladine, has been compared with that elicited by papaverine hydrochloride at equivalent dose levels and cedricladine has been found to be nearly equipotent with papaverine.
Cedrus deodara (Roxb.) Loud.
Fam: Pinaceae

Significant spasmolytic activity was found in the hexane soluble fraction of the alcoholic extract of the stem wood of this plant. The isolation of the active constituents was rather difficult because they appear in close proximity on the T.L.C. The isolation of the individual constituents was facilitated by a close coupling of the chromatographic procedures with NMR spectroscopy and assessment of biological activity. This has resulted in the isolation of the spasmolytic constituents which were identified as himachalol, allo-himachalol, a new isomer, substance A, and two new sesquiterpene diols, substances B and C.[15] The structures of himachalol and allo-himachalol were elaborated earlier by Dr. Sukh Dev and his colleagues.[16]
The relative *in vitro* spasmolytic activities of all the constituents isolated and of papaverine hydrochloride, used as the test standard, are given hereunder.
Himachalol, the major constituent, present in an yield of ca. 0.5⁰/o has

	Dosage μg/ml	Anti-acetyl-choline	Antihistamine	Antinicotine	Antiserotonin
Himachalol	1	23	69	84	77
Allo-himachalol	1	90	81	77	80
Substance A	5	30	29	49	23
Substance B	10	74	63	80	40
Substance C	5	31	56	22	0
Papaverine HCl	1	4	27	17	10
	5	5	63	70	76

been pharmacologically evaluated *in vitro* as well as in *in vivo* systems with papaverine HCl as the standard. Its LD_{50} was found to be 265 mg/kg, i.p. in mice, while LD_{50} of papaverine HCl was 116 mg/kg.

In guinea pig ileum and rabbit jejunum preparations *(in vitro)*, himachalol was found to be 2—3 times as active as papaverine HCl in antagonizing the contractions induced by various spasmogens like acetylcholine, histamine, serotonin and nicotine etc. In *in vivo* studies, it was found to be equipotent with papaverine HCl in inhibiting gastrointestinal propulsion of charcoal meal in rats and mice. However, the inhibition of carbachol-induced spasm of intestine, in immobilized but conscious cats, was found to be a little less for himachalol than for papaverine HCl.

Further, when administered i.p., in rats 30 minutes before the test meal, papaverine HCl was found to be superior to himachalol. This interesting investigation did not, however, lead to a place for himachalol as a new drug because it is less easily absorbed from the GI tract and has lower order of bronchial and vasodilatory effects as compared to those elicited by papaverine hydrochloride. Since, however, both himachalol and allo-himachalol are water-insoluble, attempts have been made to synthesise their water-soluble derivatives. One such product, sodium salt of succinyl monoester of allo-himachalol is now under study.

References

1 Dhar, M. L., M. M. Dhar, B. N. Dhawan, B. N. Mehrotra *et al.*: Part I, Ind. J. Exp. Biol., 6, 232 (1968). Part II, ibid, 7, 250 (1969). Part III, ibid, 9, 91 (1971). Part IV, ibid, 11, 43 (1973).

2 Shukla, Y. N., J. S. Tandon, D. S. Bhakuni, and M. M. Dhar: Experientia, 25, 35 (1969); idem., Phytochem., 10, 1909 (1971). Y. N. Shukla, J. S. Tandon, and M. M. Dhar: Indian J. Chem., 11, 528 (1973).

3 Agarwal, J. S., and R. P. Rastogi: Ind. J. Chem., II, 969 (1913).

4 Singh, S., N. M. Khanna, and M. M. Dhar: Phytochem., 13, 2020 (1974).

5 Agarwal, S. K., and R. P. Rastogi: Indian J. Chem., 12, 304 (1974).

6 Sarin, J. P., H. S. Garg, N. M. Khanna, and M. M. Dhar: Phytochem., 12, 2461 (1973).

7 Palstet, R., D. K. Kulshreshtha, and R. P. Rastogi: (Unpublished work).

8 Bhakuni, D. S., N. C. Gupta, and M. M. Dhar: Experientia, 26, 12 (1970); ibid., 26, 241 (1970).

9 Bhakuni, D. S., S. Tewari, and M. M. Dhar: Phytochem., 11, 1819 (1972).

10 Tandon, J. S., and M. M. Dhar: (Unpublished work).

11 Upreti, Hema, R. S. Kapil, D. S. Bhakuni, and M. M. Dhar: (Unpublished work).

12 Sarin, J. P., H. S. Garg, and N. M. Khanna: (Unpublished work).

13 Govindachari, T. R., and P. C. Parthasarathy: Indian J. Chem., 8, 567 (1970); idem, Tetrahedron, 27, 1013 (1971). T. R. Govindachari, P. C. Parthasarathy, and H. K. Desai: Indian J. Chem., 9, 1421 (1971); ibid., 10, 1117 (1972).

14 Shukla, Y. N., J. S. Tandon, and M. M. Dhar: (Unpublished work).

15 Kulshreshtha, D. K., K. Kar, B. N. Dhawan, and R. P. Rastogi: (Unpublished work).

16 Bisarya, S. C., and Sukh Dev: Tetrahedron, 24, 3861 (1968).

Long-Lived Radionuclides in Food: ^{90}Sr and ^{137}Cs in the Israel Diet *)

By Y. Feige, A. Eisenberg†, Y. Proulov and J. Klopfer

Radiation Safety Department, Soreq Nuclear Research Center, Yavne, Israel
Ministry of Health, Jerusalem, Israel

Abstract

Routine monitoring of Strontium-90 and Cesium-137 levels in the main classes of food consumed in Israel has been carried out for a number of years. Calculations have been made of the intake of these isotopes by various groups in the Israel population using Food Consumption Survey data made available by the Israel Central Bureau of Statistics. More detailed analyses have been carried out to calculate the exposure of groups selected on the basis of demographic, economic and family-size criteria. This was carried out in an attempt to detect "critical groups" in the population who might be overly exposed.

Results of these data indicate that in the case of Israel the monitoring of a single food item such as milk or the extrapolation of local fallout data cannot be relied on to give an estimate of radionuclide exposure through food.

The relation between high consumption groups and the intake of "average" consumers is discussed in detail and the use of the statistical models developed in these studies for the estimates of radionuclide exposure during periods of high fallout levels which occured in the past and possible applica-

tions with regard to other environmental contaminants is discussed.

Introduction

Routine monitoring of ^{90}Sr and ^{137}Cs in the main classes of food consumed in Israel has been carried out for the last 10 years. Although the levels encountered at present are low, studies of radionuclides in food may assist in predicting the possible exposure levels to the population during periods of high radioactivity in the environment such as might be experienced during accidental releases.

At present atmospheric testing of nuclear devices is still being carried out by China in the Northern Hemisphere and by France in the Southern Hemisphere. Fallout deposition rates in the world decreased markedly a few years after the cessation of large scale nuclear testing in 1961—62. However since 1968 this trend has not continued at the rate observed previously and levels though relatively low have been stable.(1, 2) The transfer of these long-lived radionuclides through food-chains to man can be studied more conveniently than during periods of higher levels associated with large periodic fluctuations.

This work describes an attempt to use Food Consumption Survey data supplied by the Israel Central Bureau of Statistics to estimate the exposure of groups in the population through food. The groups were selected by the Bu-

*) Published in Comparative Studies of Food and Environmental Contamination (Proc. FAO/IAEA/WHO Symp. Otaniemi, 1973), IAEA Vienna (1974) 3—22.

reau on the basis of demographic, economic and family-size criteria. A detailed analysis of the intake patterns was carried out for these groups in order to estimate the "average" population exposure which can be compared with data published for other countries, but which was found to be of limited significance. Of more interest was the attempt to study "high consumption" groups in the population who might be subjected to a higher degree of exposure to radionuclides as a result of their dietary habits.

These studies have also examined the practicability in Israel of extrapolating radionuclide exposure through food from routine monitoring of atmospheric deposition data, or by the analysis of a single food item such as milk which has been suggested as a reliable criterion in other countries.

The behaviour of long-lived radionuclides in the environment and their transfer to food can serve as a model for understanding problems connected with other contaminants, such as heavy metals, which have become of great interest in the past years.

Sampling and Method of Analysis

The food items tested were purchased periodically in Rehovoth and were selected to cover the common foods consumed in Israel. Results are reported relative to the gross weight of each food item as purchased (i.e. including bones, peels etc. although the radiochemical analysis was carried out on the edible portion only.

Fifteen to twenty food samples were analyzed each month and classified into ten main groups for which food consumption data for the Israel population are available.

Radiochemical Procedures

The method used is based on beta counting. For ^{137}Cs, this ensures a higher total efficiency than the simple gamma determination. For ^{90}Sr, by counting the ^{90}Y daughter instead of the parent directly, higher efficiencies are obtained and the results can be checked by following the decay of ^{90}Y. Moreover, the laborious separation of ^{90}Sr from Ca is avoided.

The main laboratory steps are as follows. The sample is incinerated at 450° C and the ash dissolved in HCl. Sr is precipitated as carbonate. From the remaining solution Cs is precipitated as perchlorate and beta-counted. The chemical yield of Cs is determined with an atomic absorption spectrophotometer.

The carbonates containing Sr are dissolved in HCl and, after several purification stages, Sr is obtained (together with Ca) in the HCl solution. Yttrium carrier is then added and the solution is kept for about two weeks to allow the ^{90}Y daughter to achieve equilibrium with the ^{90}Sr. The Y is milked as hydroxide and transformed to oxalate for beta-counting. The chemical yield of Y is determined by converting the oxalate to Y_2O_3 and weighing.

The ^{90}Sr ist then calculated taking into account appropriate correction factors connected with counting efficiency and the chemical yields of Y and Sr. The latter chemical yield is determined by an atomic absorption spectrophotometer.

Results

Estimates of the National Average Intakes of ^{90}Sr and ^{137}Cs for the years 1968—69 are presented in Table I. The

Table 1   Long-Lived Radionuclides in the Diet in Israel

Food	Average daily consumption (gross weight, as purchased) g/day/person	^{90}Sr		^{137}Cs	
		Concentration pCi/kg	Intake pCi/day/person	Concentration pCi/kg	Intake pCi/day/person
Fresh fruit	260	3.0	0.78	5.4	1.40
Fresh vegetables	200	2.6	0.52	5.5	1.10
Milk	130	1.9	0.25	6.1	0.79
Milk products	45	9.9	0.45	8.5	0.38
Eggs	40	2.1	0.08	5.9	0.24
Meat, poultry, fish	135	4.8	0.65	18.0	2.43
Vegetable oils	24	4.5	0.11	6.3	0.16
Bread	210	6.4	1.34	18.7	3.93
Grain products	54	7.3	0.40	19.3	1.04
Total	1098		4.58		11.47
Miscellaneous: (Tea, Cocoa, coffee, Nuts etc.)*)	15		0.30		2.50

*) This item is not included in the total because of the large variation in consumption of the individual components.

calculations were made for this period to coincide with the Food Consumption Survey. The highest contribution of long-lived radionuclides to the diet was found to be due to the food group including bread and grain products. Because of the substantial wheat imports from North America during this period (during 1968—69 about 40% of the grains were imported) these levels were higher than if locally produced grains were consumed exclusively. Fresh fruit and vegetables have relatively low activity, but because of the quantities consumed by the Israel population they become significant. Animal proteins are not consumed in quantities comparable to Western countries but because of the high ^{137}Cs levels in these foods the radionuclide intake is important. The relatively low levels contributed in general by milk and milk products has been reported previously.(3) The intake from the miscellaneous group which includes tea, coffee, chocolate, nuts, etc. cannot be evaluated accurately with the data available at present. Some of the values obtained were very high and the spread was very large. In spite of the fact that average daily intakes are very small in some individuals these food items may conceivably become a major source of ingested radioisotopes as a result of dietary habits.

From the data presented it is clear that for the period under discussion the intake of ^{90}Sr and ^{137}Cs through food in Israel was strongly dependent on the radioactive levels of wheat grown abroad. Thus intake cannot be predicted solely on the basis of correlations with local fallout deposition levels.

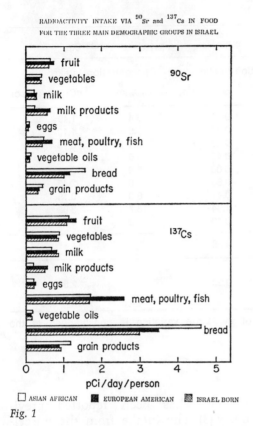

RADIOACTIVITY INTAKE VIA ^{90}Sr and ^{137}Cs IN FOOD
FOR THE THREE MAIN DEMOGRAPHIC GROUPS IN ISRAEL

□ ASIAN AFRICAN   ■ EUROPEAN AMERICAN   ▨ ISRAEL BORN

Fig. 1

Variations in ^{90}Sr and ^{137}Cs Ingested by Different Groups in the Population

Dietary habits vary significantly among the different sections of the population. This may lead to significant differences in radionuclide intake. Calculations were made of the radionuclide intakes for the different demographic, economic and family-size groups covered by the survey.

Food Consumption data were derived from the Consumption Survey carried out in 1968–69 by the Israel Central Bureau of Statistics. The survey covered a representative sample of urban households where all food purchases were recorded. Food consumed outside the home was not included. The demographic groupings were based on the country of birth of the head of the family. A computer program was prepared to convert raw food consumption survey data to radionuclide intakes via the various foods. Ten levels of con-

Table 2   Average Radionuclide Intake on the Basis of Income (picocurie / person /day)

Food income	Sr - 90			Cs - 137		
	H	M	L	H	M	L
Fresh fruit	.70	.82	.90	1.26	1.48	1.61
Fresh vegetables	.51	.49	.43	1.08	1.03	.94
Milk	.25	.26	.25	.82	.86	.81
Milk products	.63	.48	.37	.54	.41	.32
Eggs	.47	.44	.36	1.32	1.24	1.01
Meat, poultry, fish	.65	.59	.55	2.42	2.23	2.09
Vegetable oils	.12	.13	.13	.18	.19	.17
Bread	1.20	1.54	1.61	3.51	4.50	4.70
Grain products	.44	.47	.46	1.16	1.25	1.18
Total	4.97	5.22	5.06	12.29	13.19	12.83

H = high income
M = medium income
L = low income

Table 3 Average Radionuclide Intake on the Basis of Family-Size (Picocurie/Person/Day)

Family of	Sr − 90						Cs − 137					
	1	2	3	4	5	6+	1	2	3	4	5	6+
Food												
Fresh fruit	.79	.94	.83	.75	.75	.65	1.43	1.71	1.50	1.36	1.35	1.17
Fresh vegetables	.57	.63	.53	.48	.43	.44	1.18	1.33	1.12	1.01	.92	.93
Milk	.32	.31	.27	.29	.27	.21	1.03	.98	.88	.92	.88	.70
Milk products	.85	.82	.65	.61	.46	.26	.76	.73	.58	.54	.40	.23
Eggs	.39	.41	.41	.47	.44	.50	1.10	1.14	1.15	1.31	1.25	1.40
Meat, poultry, fish	.78	.89	.71	.61	.58	.49	2.90	3.35	2.66	2.29	2.18	1.85
Vegetable oils	.11	.15	.13	.12	.13	.13	.16	.21	.18	.17	.17	.18
Bread	1.41	1.59	1.34	1.26	1.54	1.10	4.12	4.64	3.91	3.69	4.50	3.20
Grain products	.41	.50	.46	.41	.44	.50	1.07	1.33	1.21	1.08	1.16	1.34
Total	5.63	6.24	5.33	5.00	5.04	4.28	13.75	15.42	13.19	12.37	12.81	11.00

sumption were calculated for each food group but only some of these levels have been used in the calculations presented. In Fig. 1 the 5'th decile intakes of ^{90}Sr and ^{137}Cs by the three main demographic groups are presented.[4] This diagram shows in graphic form the contribution of the various food items and their variation among the different demographic groups.

Table II shows the average consumption of these radionuclides by Economic grouping and Table III shows the average radionuclide intakes calculated for the different Family-Size groups.

From our data we computed the consumption for "low consumers" (the 2'nd decile), "high consumers" (the 9'th decile) and "very high consumers" (the 10'th decile) normalized to the average (1/2.5'th and 1/2.6'th deciles).

Table 4 Variations in Food Consumption for Different Income Level Groups
[The Intakes of the Different Groups are Normalised to the Average]

Income	High			Middle			Low		
Food	A	B	C	A	B	C	A	B	C
Fresh fruit	.41	2.1	3.0	.44	2.2	2.9	.49	1.7	2.0
Fresh vegetables	.53	1.7	2.5	.49	1.7	2.4	.44	1.9	3.0
Milk	.28	1.9	2.8	.35	1.8	2.4	.24	1.8	2.8
Milk products	.41	2.1	3.1	.33	2.3	3.9	.11	2.6	4.9
Eggs	.49	1.7	3.2	.45	1.7	3.9	.39	1.9	4.1
Meat, poultry, fish	.55	1.7	2.5	.58	1.7	2.4	.49	1.9	2.8
Vegetables	.42	1.8	2.7	.38	1.9	2.7	.31	2.0	3.2
Bread	.51	1.5	2.2	.60	1.5	1.9	.39	1.8	2.0
Grain products	.48	1.8	2.4	.45	1.7	2.1	.39	2.2	3.4

Values are calculated for food intake. These same ratios apply to radionuclide intake.
A = 2nd Decile / Average
B = 9th Decile / Average
C = 10th Decile / Average

Table 5 Variations in Food Consumption for Different Family-Size Groups
[The Intakes of the Different Groups are Normalised to the Average]

Family of	1			2			3			4			5			6+		
Food	A	B	C	A	B	C	A	B	C	A	B	C	A	B	C	A	B	C
Fresh fruit	.19	2.8	3.8	.40	2.3	2.5	.48	1.8	2.6	.48	2.0	2.9	.50	1.7	2.4	.43	1.7	2.5
Fresh vegetables	.10	3.0	3.0	.46	2.2	2.3	.53	1.8	2.4	.48	1.7	2.3	.57	1.6	2.3	.54	1.7	2.1
Milk	.01	2.5	2.5	.26	1.7	2.4	.26	1.9	2.7	.43	1.7	2.4	.24	2.0	2.4	.35	1.7	2.4
Milk products	.13	2.9	2.9	.37	2.6	2.6	.42	2.0	3.0	.42	2.0	2.9	.38	2.0	3.3	.22	2.6	4.2
Eggs	.07	2.7	4.0	.42	1.7	3.1	.37	1.8	3.3	.53	1.6	3.0	.37	1.8	2.6	.47	1.6	2.5
Meat, poultry, fish	.12	2.0	2.4	.50	2.0	2.0	.52	1.7	2.2	.57	1.6	2.1	.60	1.6	2.0	.58	1.5	2.3
Vegetable oils	.04	4.0	5.3	.35	2.1	2.6	.38	1.9	2.8	.43	1.8	2.4	.43	1.9	2.8	.41	1.8	2.7
Bread	.30	2.2	2.7	.58	1.7	2.0	.55	1.6	2.2	.59	1.6	2.1	.56	1.5	2.0	.58	1.6	1.8
Grain products	.13	3.7	3.5	.42	2.2	2.8	.41	1.7	2.9	.48	1.7	2.6	.52	1.7	2.6	.48	1.7	2.8

Values are calculated for food intake. These same ratios apply to radionuclide intake.

A =   2nd decile / average
B =   9th decile / average
C =   10 th decile / average

These intake ratios are presented in Table IV for the Economic groups and in Table V for the Family-Size groups.
In these tables a number of points regarding the dietary habits of these groups and the ways in which it was estimated stand out:
1) The average intakes of the sub-groups of the Economic and Family-Size groups vary to a much lesser extent than the variations within the deciles comprising each group. Thus if there is a "critical group" to be detected it will be difficult to define them using this type of data, but it will be possible to estimate the *intake* of such a group as expressed by the "very high consumers" category.
2) The intakes recorded for the one person family-size may be misleading because the survey recorded only home food purchases. The data appearing in Table III where the per capita intake of the 2 person family is consistently higher than the single person who takes part of his meals in restaurants etc. may reflect this type of error in esti-mating exposure through food ingested.

Discussion

The estimation of population exposure to radiochemical residues in food has been studied for an extensive period and found to be very complex.
Figures based on National Average Food Intakes are at best rough estimates which may camouflage the main point of interest, namely the possibility of predicting the extent and nature of high exposure of groups in the population to a particular contaminant.
Using the findings of the Israel Food Consumption Survey we have attempted to arrive at better understanding of the factors in operation and to formulate guidelines to help estimate possible high exposures using data normally available.
For this purpose we have calculated the ratio of food intake for „high consumers" and for „very high consumers" as compared to the „average consumers". For the 9'th decile these ratios generally approached 2 and occasionally were twice this value. The intakes of the „very high consumer" group as characterized by the 10'th decile were even higher. What is the significance of these ratios? In the first case it indicates that at least 15% of the population has an intake of the contaminant under study which ex-

ceeds the average intake by this factor and in the case of „very high consumers" at least 5% of the population exceeds the average by the appropriate ratio for the particular food item.

Using this line of reasoning it is of interest to estimate the exposure of populations during periods of high radioactive levels in food which have been reported in the past.

V. A. Knizhinikov et al. (5) presented data on the „Average dietary intake of ^{90}Sr in the Soviet Union" for the period 1963 to 1971. The highest values reported were for the rural population during 1964 and estimated at 74 pCi/day/person. Flour and bakery products comprised 60.4% of the diet and correcting these figures with the ratios found in our study for „very high consumers" we see that it is reasonable to project that very high consumers in the population consumed at least twice that of the average for bread and flour products, and similarily, the factor emerging from our data for milk and milk products is even higher and can conservatively be set at 3. Assuming that individuals were „very high consumers" for both of these food items we arrive at an estimated exposure for some individuals in the population of the order of 140 pCi/day/person. This value is still below the level set by National Authorities (6) but appreciably higher than the value reported for the average intake.

On the basis of even detailed Food Consumption data it is very difficult to pinpoint „critical groups" in the population. However, from the experience gained in these studies it is possible to set a „Safe National Average Level" which will take into account the exposure to be expected in the extreme groups as characterized by their dietary habits.

Of general interest is the finding in these studies that the ratio of „very high consumers" to the „average consumers" does not exceed an order of magnitude, an assumption which has been commonly made by toxicologists in estimating the „Food Factor" in population exposure studies.

Acknowledgement:

We would like to thank Prof. Y. Nishiwaki, IAEA, Vienna, for the interest and encouragement he has shown in this project.

References

1 Ionizing Radiation: Levels and Effects. UNSCEAR 1, UN New York (1972), 39–47.

2 Feige, Y., I. Proulov, P. Karp, A. Eisenberg, E. Shalmon, E. Asculai, M. Gaaton: Radioactivity in the environment in Israel, July 1966–June 1968, Israel AEC Rep. IA-1205 (1972).

3 Feige, Y., A. Eisenberg, N. Passy, M. Stiller: The estimation of ^{90}Sr in the Israeli diet during 1962–63 as based on the radioanalysis of milk. Health Physics 11, 629 (1965).

4 Eisenberg, A., Y. Feige, Y. Proulov: Long-lived radionuclides in food and their significance for different groups in the Israel population. Israel AEC Rep. IA-1262, 157(1972).

5 Knizhnikov, V. A., E. V. Petukhova, R. M. Barkhudarov: ^{90}Sr Intake in food by the population of the Soviet Union, 1963–1971. Second International Conference on Strontium Metabolism, Glasgow, Strontian, 477 (1972).

6 Loutit, J. F., R. Scott Russell: 'Criteria for Radiation Protection', "Radioactivity and human diet" 40, Pergamon Press (1966).

Selenium Content of Some Foodstuffs and Other Environmental Samples in a Mineralized Area of Italy*)

By **M. A. Bombace†, L. Cigna Rossi, G. F. Clemente**

Laboratorio per lo Studio della Radioattività Ambientale, CNEN - CSN Casaccia, S. Maria di Galeria — Roma (Italy)

Abstract

No data are available on Selenium concentration in the Italian environment in spite of its importance as an essential and toxic element.

In order to get some information about the concentration and distribution of this element in the environment and in the food-chain, a survey has been made in the Monte Amiata area (Toscana, Italy). Such an area was selected for this study owing to its large mineralization due to many elements (e.g. Fe, Cu, Ag, Sb and particularly Hg). The selenium content has been determined by thermal neutron activaltion analysis and a large volume high-resolution Ge-Li gamma ray spectrometer connected on line to a DEC PDP 8/L computer.

The instrumental method requires neither a chemical separation technique nor a pre-or post-concentration of Selenium.

The Selenium concentration for wet weight in foodstuffs ranges between a value of 0.008 ppm in milk and a value of 0.2 ppm in chicken.

The average Se intake for the population living in the area under investigation has been also estimated. A critical discussion of the data is reported together with a comparison with the Se concentration values referring to other countries.

1. Introduction

Since the discovery by Schwarz (1) of the importance of Se as an essential trace element for rats, an increasing interest has been directed toward the investigation of Se content in the animal and human diet. Many papers (2—5) have shown that Se deficiency can affect the health of various animals. Further more in 1965 the Se content in the human diet was involved with the occurence of *Kwashiorkor*, a protein-calorie deficiency disease. (6) More recently it has been found that Se level in whole blood and plasma of Guatemalan children suffering from *Kwashiorkor* (7) and in placentas of under-nourished Guatemalan women (8) is lower than in normal controls.

In its biological effects, Se is intimately related to two important dietary components, namely vitamin E and sulfur aminoacids (9).

In spite of its importance as an essential element influencing the health status of man, there exist few data concerning amounts of Se in foodstuffs consumed by the human population. Moreover the greatest part of all the existing data are referred only to few countries (13—17) and no data are available on the Se content in the Italian foods.

*) Published in Comparative Studies of Food and Environmental Contamination (Proc. FAO/IAEA/WHO Symp. Otaniemi, 1973), IAEA, Vienna (1974) 3–22.

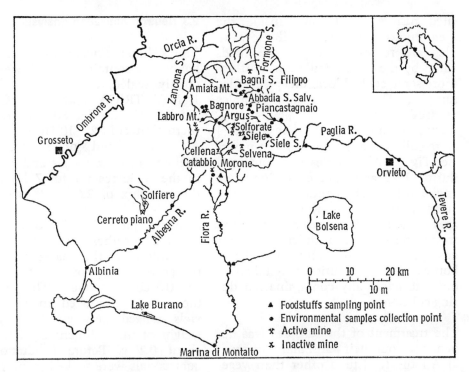

Fig. 1

For these reasons a study was planned in order to obtain some information about the concentration and distribution of this element in the environment and in the food-chain in a well defined area of Italy, where a large mineralization of Hg, Fe, Cu, Ag and Sb exists. In a previous paper (10) the Hg content and distribution in the diet and in the environment referred to this area were reported; the levels found for Hg are comparable with those from other countries. Furthermore there are no particular reasons to suspect this area far from having a normal Se content. Only two very scarce Se minerals are present in this area: the tiemannite (HgSe) and the onofrite (Hg[S,Se]).

2. Material and Methods
2.1 The Amiata Mt. area.

The area under investigation is in the Center of Italy (Toscana) and is delim-

ited by Marina di Montalto, Albinia, Amiata Mt. and Orvieto (see Fig. 1). Inside the region there are the basins of three main rivers: the Paglia, Fiora and Albegna which are also used for irrigation purposes. The area can be devided in two parts: the northern part which is principally mountain and the southern one which is a large flat land nearby the sea. In the northern part there are many hills about 200—700 meters high and two mountains: the Amiata Mt. (1738 meter) and the Labbro Mt. (1193 meter). In Fig. 1 are also reported the principal mines; the mineralization is present particularly for Hg in the northern part, whilst some Sb mines are existing in the southern part.

2.2 Sample Collection and Preparation

The foodstuffs (see Table 1) were collected at five different sampling points sited nearby small town inside the area studied (see Fig. 1); all the samples are representative of the foods eaten by the local population. Furthermore some cooked local diets were also sampled. The environmental samples as fresh water, aquatic plants sediments, pasture grass etc. (see Table 3) were collected at many different points (see Fig. 1). All samples were obtained during 1972 and after collection were enclosed in plastic containers carefully cleaned; 1% in volume of pure nitric acid was added immediately to the freshwater samples in order to prevent adsorption from the walls of the plastic bottles. Every precaution was taken in order to avoid any contamination during collection.

The treatment of the samples was dependent from their particular nature:

a) All the foodstuffs other than vegetables, were irradiated without any previous treatment.

b) The blackberry and the human hair were irradiated after a careful and repeated washing in bi-distilled water.

c) Vegetables, chestnut, pasture grass, aquatic plants, aquatic moss and filamentous algae were carefully washed with bi-distilled water and successively dried in air in a clean room at 24° C for fifteen days. After such a period the weight of the samples was constant and a complete loss of water was consequently assumed.

d) The sediments were only dried as the vegetables and after irradiated.

e) The freshwater samples were irradiated after a sedimentation for 24 hours without any other previous treatment.

All the aliquots irradiated were weighted with a precision balance.

2.3 Neutron Activation Analysis

The sediments were enclosed in polyethylene containers which were thoroughly sealed, and irradiated in the 1 MW TRIGA reactor of Casaccia Nuclear Centre for 30 minutes in a thermal flux of $1.25 \cdot 10^{13}$ n $\cdot$ cm$^{-2} \cdot$ sec^{-1}. All other kind of samples were enclosed in quartz vials and irradiated in the same reactor for 7 hours in a thermal flux of $2.6 \cdot 10^{12}$ n $\cdot$ cm$^{-2} \cdot$ sec^{-1}.

Standard pure Se solutions were irradiated together with the samples. The volume of the quartz vials and of the polyethylene containers was 10 cc and 0.3 cc respectively. The weight of the sample irradiated in the quartz vials ranged between 1–10 g; the weight of the sediment irradiated was about 0.2 g. Before irradiation the quartz vials were soaked in pure nitric acid for 12 hours and then washed with bi-distilled water, in order to avoid any contamination of the sample. The polyethylene containers were carefully washed with bi-distilled water.

The gamma measurements were performed in both case without removing the sample from the container irradiated. A careful testing of the quartz vials and of the polyethylene containers show in fact the complete absence of Se in these materials.

The gamma spectrometer was composed by two Ge(Li) true coaxial detectors of about 50 cm^3, 10% efficiency and 2.5 keV resolution, together with a 4096 channel analyzer (2048 channels for each detector) connected on-line to a DEC PDP 8/L computer. The Se content was evaluated on the 265 keV gamma-line, emitted by the ^{75}Se obtained with the reaction ^{74}Se (n, γ) ^{75}Se.

The spectra were automatically processed on the PDP 8/L computer by means of the Asper (11) program. The samples were measured 25–30 days after the irradiation; the counting time ranged between 400 and 1000 minutes. The minimum detectable amount of Se was 0.05 μg in sediments and 0.02 μg in all other samples. The accuracy and precision of the method, for many different elements, were given in a previous paper.(12) Moreover, to detect any possible loss and contamination

Table 1   Selenium Content of Foodstuffs in the Amiata Mt. Area (μg/g wet weight)

Foodstuffs	Se content ± 2σ*) µg/g wet weight	Number of samples with Se content above detection limit	Total Number of samples analyzed
Chicken			
kidney	0.21 ± 0.02	1	1
liver	0.21 ± 0.03	1	1
feathers	0.10 ± 0.07	1	1
skin	0.08 ± 0.01	1	1
flesh (breast)	0.06 ± 0.02	1	1
Fish (*salmo trutta*)		1	1
gills	0.11 ± 0.03	1	1
skin	0.06 ± 0.03	1	1
flesh	0.06 ± 0.01	1	1
Egg			
white	0.036 ± 0.001	5	6
yolk	0.068 ± 0.003	6	6
shell	0.020 ± 0.007	1	6
Cheese			
soft	0.010 ± 0.005	2	2
hard	0.030 ± 0.005	2	2
Pork			
sausages	0.03 ± 0.01	1	1
Bread	0.012 ± 0.006	1	2
Milk, Cow	0.008 ± 0.004	1	3
Mixed Diet (cooked)	0.05 ± 0.007	2	7
Vegetable			
salad (radish, *Cichorium* sp.)	N. D.		3
cabbage	N. D.		1
cauliflower	N. D.		1
tomatoes	N. D.		2
pepper, yellow	N. D.		2
parsley	N. D.		1
celery	N. D.		1
carrots	N. D.		1
mushroom (*Boletus* sp.)	0.13 ± 0.02	1	1
id. stalk only	0.13 ± 0.02	1	1
Fruits			
chestnut	N. D.		1
blackberry	N. D.		1

*) The σ represents the analytical error on the single value, when only one sample was measured, or the standard deviation of the mean value when more than one sample was considered.

due to the method, a comparison has been made between the data obtained for Hg on the same samples by neutron activation and atomic absorption analysis (10); the good agreement between the results obtained confirms the validity of the analytical technique referring to Hg. Thus it is not unreasonable to suppose that the accuracy referring to Se, which was not directly tested, is also correct, owing to the absence of any interference from other elements.

3. Results and Discussion

In Table 1 are reported the Se concentrations ($\pm$ 2 σ) found in the foodstuffs collected in the area under investigation. In the same table the number of samples, where the Se content was detectable, and the total number of samples analyzed are reported. The average Se content in the foodstuffs is given when the Se was measured in more than one sample (e.g. Egg, yolk).

The Se content was above detection limits only in a mushroom among all vegetable samples. The very low Se content in vegetables and fruits is in good agreement with the results reported by Morris and Levander (13) for the American Foods.

The small number of fodds samples analyzed is principally due to the difficulty to collect local foodstuffs from the farmers.

In Table 2 a comparison is given, for some food, between the Se concentrations found in USA (13) and those in the area under investigation. The Se content in chicken, egg, cheese and milk seems to be lower in Amiata Mt. Area than in USA by a factor 2 or 3. The big difference observed in the case of the pork meat is mainly due to the dif-

Table 2   Comparison of Values Reported for Selenium in Some Foodstuffs in Italy and USA

Foodstuffs	Italy Se concentration μg/g wet weight	USA (13) Se concentration μg/g wet weight
Chicken		
breast	0.06	0.12
skin	0.08	0.15
Egg		
yolk	0.068	0.18
white	0.036	0.05
Pork	0.03	0.24
Cheese		
soft	0.01	0.05
hard	0.03	0.1
Milk	0.008	0.013

ferences in the sampling; in our case the Se content for pork was measured in the sausage, whilst in ref. 13 was measured in chops. Furthermore great differences in the Se concentrations in different parts of the same animal were observed in this paper for chicken (Table 1) and in ref. 13 for beef, pork and lamb.

The Se concentration in milk in the Amiata Mt. area was found to be lower than those given in ref. 14 for Germany and in ref. 16 for USA.

The data referred to the environmental samples are given in Table 3. The Se content was undetectable in freshwaters; the limit of detection for Se in water samples is 0.002 mg/l well below the value (0.05 mg/l) of the maximum permissible Se content in drinkable water suggested by the W.H.O.(18) The macrophytes, aquatic moss and filamentous algae seems to show high concentration factors for Se with reference to water; in fact all the fresh waters where those plants were living showed a Se concentration less than 0.002 mg/l. The undetectability of Se on pasture grass is in good ac-

Table 3   Selenium Content of Environmental Samples in the Amiata Mt. Area (μ/g dry weight)

Sample	Se content $\pm 2\sigma$*) μg/g dry weight	μg/g wet weight	Number of samples with Se content above detection limit	Total number of samples analyzed
Freshwater				
sub surface water		N. D.		7
surface water		N. D.		32
Sediment	0.57 ± 0.4		2	25
Aquatic plants (Macrophytes)	0.11 ± 0.06		5	8
Aquatic moss	0.75 ± 0.10		2	4
Filamentous algae	0.58 ± 0.19		2	5
Pasture grass	N. D.			12
Human hair (composite sample)	0.23 ± 0.09		1	1

*) See table 1

cordance with the results reported for vegetables in Table 1.

No comparison can be made among the Se concentration in environmental samples in other countries in reason of the absence of any data.

Se seems to be present in the human hair (see Table 3) and it could be possible that the Se hair concentration in humans is dependent from the Se intake with the diet. In such a case the Se in human hair could be a good indicator of Se level in the diet.

The daily Se intake for the population living inlide the area under investigation is reported in Table 4. More than 50% of Se intake is due to the animal products (meat, eggs) and to the cereal products.

Thompson and Scott (19) have shown that 0.04 to 0.10 ppm (part per millions) of Se are needed in the diet to prevent Se deficiency in chickens. A flat extrapolation of these limits to humans is not possible; in any case the comparison, reported in Table 2 be-

Table 4   Daily Se Intake Through Raw Foods for the Population of Amiata Mt. Area

Foodstuff	Average Daily Intake Pro capite (g)	Average Se concentration (μg/g wet weight)	Se Daily Intake Pro capite (μg)
Chicken (meat)	69	0.06	4.14
Fish	20	0.06	1.20
Vegetables	107	0.002	0.2
Egg	37	0.05	1.85
Cereal products (bread)	288	0.012	3.45
Cheese	29	0.02	0.58
Milk	144	0.008	1.15
Total	694		12.57

Maximum Se daily Intake pro capite = 12.57 μg = A
Total Daily Intake of Foods pro capite = 694 g = B
Average Se concentration in Foods = 0.018 ppm = A/B

tween the American and Amiata Mt. area data, and the results reported in Table 4, suggest that the Se level in the diet of the population of the Amiata Mt. area could be not sufficient for good nutrition.

Furthermore the maximum average Se content in the cooked diets (see Table 1), considering all the samples analyzed and assuming a value of 0.002 ppm as the Se content of the diet samples where the Se was below the detection limits, is 0.016 ppm. Such a value is in a fairly good agreement with the estimate of the average Se concentration in raw foods reported in Table 4.

4. Conclusions

The data, reported in this paper, may suggest that the diet of the population, living in Amiata Mt. region, is poorer in Se, as compared to other countries. However, some important considerations should be made, before to flatly assume that the population under control is receiving a low Se diet. First of all, there may be certain local areas of low Se soils which could contribute to a possible Se deficiency; the use of interregional foods shipments should in any case minimize any geographical Se deficiency.

Secondly there is a wide variation in the ability of various Se compounds to prevent liver necrosis in rats [20] and it is not well known in what active form the Se is normally present in the various foods. Therefore the total dietary Se content may not be a valid indicator of their nutritional value.

In any case further studies are planned in the area under investigation, in order to assess, from the epidemiological point of view, the health status of the population.

Furthermore the need of more studies on the Se environmental content in other countries is evident; in fact environmental factors may cause the Se concentrations in foods from different geographical areas to vary widely.

In Oregon, for example, one of the Se deficient area of the USA, the difference between two samples of eggs and milk from two areas within the state was tenfold.[21] It is consequently clear the importance of the knowledge of some environmental indicators of the Se deficiency or abundancy in the human diet. Another important consideration, made on the basis of the results reported in this paper, is the need of a good correlation between the Se dietary intake and the Se concentration in man. The hair Se content seems to be very useful for an easy check of the Se nutritional status of large groups of population. Also in this case further studies are planned, in order to control the constancy, in other Italian regions, of the ratio between Se dietary concentration and Se hair concentration.

References

1 Schwarz K., C. M. Foltz: J. Amer. Chem. Soc. 79 (1957) 3292.
2 Hartley, W. J., A. B. Grant: Federation Proc. 20, (1961) 679.
3 Nesheim, M. C., M. L. Scott: Federation Proc. 20 (1961) 674.
4 Schubert, J. R., O. H. Muth, J. E. Oldfield, L. F. Remmert: Federation Proc. 20 (1961) 665.
5 Kuttler, K. L., D. W. Marble, C. Blincoe: Amer. J. Vet. Res. 22, (1961) 422.
6 Schwarz, K.: Lancet, 1 (1965) 1335.
7 Burk, R. F., W. N. Pearson, R. P. Wood, F. Viteri: Amer. J. Clin. Nutr. 20 (1967) 723.
8 Dayton, D. H., L. J. Filer, C. Canosa: Federation Proc., 28, (1969) 488.
9 Schwarz, K.: Federation Proc. 24 (1965) 58.
10 Bombace, M. A., L. Cigna-Rossi, G. F. Clemente, G. Zuccaro Labellarte, M. Allegrini,

E. Lanzola: "Recherche écologique sur le Mèrcure dans la region du Monte Amiata (Toscana, Italia)" to be published in Proceedings of the Colloque Européen sur les Problèmes posès par la contamination de l' homme et de son milieu par le Mèrcure et le Cadmium, Luxembourg, 3–5 juillet 1973.

11 Giorcelli F., G. Zuccaro Labellarte: Italian Health Physics Association: AIFSPR (Proc. XVI Congr. Florence 1970) Florence (1971) 290.

12 Clemente, G. F., G. Mastinu: J. Radioanal. chem. 20 (1974) 707.

13 Morris, V. C., O. A. Levander: J. Nutr. 100 (1970) 1383.

14 Kiermeier, F.: Nahrung 12 (1968) 565.

15 Oelshlaeger, W., K. H. Menke: Ernaehrungswiss Z. 9 (1969) 208.

Bioaccumulation of ^{14}C-DDT in a Small Pond *)

By **Laina Salonen** and **H. A. Vaajakorpi**

Department of Radiochemistry and Zoological Museum, University of Helsinki

Abstract

^{14}C ring-labelled DDT, together with inactive carrier-DDT, were pumped and mixed into a small polyhumous pond located in Southern Finland. At the beginning of the experiment, DDT concentration in the pond water was 1 μg/l and ^{14}C activity 1 nCi/l. Initial samples were collected 4 hours and final samples 2 months after application. During this period, DDT concentrations were determined in filtered water, suspended material sediment, fish, newt and certain invertebrates and plants. The ^{14}C-activities of all samples were determined utilizing liquid scintillation counting. With the exception of the filtered water, all samples were dried and prepared for counting by means of a combustion technique, which converted all samples into the same chemical form and gave high counting efficiencies.

Results demonstrated that the DDT concentration in filtered water decreased quickly and that the uptake of DDT by suspended material was rapid. Accumulation and elimination of DDT varied considerably in different fish tissues. The highest concentration (23.8 mg DDT/kg) was noted in the mesenteric adipose of perch *(Perca fluviatilis L.)* at the end of the test period. This concentration was clearly higher than in other tissues (liver, 16.1 mg/kg; gills, 3.7 mg/kg; muscle, 0.4 mg/kg). Accumulation in perch liver was rapid, with the maximum value being reached after only 1 day. DDT residues determined in Crucian carp *(Cyprinus carassius L.)* were smaller than in perch. Maximum concentrations of DDT in the newt *(Triturus vulgaris L.)*, bivalve mollusk *(Sphaerium lacustre L)*, caddisfly larvae *(Phryganea sp.)* and backswimmer *(Notonecta glauca L.)* were attained in 2–4 days. The highest residue (3.0 mg/kg) determined among the invertebrates was observed in the bivalve mollusk after 2 days. Among the aquatic plants examined, the bladderwort *(Utricularia minor L.)* yielded higher residues than the moss *(Drepanocladus sp.)*. Residue elimination, after the attainment of the maximum concentration, was approximately exponential in some of the samples.

The application of ^{14}C ring-labelled DDT and the liquid scintillation counting technique rendered possible the observation of DDT bioaccumulation in the typical Scandinavian polyhumous pond utilizing small amounts of DDT. This amount did not cause detectable symptoms of poisoning over the course of the test period. The metabolites of DDT were not studied in this experiment.

*) Published in Comparative Studies of Food and Environmental Contamination (Proc. FAO/IAEA/WHO Symp. Otaniemi, 1973), IAEA, Vienna (1974) 201–211.

1. Introduction

DDT (1,1,1-trichloro-2.2 bis [p-chlorophenyl] ethane) is the most widely used of the synthetic organochlorine pesticides. Millions of tons of it have been used as an effective and toxic insecticide. Due to the spreading techniques and transportability of DDT, it is distributed to all kinds of environments, and into waters, where it has a great affinity for living organisms.

In the natural aquatic ecosystems many of the persistent pesticides, such as DDT, are considered hazardous because they kill water organisms even at low concentrations and are enriched in these ecosystems.[1] DDT's solubility characteristics and its high resistance to chemical, physical and microbiological degradation explain its enrichment. DDT is almost insoluble in water (1.2 μg/l = 1.2 ppb at 25° C) but quite soluble in organic solvents and in lipids (ca 100 g/l). Therefore its concentration in fatty tissues may become very high.[2] The DDT levels need not always rise with progress from one trophic level to another, for even at the lowest trophic levels bioaccumulation is very fast and effective.[3]

The amount of DDT used in Finland is small compared to many other countries. From the beginning of 1946 to the end of 1971 304 tons of DDT were sold in Finland.[4] Attempts have been made to limit the use of DDT and since the beginning of 1971 its use is permitted only for treatment of saplings in forest nurseries.[4] The determinations done during 1970–72 in one of our largest lakes, Lake Päijänne, show that

the residues of DDT in fish (average 0.008 ppm in wet tissue) are low compared to values obtained in other studies around the world.[5]

However, even the small amounts used, together with the global distribution of DDT, may have an effect on our water courses which due to their small volumes are especially sensitive to pollution. The purpose of this study was to observe the bioaccumulation and translocation of DDT in a natural aquatic ecosystem, a small polyhumous pond typical to Scandinavia. The bioaccumulation studies performed with ^{36}Cl-DDT in freshwater marsh ecosystems in Ohio, USA, and a simulation study based on these field experiments offer a basis for comparison.[6, 7, 8]

2. Test Bond

The small pond used in this study is situated in Southern Finland. It is in an old gravel pit, not used for 20–30 years. The pond drains its water supply mainly from a nearby marsh, but during the summer they are not connected. The area of the pond is 1100 m^2, and the mean depth 0.45 m, the maximum being 0.75 m. The pond water contains plenty of suspended humus matter and the water is typically brown. The ratio of plankton to dead suspended matter was not deter-

Table 1   Water Analyses.

Date	4.8.	14.8.	23.8.	31.8.	7.9.	13.9.	13.10.
Temp. °C	21.0	18.0	16.5	16.5	12.0	11.5	4.7
Dissolved O₂ mg/1	7.4	7.6	8.0	8.8	9.6	8.8	10.8
pH	6.5	6.5	6.7	6.7	6.7	6.4	7.0
Conductivity, μS/cm	35	33	30	31	30	30	30
Alcalinity, mval/1	0.16	0.10	0.03	0.04	0.08	0.10	0.17
Colour, mg Pt/1	281						
Productivity, mg C/1	65.5						

mined — the combination is called "seston" for short.

The characteristics of the water were determined as usual. Temperature, specific conductivity and the amount of dissolved oxygen were observed throughout the test period (Table 1). The amount of bottom sediment varied in the different parts of the pond. Where the amount of sediment was high, there were bottom animals and higher water plants in abundance. Among the bottom animals, caddisfly, isopod as well as various genera of *Odonata* were most abundant. The most common vertebrates were tadpoles and newt larvaes. The metamorphosis of frog tadpoles was completed before the start of the experiment and they were not included in this test. The original Crucian carp in this pond was rather small, of so-called pond type. About 50 bigger Crucian carps and 15 perches were brought to the pond some weeks before the labelling.

Bladderwort and moss were the most common plant species. In addition, members of the *Sparganum* and *Potamogeton* families were common.

In addition to filtered water, sediment and seston, the following species were analyzed:

Bladderwort, *Utricularia minor L.*
Moss, *Drepanocladus sp.*
Dragonfly larvae, *Libellullidae*
Caddisfly larvae, *Phryganea sp.*
Backswimmer, *Notonecta glauca L.*
Bivalve, *Sphaerium lacustre L.*
Newt, *Triturus vulgaris L.*
Perch, *Perca fluviatilis L.*
Crucian carp, *Cyprinus carassius L.*

3. Methods

The ^{14}C ring-labelled DDT (0.5 mCi) was obtained from the Radiochemical Centre at Amersham, England. Its specific activity was 67 μCi/mg. For labelling, the radioactive DDT (7.43 mg) was solubilized together with technical grade (carrier) DDT (492.6 mg) into acetone (1 l) and then mixed with the pond water (50 l). This solution was gradually pumped and mixed into the pond. Thus, the DDT concentration in the pond water was 1 μg/l and the ^{14}C activity 1 nCi/l at the beginning. Mixing up of the bottom sediments could not be avoided entirely which must have had some effect on the rate of bioaccumulation. The test period was the two months following the labelling of the pond (August 15, 1970).

Most of the samples were combined from a number of subsamples collected from various parts of the pond. The samples were always taken in the same order: water — sediment — plants — bottom animals and fish. Water samples were collected into plastic bottles (0.5 l.). Seston was filtered from these samples on tarred Millipore filters (first 8 μ and then 3 μ). Bottom sediment was collected from the top of the bottom mud. From the fish, samples of muscle, gills, liver and — from perch — mesenteric adipose, were prepared separately. The samples, dried at 95° C, were stored in closed plastic containers.

Liquid scintillation counting was considered the best method for radioactivity determinations because of the soft beta particles of ^{14}C and the low activities of some samples. The tracer technique applied was practical for analyzing the total-DDT residues of various types of biological samples. Without the use of radioisotope DDT-, and its metabolites DDE and DDD would have required the use of a number of separation, cleaning and quantitation techniques, often various for different samples. These methods with their recovery experiments are quite complicated and tedious for this type of bioaccumulation study and for the amount of DDT used in this experiment.

For liquid scintillation countig, the combustion method described below was used to prepare the dried samples. For the water samples an extraction method was used.

Combustion method

Furnace combustion was the most practical method for our purposes. The combustion furnace used had an inner silica tube which contained the catalyst, copper oxide and platinum, at a temperature of 850° C to complete the oxidation. A dried sample was inserted in a silica sample boat into the furnace tube and combusted in an oxygen stream. Combustion of a sample took a few minutes and then, the possible residue was ignited to insure the complete combustion of all organic matter.

Products of combustion were carried by the oxygen stream into an absorption solution through the copper oxide catalyst where the complete conversion of carbon to CO_2 was achieved. After a 10-minute collection time the oxygen stream was replaced by nitrogen in order to eliminate the oxygen quenching in the counting solution.

The mixture of redistilled β-phenylethylamine and methanol (1:1) was used for trapping of the CO_2 in two absorbers coupled in series.(9) The phenylethylamine was an effective absorbent for CO_2 with which it formed a carbamate. In order to introduce this carbamate into the toluene based scintillation solution methanol was used as a secondary solvent. The deposition of carbamate in the first absorber was pronounced when the sample weighed over 200 mg. In these cases, the carbamate was dissolved by warming the absorber in a water bath (ca 50° C). The absorption solutions (10 ml) from both absorbers were placed into their own polyethylene counting vials (20 ml). Both absorbers were washed with 8.5 ml of scintillation solution which were added into the counting vials. Usually the activity of the second absorber was of the background level. The final composition of the scintillation solvent was (10):

27 ml β-phenylethylamine
27 ml methanol
46 ml toluene
500 mg PPO (2.5-diphenyloxazole)
and
10 mg POPOP (1.4-bis-[5-phenyl-oxazol-2-yl] benzene).

The ^{14}C-activities of the homogeneous samples were measured at a temperature of 10° C with a model Decem-NTL^{314} Wallac liquid scintillation counter, provided with a refrigerated automatic sample changer. The counter had three separate counting channels, one of which was adjusted for ^{14}C, and in addition two standardization channels. The external standardization method was used for determining counting efficiencies. Every sample was standardized separately because the counting efficiency varied from sample to sample. The counting efficiency for most samples was good (ca 80%). Only the samples prepared from liver and muscle tissues were colored and quite quenched, which lowered the counting efficiency even down to 10%. The background varied between 45—55 cpm depending on the type and the counting efficiency of a sample.

The recovery determinations for this combustion method were performed by burning inactive samples of fish muscle and liver, grass, and sucrose, spiked with a known amount of ^{14}C-hexadecane standard solution. The average recovery was found to be 91.8% with a standard deviation of 1.3%. The recovery did not depend on the carrier type but on the amount of the carrier. When this exceeded 400 mg, the amount of carbamate precipitate was so great that it was difficult to keep it in the scintillation solution during the counting.

The weight of the dried samples used for combustion varied from 30 to 400 mg except for seston and bottom sediment. The dry weigths of seston samples were between 0.1 to 7.2 mg, obtained from the 0.5 l water samples. The Millipore filters became clogged, when more water was used. Since the bottom sediment contained also a lot of inorganic material, samples of two gram weight could be combusted.

Extraction method

It was necessary to concentrate DDT from the water samples because the ^{14}C-activity of the water was even at the beginning only 1 nCi/l, and only small water samples (2–3 ml) can be mixed with dioxane or Triton X based scintillation solutions. Extraction of DDT into toluene was the most practical method for liquid scintillation counting.

One ppm of inactive carrier DDT was added into the water bottle. After two days an aliquot of 150 ml was taken for extraction. DDT was extracted into 20 ml of toluene, and an aliquot (10 ml) mixed with toluene based scintillation solution was measured. The final composition of the scintillation solvent was:

    1   l toluene
    5   g PPO and
    0.5 g POPOP

Because only low activities were found in the water, 0.5 l of toluene and 1 ppm of inactive DDT was added into an empty plastic bottle. After two days this toluene solution was placed into a vial to evaporate the toluene to dryness. The residue was soluted to toluene and measured. It was cheched using ^{14}C-DDT did not evaporate with toluene. Most of the activity (ca 95%) was found on the walls of the bottle.

Extraction of DDT from water to toluene was also checked by employing a known amount of ^{14}C-DDT. Only one extraction was necessary for a 100% recovery.

4. Results

The results are given in Table 2 where the total DDT concentrations of the samples are presented as a function of time elapsed from the application of DDT into the pond. To eliminate the effects of moisture and inorganic matter in some samples, data are expressed in ppm of DDT in organic matter. For most samples the dry weight and the amount of organic matter has the same value. The specific activities of the samples (pCi/g of org.mat.) have been converted to ppm.

The errors calculated consist of the statistical deviations of the counting rate and efficiency and from the standard deviation of the percentual recovery. Also the accuracy of weighing the sample has been included. In the case of some seston samples this was the principal error.

Water, seston and bottom sediment

In filtered water the concentrations varied between 0.86–0.01 ppb (12 h and 14 d). Even the first values were

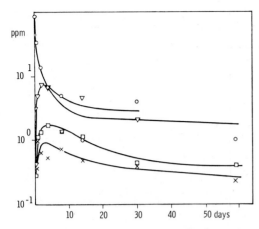

Fig. 1 DDT concentrations in filtered water (○, 10⁻⁵ x ppm indicated), in bladderwort – (△), in moss (✕) and in caddisfly larvae (□) as a function of time after labelling the pond.

and non-living particles in the water, are known to be efficient.[3, 11] The rapid decline of the DDT concentration in water is proof of the effectiveness of these mechanisms. In this test the adsorption by the non-living particles of seston was probably the decisive factor. The colour determination of the water of the test pond indicates that the proportion of suspended particles of various sizes and finely divided humus colloids was markedly high compared to live plankton (Table 1).

The DDT bound to total seston mass in the test pond can be roughly estimated when the total amount of seston and the seston-bound DDT in a given water volume is measured after filtering. At the 24-hour accumulation maximum seston-bound DDT is ca. 34 mg, which is about 7% of the total DDT added. Considering the sources of errors this estimate is rather approximate and should not lead to further conclusions. After the maximum, the level in seston decreased rapidly, the DDT being depleted by both bioaccumulation and sedimentation.

Some of the biosides in the water are transported before long into the bottom sediment, from where they can be again taken up into the active cycle under suitable conditions.[12] In the experiment carried out, the bottom sediment offered an effective adsorption base when it was mixed with the water during labelling and a considerable amount of DDT was taken up thereby.

The role of the bottom sediment in the DDT-balance in the pond is not clearly evident. The concentrations varied from 0.41 to 1.12 ppm. The regression coefficient calculated from the values obtained 2 days after labelling (−0.0026) shows slight negative correlation between the change in concen-

below the theoretical value of 1 ppb, and thereafter the concentrations remained at a fairly low level (Fig. 1). Towards the end of the test period when the concentrations were close to the detection limit, the direction of the change was not clearly evident. It does seem obvious, however, that no sharp decline took place, but a plateau was reached from where the decrease was very slow.

The accuracy of the determinations for seston was fairly poor. This was due to the clogging of the filter when the sample volume reached a certain size. Therefore it can be stated only with certain reservations that the 3 μ seston (seston that remained on the 3 μ-membrane filter) reached the highest concentration of the test, 39.8 ppm, 1 day following the labelling, while 8 μ seston reached its maximum, 20.6 ppm, in 12 hours. The differences in the 3 μ and 8 μ seston accumulation are mainly due to the difference in the surface area/volume ratio.

Of the mechanisms binding DDT, absorption and adsorption by the living

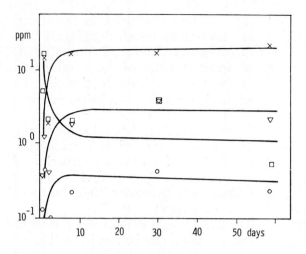

Fig. 2   DDT concentrations in various perch tissues ($\bigcirc$ muscle, $\triangle$ gills, $\square$ liver and $\times$ mecentric aripose) as a function of time after labelling the pond.

tration and time. It can be assumed, however, that when DDT is removed from water through accumulation, some DDT has been redissolved from the bottom sediment into water.

Plants

Two physiologically different types of plants, bladderwort and moss, that were also the most common plant species in the pond, were chosen for analysis. The results are interesting, because bladderwort which uses mostly its bladders to obtain nutrients, differs greatly from the prototype that moss represents. Bladderwort with its branches forms a rather limited adsorption surface compared to moss but has instead plenty of bladders.

In bladderwort, DDT varied from 0.40 to 7.25 ppm and in moss from 0.22 to 1.00 (Fig. 1). On the average, the levels in bladderwort were seven-fold compared to moss. This is evidently due to its efficient particle-capture capacity.

Invertebrates

Dragonfly larvae of numerous species were found in the pond, but to avoid dispersion we tried to limit the types of larvae analyzed to a few most common species. The larvae collected were mostly of the suborder *Anisoptera*, principally of families *Cordulidae* and *Libellulidae*. The larvae were mostly of the bottom dwelling types that feed on plankton organisms and on small bottom animals. The accumulation maximum (0.54 ppm) was reached fairly late, on the 14th day after labelling. In 30 days the level had already decreased to the level of the earliest samples.

The caddisfly larvae were mostly members of the genus *Phryganea*. The crust surrounding the larva was removed and only the main body was used for analysis. The accumulation maximum (1.71 ppm) was observed at 4 days (Fig. 1). On the average, the levels were five-fold compared to the dragonfly larvae, and the maximum was observed earlier (14 d for dragonfly).

The maximum of accumulation in the backswimmer was at 3 days. The concentration was the lowest measured for the invertebrates. The backswimmer differes from the others in its breathing mechanism.

Table 2   Total DDT Concentrations (ppm in org. matter) in Samples Collected from the Pond.

	4 h	12 h	1 day	2 days	4 days	8 days	14 days	30 days	59 days
Filtered water*)	0.46	0.86	0.33	0.14	0.067	0.05	0.01	0.04	0.01
Seston, 8 μ	9.0±4.0	20.6±8.0	11.2±6.1	3.9±1.5	2.3±0.9	2.9±1.3	1.2±1.1	0.0	0.0
Seston, 3 μ	36.1±62.7	25.4±19.9	39.8±50.2	15.9±42.7	5.6±29.5	3.7±17.7	3.7±17.3	0.0	0.0
Sediment	0.74±0.03	0.41±0.02	1.12±0.04	1.09±0.04	0.74±0.02	0.94±0.03	1.08±0.03	0.69±0.02	0.71±0.02
Bladderwort	0.40±0.02	3.09±0.07	4.66±0.10	7.25±0.15	6.84±0.14	1.35±0.04	4.64±0.10	1.98±0.06	—
Moss	0.36±0.01	0.36±0.02	1.00±0.03	0.62±0.02	0.52±0.02	0.73±0.02	0.48±0.02	0.38±0.01	0.22±0.01
Dragonfly larvae	—	0.22±0.02	0.22±0.02	0.27±0.02	0.17±0.02	0.25±0.02	0.54±0.04	0.20±0.02	0.17±0.01
Caddisfly larvae	0.27±0.05	1.01±0.08	1.09±0.09	1.31±0.09	1.71±0.08	1.33±0.08	1.11±0.05	0.43±0.02	0.38±0.02
Backswimmer	0.08±0.01	0.12±0.01	0.20±0.02	0.30±0.02	0.30±0.02	0.21±0.02	0.16±0.01	0.24±0.02	—
Bivalve mollusc	0.23±0.05	2.02±0.07	1.70±0.09	3.00±0.11	2.31±0.09	1.02±0.09	0.65±0.09	0.24±0.03	0.17±0.03
Newt	0.07±0.02	0.29±0.03	0.32±0.02	0.98±0.05	2.01±0.07	1.17±0.05	1.55±0.06	0.58±0.03	—
Perch; muscle	0.19±0.03	0.13±0.01	0.43±0.02	0.09±0.01	0.26±0.01	0.22±0.02		0.44±0.05	0.24±0.03
gills	1.67±0.06	0.37±0.04	1.16±0.05	0.39±0.04	—	1.72±0.06	—	3.73±0.19	2.15±0.08
liver	5.63±0.16	5.02±1.47	16.10±4.40	2.07±0.06	—	2.00±0.06	—	3.79±1.02	0.52±0.02
mes. adip.	0.39±0.09	1.19±0.35	14.30±4.10	1.82±0.49	—	16.80±0.40	—	17.20±4.60	23.80±6.60
Crucian carp muscle	0.15±0.03	0.19±0.03	0.10±0.01	0.11±0.02	0.26±0.01	0.08±0.01	0.09±0.01	0.10±0.01	0.37±0.04
gills	0.16±0.03	0.33±0.02	0.15±0.02	0.26±0.02	0.78±0.05	0.39±0.04	0.37±0.02	0.33±0.02	0.83±0.03
liver	0.06±0.02	0.09±0.01	0.09±0.01	0.34±0.01	0.39±0.01	0.06±0.01	0.17±0.01	0.09±0.01	0.20±0.01
Crucian carp total	—	—	—	—	—	—	—	2.06±0.06	2.02±0.05

*) ppb

The bivalve molluscs were analyzed including their valves. The concentrations were the highest for the invertebrates. They were also somewhat higher than those for newts and Crucian carps. Accumulation was rapid, the maximum (3.00 ppm) being reached at 2 days. This is without doubt due to the highly efficient filtering capacity of the bivalves.

Vertebrates

The newts were all small (3—5 cm) with external gills and at the larvae phase, and were therefore in the water during the whole test period. DDT accumulation could therefore have taken place both through the gills and the food. The accumulation maximum was at 4 days.

In fish, concentrations were determined for muscle, gills, liver and, in perch, also for the mesenteric adipose (Fig. 2). The levels in perch were considerably higher than in Crucian carps. The differences were most marked in gills and liver. The highest level in perch (23.8 ppm) was observed in the mesenteric adipose at the end of the test period. The Crucian carp has no uniform mesenteric adipose and no comparisons can therefore be made. For both species, the levels in the muscle tissue are the lowest. The highest concentration (16.1 ppm) in the perch liver was measured in the 1 day sample. It declined thereafter fairly rapidly, showing a fast response of the tissue to DDT added to the water. The highest concentration in the liver of the Crucian carp (0.39 ppm) was measured at 4 days. This is below the lowest value of perch liver (0.52 ppm), measured at the end of the test period.

The differences are probably caused by the physiological differences on the species and their different means of obtaining nutrients. The perch belongs to the group that has a more efficient metabolic system and a higher oxygen demand. The metabolically less efficient species, such as the Cyprinids have been observed to accumulate less of toxins in water, than the species with an efficient metabolism.[13]

The DDT concentration was also determined for the original Crucian carp in the pond at the end of the study. It was analyzed as a whole. The concentrations obtained (30 and 59 days) were of the order of 2 ppm and therefore higher at a corresponding point of time than those in the tissues of the bigger type of Crucian carp used in the experiment. This may be due to species specific or individual differences or due to the fact that in the Crucian carps analyzed as a whole there was an average increase caused by some other tissues. For instance, in goldfish levels higher than those in liver have been found e.g., in adipose tissue, nerves, brain and feces.[2]

5. Discussion

The results of this bioaccumulation study are in good agreement with those of the freshwater marsh experiments carried out by Peterle and Meeks in Ohio, USA. In Peterle's and Meeks' studies the labelled DDT mixed with inert granules was applied by helicopter to an enclosed four-acre marsh at the rate of 0.2 lb per acre (50 times our dose per m^2). The DDT released into the water from granules was rapidly removed by plankton and larger organisms. Producer organisms contained their maximum residues between 1—3 days and exceeded the environmental level by a factor of 3125. In our study the amount of DDT was small enough to be completely dissolved during the application. The maximum concentration in seston was attained in 1 day. The concentration factor at the maximum was 3600 for the 3 μ seston and 1800 for the 8 μ seston.

Many species of plants analyzed in Meeks' study showed that the accumulation and loss process was essentially the same in all plants — with the differences mostly in magnitude. The bladderwort reached its maximum value 10.8 ppm (range 4.8 to 15.4) in 3 days. Concentrations measured in individual plants were extremely variable, showing that this species is a poor indicator for DDT. In our study the maximum, 7.3 ppm, was attained also in 3 days.

Our invertebrates reached their maximum values in 4 days, with the exception of the dragonfly larvae (14 d) and bivalve molluscs (2 d). In Meeks' marsh studies the invertebrates attained their maximum DDT content several days later than plants. Rapid accumulation of DDT was observed with ramshorn snail (Planorbidae).

Accumulation and loss of DDT in fish tissues observed in our study agrees, on the whole, with studies performed with other fish species.(2, 14) That there are species differences and also individual variance was brought out also by our study. Since the natural state of the pond had to be preserved, neither many fish species foreign to it nor a large number of fish could be brought in. The variance and the short test period did not allow mathematical calculations of elimination rates.

The results and observations on the bioaccumulation of DDT in this pond cannot be compared as such with other studies because of the many complex factors affecting the accumulation of DDT by aquatic organisms. For instance, the accumulation is related to the concentration of DDT in the water and increases with higher concentrations and temperatures. In this polyhumous small pond the abundance of suspended material, with its large surface area that made it suitable for adsorption of DDT, affected strongly the course of bioaccumulation. Also, the effects of disturbing the sediment during the application, and of the late season are not easily estimated. However, this study has enabled us to elucidate some of the factors involved in bioaccumulation of small sublethal amounts of DDT in a natural pond representing te polyhumous lake type.

Acknowledgement

We are grateful to Prof. Jorma K. Miettinen (Department of Radiochemistry, University of Helsinki) and Dos. Pekka Nuorteva (Zoological Museum, University of Helsinki) for advice and to Dr. Esko H. Koskinen (Isotope Laboratory, Faculty of Agriculture and Forestry, University of Helsinki) for making the scintillation counter available to us.

References

1 Eisler, R.: "Pesticide-Induced Stress Profiles", Marine Pollution and Sea Life (Ruivo, M.), FAO, Fishing News LTD, London (1972) 229.

2 Grzenda, A. R., F. D. Paris, W. J. Taylor: Trans. Am. Fish. Soc. 99(1970) 385.

3 Södergren, A.: Oikos 19 (1968) 126.

4 Markkula, M.: Kemian Teollisuus 29 (1972) 534.

5 Hattula, M. L.: Analysis of DDT- and BCB-type Compounds at Low Level in Fish with Reference to Pike, Perch and Bream in Lake Päijänne, University of Helsinki, Institute of Food Chemistry and Technology, EKT Series 301, Jyväskylä (1973).

6 Peterle, T.: Translocation and Bioaccumulation of Cl–36 DDT in Freshwater Marsh, VIIᵉ Congres Des Biologistes Du Gibier, Beograd-Ljubljana (1967) 297.

7 Meeks, R. L.: Jour. Wildl. Mngt. 32 (1968) 376.

8 Eberhardt, L. L., R. L. Meeks, T. J. Peterle: DDT in a Freshwater Marsh — a Simulation Study, AEC Research 8 Development Report, BNWL-1297, UC-48, Washington (1970).

9 Griffiths, M. H., A. Mallinson: Anal. Biochem. 22 (1968) 465.

10 Woeller, F. H.: Anal. Biochem. 2 (1961) 508.

11 Poirrier, A., B. R. Bordelon, J. L. Laseter: Environ. Sci. Technol., 6 (1972) 1033.

12 Albone, E. S., G. Englinton, N. C. Evans, J. M. Hunter, M. M. Rhead: Environ. Sci. Technol. 6 (1972) 914.

13 Bauer, K.: Studien über Nebenwirkungen von Pflanzenschutzmitteln auf Fische und Fischnährtiere, Mitteilungen aus der Biologischen Bundesanstalt für Land- und Forst-wirtschaft Berlin-Dahlem, Heft 105, Berlin (1961).

14 Macek, K. J., C. R. Rodgers, D. L. Stalling, S. Korn: Trans. Amer. Fish. Soc. *99* (1970) 689.

Laboratory Model Ecosystem Evaluation of the Chemical and Biological Behavior of Radiolabeled Micropollutants *)

By **R. L. Metcalf**

University of Illinois, and Illinois Natural History Survey, Urbana-Champaign, Illinois USA

Abstract

A standardized laboratory model ecosystem has been used to evaluate the comparative behavior of radiolabeled micropollutants including organochlorine, organophosphorus, carbamate, and hormone-mimic insecticides; herbicides; important industrial organic compounds including phthalate esters and PCB's; and specific pollutants such as TCBD and hexachlorobenzene. The compounds have been studied for ecological effects over a terrestrial-aquatic interface and through several food chains, especially with regard to the development of quantitative information on ecological magnification and biodegradability index. The pathways of chemical degradation in the various organisms have been evaluated and the ecological effects of degradation products investigated.

Illustrations are given of the use of the model ecosystem technology for screening new candidate pesticides for effects on environmental quality; for evaluating the hazards of environmental pollution by industrial waste effluents; and for fundamental studies of the principles of biodegradability of organic chemicals in a variety of organisms.

---

*) Published in Comparative Studies of Food and Environmental Contamination (Proc. FAO/IAEA/WHO Symp. Otaniemi, 1973), IAEA, Vienna (1974) 3–22.

I. Introduction

The United States annual production of organic chemicals has increased from 10×10^9 lb. $(4.5 \times 10^9$ kg.) in 1943, to 140×10^9 lb. $(64 \times 10^9$ kg.) in 1972.[1] The growth rate is exponential and the increase was 19.7% from 1971 to 1972.[2] The present rate of production is equivalent to 676 lb. (307 kg.) per capita per annum. Other highly industrialized nations are not greatly different [3] and it is estimated that the annual world production of organic chemicals is about 450×10^9 lb. $(209 \times 10^9$ kg.) and is growing at about 10% per annum.[3]

The diversification of organic chemicals is seemingly endless. There are more than 10,000 organic chemicals in commercial use as intermediates in synthesis, or as fuels, plastics, elastomers, plasticizers, brighteners, rubber chemicals, paints, lacquers, solvents, fibers, detergents, fuel additives, coolants, hydraulic fluids, pesticides, fertilizers, food additives, and pharmaceuticals.[4] These enter the environment *accidentally* from industrial effluents, household applications, or as by-products of transportation; or *purposefully* from applications of agricultural chemicals, protective coatings, and preservatives. It has been variously estimated that 500 to 700 new organic chemicals are produced each year on a scale large enough so that traces of them enter directly into the environment.[5] As micropollutants of air, water, soil, and food these chemicals may be harmless, toxic, carcinogenic, mutagenic, or teratogenic. It has been suggested recently that 80% of human cancers are caused by chemicals that humans encounter in their daily lives.[6] Thus the potential problems of toxic chemicals in the environment are enormous and will become steadily more awesome because of the expanding production, estimated to reach 1700 × 19^9 lb. by 1985. It is of great importance to determine the distribution, fate, and potentially hazardous effects of these chemical contaminants so that they may be used properly and with due regard to environmental quality. The ultimate aim must be the establishment of criteria for permissible levels of chemical contamination in soil, water, air, and food.

The toxicological problems posed by environmental micropollutants involve experimental complexities beyond those of conventional investigations with laboratory animals. These substances as defined by Warner [7] (a) are present in ppt to ppm quantities (b) are water insoluble and lipid soluble, (c) exhibit biological concentration in organisms of the trophic web, (d) show delayed onset of biological activity, and (e) are toxic, carcinogenic, mutagenic, or teratogenic. We report here on the development of a simple, reproducible laboratory system to screen and evaluate the micropollutant properties of chemicals. This system involves the transfer of the compound from terrestrial to aquatic situation at micropollutant levels and incorporates a food web of both phytophagous and carnivorous elements (Fig. 1A). Using radiolabeled compounds this system produces data on the biological activity of the compound and its degradation products to a variety of organisms, demonstrates pathways of environmental degradation, measures ecological magnification, and determines a quantitative biodegradability index. The system can be adapted to mimic several of the ways in which micropollutants enter the aquatic environment, i.e. as applications to plant foliage, to seeds, by leaching from soil, and by direct application.

II. Methodology

Fig. 1a

Radiotracer methodology is probably the only practicable way of evaluating the behavior and fate organic environmental micropollutants.[8] We have sought to develop, over the past seven years, a standardized model ecosystem (Fig. 1 A) in which radiolabeled molecules could be followed under simulated environmental conditions, for their transfer from the terrestrial to the aquatic environment and for their uptake in a variety of food chain organisms, their toxicity, biodegradation, and environmental fate. The techniques involved are extraction, thin-layer chromatography, autoradiography, liquid scintillation counting, and identification by microchemical techniques and mass spectrometry.(9, 10) The model ecosystem methodology has been described elsewhere (9, 10) and will be but briefly summarized here. The basic unit is a glass aquarium 25 × 30 × 46 cm. with a plexiglass top incorporating a screen wire portion, which serves to retard evaporation. A sloping shelf of 15 kg of washed white sand is bisected by a "lake" of seven l. of standard reference water.(11) On the terrestrial shelf are grown 50 *Sorghum vulgare* plants to comprise a "farm". In typical experiments with candidate pesticides the radiolabeled compound is applied quantitatively to the leaves of the *Sorghum* with a micropipette, using a dosage of 1.0 to 5.0 mg. This is equivalent to 0.2 to 1 kg. per hc. Alternatively the radiolabeled compound can be incorporated into the sand, applied in soil, applied as a seed dressing, applied directly to the water, or applied in animal waste. Typically, 10 5th instar salt marsh caterpillar larvae, *Estigmene acrea*, are added to the system after treatment to consume the treated leaves and to serve as the dispersing agent for the radiolabeled pollutant.

The lake contains a complement of microorganisms, plankton, 30 *Daphnia magna*, 10 *Physa* snails, and a clump of alga *Oedogonium cardiacum*. After the model ecosystem has equilibrated for 26 days in an environmental plant growth chamber at 26° C (80° F) with a 12-hour daylight exposure to 5000 ft. candles (54,000 lux); *Culex fatigans* mosquito larvae are added, and after 30 days, three *Gambusia affinis* fish. The experiment is terminated after 33 days when weighed samples of the organisms are removed for radioassay. Measurements which are normally made include rate of radioactive contamination of the water phase, toxic effects on the organisms of the system, chemical changes in the nature of the radiolabeled products in water and the various organisms, ecological magnification (E.M.) or ratio of amount of parent compound or specific degradation products in organisms/amount in water; and biodegradability index (B.I.) or amount of polar radioactivity in organism/amount of non-polar radioactivity.(12)

Results

Over the past seven years we have evaluated quantitatively more than 50 organic compounds for their environmental behavior in the model ecosystem. These include a variety of established and experimental insecticides and herbicides, growth regulators, plasticizers, heat transfer agents (PCB's) toxic impurities (e.g. TCBD, DDE), and miscellaneous industrial chemicals. The use of the model ecosystem to evaluate micropollutants has been well received by regulatory agencies in the United States and has been selected by the Division of Vector Biology and Control, World Health Organization as an essential preview for use of new pesticides proposed for vector control. A number of industrial firms have collaborated in studying new products. We have found the model ecosystem especially useful in a research program to study the principles of biodegradability as applied to the DDT-type molecule.(12, 13, 14) Examples of the results obtained in these various types of investigations will be given in this brief review.

A. Organochlorine Pesticides

These compounds applied to the environment as deliberate contaminents have become the prototype microchemical pollutants. Despite efforts to regulate and control them large quantities are still being applied in the United States and elsewhere in the world. The cumulative U. S. production over the past 25 years, for these very stable environmental pollutants, is estimated as: DDT 2,000 million lb., BHC 800 million lb., cyclodienes and toxaphene 1,750 million lb. As a result micropollution has become ubiquitous and these compounds and their degradation products are found in surface waters, food, wildlife, and human fat.

Detailed model ecosystem studies have been reported with DDT, DDD, DDE (9), and aldrin, dieldrin endrin, mirex, lindane, and hexachlorobenzene.(15) The results with DDT are described in Section D. Aldrin and endrin will be described briefly here.

Aldrin or 1,2,3,4,10,10-hexachloro-1,4,4a,5,8,8a-hexahydro-1,4-*endo*, *exo*-5,8-dimethanonaphthalene, ring U-labeled at 13.6 m Ci per m mole, was applied to *Sorghum* at 5.0 mg. per

chamber. The radioautograph of Figure 1B shows very clearly the rapid conversion of aldrin to the 6,7-epoxide dieldrin. As shown in Table 1, dieldrin was stored in the organisms of the model system as 95.9% of the total ^{14}C in fish, 91.6% in snail, and 85.7% in alga (Fig. 1B). Three minor degradation products were clearly visible in the radioautograph of TLC plates developed in ether hexane 1 : 1, R_f 0.63, 0.45, and 0.34 (Figure 1B). The compound R_f 0.45 cochromatographed with an authentic sample of 9-hydroxydieldrin, and that at R_f 0.34 with 9-keto-dieldrin. The unknown R_f 0.08 is probably *trans*dihydroxydihydroaldrin. The E.M. values can be calculated for these degradation products from the data in Table 1 as: aldrin 3140 (fish), dieldrin 5957, 9-OH-dieldrin 617, and 9-C = O dieldrin 220. Thus aldrin and particularly its primary degradation product dieldrin provide the major environmental hazards from bioconcentration and the other compounds including aldrin *trans*-diol are on the degradation pathways and are slowly being eliminated.

Endrin or 1,2,3,4,10,10-hexachloro-6,7-epoxy-1,4,4a,5,6,7,8,8a-octohydro-1,4-*endo,endo*-5,8-dimethanonaphthalene is the *endo,endo*-isomer of dieldrin. A preliminary evaluation of the ^{14}C ring-U labeled product 23 m Ci m mole, at 5.0 mg produced such high kills of the organisms that the experiment could not be completed. It was repeated at 1.0 mg per chamber and repeatedly killed snails and *Daphnia* over the first month of the experiment. The aqueous concentration of 1 to 2, ppb repeatedly killed the fish in the system and delayed the termination of the experiment for twice the normal 33 days. These observations show the value of biological

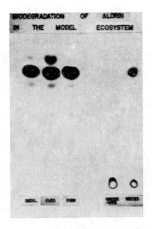

Fig. 1b

Table 1 Behavior of Aldrin and Endrin in the Model Ecosystem

| | conc. | | ppm (equivalents) | | |
	water	alga	snail	mosquito	fish
A. aldrin					
total ^{14}C	0.0117	19.70	57.20	1.13	29.21
aldrin	0.00005	1.95	2.23	—	0.157
dieldrin	0.0047	16.88	52.40	1.1	28.00
9-OH dieldrin	0.00052	0.12	0.17	—	0.322
9-C=O dieldrin	0.0004	0.079	0.217	—	0.088
Unknowns	0.00039	0.585	2.05	—	0.612
polar metabolites	0.0040	0.015	0.097	—	0.004
	E.M. 3140, B.I. 0.00014 (fish)				
B. endrin					
total ^{14}C	0.0134	13.62	150.58	all died	4.48
endrin	0.00254	11.56	125.00	—	3.40
Unknowns	0.00385	1.58	15.11	—	1.04
polar metabolites	0.00436	—	1.85	—	0.04
	E.M. 1335, B.I. 0.009 (fish)				

information gained from model ecosystem studies.

Endrin is somewhat less stable in the model ecosystem than aldrin or dieldrin and was stored as 75.8% of the total ^{14}C in fish, 82.8% in snail, and 84.9% in alga. The major degradation products were thought to be 9-OH and 9-C = O endrin but standards for comparison were not available. Table 1 shows the information gained from a radioautograph of a TLC plate from the endrin experiments, as pictured in.[15] The E.M. values (fish) were: endrin 1335, and 9-OH endrin 270.

The following values sum up the information on E.M. and B.I. for organochlorine pesticides gained from model ecosystem studies.

compound	E.M. (fish)	B. I. (fish)
DDT	84,500	0.015
DDE	27,400	0.032
aldrin	3,140	0.00014
dieldrin	2,700	0.0018
endrin	1,335	0.009
mirex	219	0.0145
lindane	560	0.091
hexachlorobenzene	287	0.46

B. Evaluation of Pesticides for Vector Control

During the past 13 years, WHO has conducted an extensive collaborative program for the "Evaluation of Insecticides for Vector Control".[16] More than 1800 compounds from the world chemical industry have been intensively screened for their utility in vector control both against DDT and cyclodiene resistant insect vectors and for unique properties in the control of vector borne diseases. This program has developed the vector control potentialities of propoxur (OMS 33), fenitrothion (OMS 43), fenthion (OMS 2), chlorpyrifos (OMS 971), diphenphos (OMS 786), and other new compounds. As new vector control programs are undertaken in developing nations, particularly those involving pesticide applications to the aquatic environment, prudent regard for environmental quality requires selection of degradable pesticides to replace the environmentally polluting organochlorines. WHO has initiated studies of the biodegradability and possible incorporation in food chains as part of its protocols for the evaluation of new insecticides using the laboratory model ecosystem as described in this report.

Larvicides for onchocerciasis control. The transmission of the filarian *Onchocerca volvulus* by the biting black fly *Simulium damnosum* has retarded the agricultural and economic development of large areas of Central Africa. In Upper Volta there are approximately 400,000 victims of the disease onchocerciasis or river blindness and in Nigeria about 340,000. Approximately one of every ten suffers from severe eye lesions and may become blind. Fear of the disease by local inhabitants has resulted in abandonment of fertile river valley land where the black flies are abundant. The only feasible control measure at present, involves treatment of streams where the black fly larvae breed with a larvicide effective enough to cause their detachment from the silken threads anchoring them to rocks. Chlorpyrifos or Dursban® or 0,0-diethyl 0-3,4,6-trichloropyridinyl phosphorothionate (OMS-971) is one of the most effective larvicides for aquatic insects yet developed. Its 0,0-dimethyl analogue chlorpyrifos-methyl (OMS 1155) appears equally effective as a larvicide and is much less toxic to mammals; rat oral LD_{50} in mg per kg, chlorpyrifos methyl 1000, chlorpyrifos 135.[16] The question arose as to which of the two organophosphorus esters would have the least effect on environmental quality when used for large scale larviciding experiments for *Simulium* control.

Both chlorpyrifos and chlorpyrifos methyl labeled with ^{14}C in the pyridinyl ring, specific activity 10.7 m Ci per m mole, were applied to the model ecosystem at 1.0 mg. because of their very high biological activity. The data in Table 2, provides a comparison of the model ecosystem behavior of the two compounds.[9]

The E.M. value for the fish were chlorpyrifos 314 and chlorpyrifos methyl 95 and the corresponding B.I. values were 1.02 and 3.95. The unextractable radioactivity in the fish was 23 per cent of the total for chlorpyrifos and 52 per cent for chlorpyrifos methyl. All of these parameters indicate the substantially greater biodegradability of chlorpyrifos methyl over chlorpyrifos and suggest that the latter is a more suitable larvicide for the preservation of environmental quality.

Table 2 Behavior of Chlorpyrifos and Chlorpyrifos Methyl in the Model Ecosystem

	water	alga	conc. ppm (equivalents) snail	mosquito	fish
A. chlorpyrifos					
total ^{14}C	0.00056	0.0261	0.158	0.058	0.0711
chlorpyrifos	0.00011	0.0079	0.076	0.005	0.0352
trichloropyridinol	0.000115	0.0051	0.051	0.022	0.0207
polar metabolites	0.00005	0.0141	0.031	0.036	0.0152
	E.M. 314, B.I. 1.02 (fish)				
B. chlorpyrifos methyl					
total ^{14}C	0.00272	0.0780	0.0882	0.220	0.0367
chlorpyrifos methyl	0.00008	0.0382	0.0435	0.15	0.0076
trichloropyridinol	0.0002	0.0087	0.0082	0.037	0.0109
polar metabolites	0.00113	0.0051	0.0067	0.033	0.0101
	E.M. 95, B.I. 3.95 (fish)				

Table 3   Behavior of ¹⁴C Propoxur in the Model Ecosystem

| | water | conc. ppm (equivalents) | | | |
		alga	snail	mosquito	fish
total ¹⁴C	0.00408	0.4167	0.3946	2.2913	0.1173
isopropoxyphenol	0.000083	—	0.0406	—	0.0252
propoxur	0.00032	0.0327	0.0963	0.4441	0.0468
isopropoxyphenyl carbamate	0.000032	—	0.0237	—	—
isopropoxyphenyl hydroxymethyl carbamate	0.000006	0.0223	—	1.1520	0.0180
Unknowns	0.000032	0.2486	0.1631	—	—
polar metabolites	0.00108	0.1131	0.0746	0.2640	0.0273

E.M. 146, B.I. 0.30 (fish)

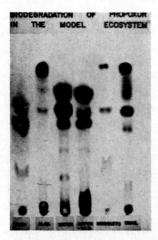

Fig. 1c

Propoxur or 2-isopropoxyphenyl N-methylcarbamate (OMS-33) is a highly effective contact insecticide which has been found in the WHO Programme (16) to be perhaps the most promising substitute for DDT in various areas of the global malaria eradication program, especially where anopheline vectors are resistant to DDT and/or dieldrin. Propoxur is also widely used for household pest control. Model ecosystem investigations were made with 5.0 mg. of 2-(isopropoxy-1,3-¹⁴C) phenyl N-methylcarbamate, 10.4 m Ci per m mole. Radioautographs of the extracts of the organisms of the system are shown in Figure 1C and the quantitative summation of results in Table 3. Trace amounts of propoxur persisted in the organisms throughout the experiment, together with 2-isopropoxyphenol and other partially degraded carbamates. The E.M. value of 146 and the B.I. of 0.30 show that propoxur is markedly more biodegradable than DDT (see Table VI).

C. Industrial Compounds

As described above, the rapidly increasing production of synthetic organ-

Table 4   Behavior of 2,5,2',4',5'-Pentachlorobiphenyl in the Model Ecosystem

| | water | conc. ppm (equivalents) | | | |
		alga	snail	mosquito	fish
total ¹⁴C	0.04340	62.4660	633.0165	181.4565	127.6945
pentachlorobiphenyl	0.00985	53.8440	587.3545	170.8480	119.7060
Unknowns	0.00450	1.5925	20.8030	4.7785	4.1925
polar metabolites	0.02055	1.6265	16.5550	2.6745	2.3610

E.M. 12,153, B.I. 0.019 (fish)

ic chemicals has resulted in substantial losses to the environment. Two examples of important world-wide micropollutants will be discussed in terms of model ecosystem evaluation.

Polychlorinated biphenyls or PCB's are used as dielectric fluids, heat transfer fluids, hydraulic fluids, plasticizers and solvents. The commercial products consist of mixtures of 20 to 30 isomers of average percent chlorination. The United States cumulative production during the past 44 years is about 800×10^6 lb.[17] PCB's, like organochlorine pesticides, have become ubiquitous yet research has been hindered by the complexities of the multicomponented mixtures. Very recently ^{14}C ring-U labeled PCB's have become available. We report here model ecosystem results with 2,5,2',4',5'-pentachlorobiphenyl 9.87 m Ci per m mole, a major constituent of Aroclor® 1254. This compound was applied to *Sorghum* at 5.0 mg. The results are shown in the radioautograph of Figure 1D and in Table 4.[18] This PCB accumulated in the organisms of the model system to relatively enormous levels and was only slightly degraded to polar products (Table 4). The parent compound comprised 93.0% of the total ^{14}C in snail, 93.9% in fish, 98.4% in mosquito, and 86.3% in algae. The unknowns represent largely other PCB

trace impurities in the labeled product. The E.M. value (fish) was 12,153 for pentachlorobiphenyl and from 2900 to 25,000 for the other PCB impurities.

Several other radiolabeled PCB isomers have been investigated in the model ecosystem and the results will be reported elsewhere.[18] They demonstrate a surprisingly individualistic behavior for each isomer.

Dialkyl phthalate plasticizers have recently become suspect as environmental pollutants because of their very large usage as components of vinyl resins. More than 4×10^9 lb. of these esters have been produced in the United States since 1943.[19] They have been found in a variety of human tissues [20] and are known to be teratogenic in rats.[21] When ^{14}C-O labeled di-2-ethylhexyl phthalate (DEHP) was applied to the *Sorghum* of the model ecosystem, the distribution of radioactivity was that summarized in Table 5.[22] The radioautographs of TLC plates from extracts of the biological components of the system showed that DEHP is a fairly stable environmental pollutant which is stored intact in the tissues of organisms, with E.M. values of: alga 53,890, snail 21,480, and mosquito 107,670. The fish, E.M. 130, was more successful in degrading DEHP and contained substantial amounts of mono-ethylhexyl

Table 5   Behavior of di-2-ethylhexyl phthalate (DEHP) in the Model Ecosystem

	conc.		ppm (equivalents)		
	water	alga	snail	mosquito	fish
total ^{14}C	0.0078	19.105	20.325	36.609	0.206
DEHP	0.00034	18.322	7.302	36.609	0.044
MEHP	0.00099	0.325	2.541	—	0.021
phthalic anhydride?	0.00363	0.180	5.772	—	0.113
phthalic acid	0.00077	0.094	2.724	—	0.018
Unknowns	0.00190	0.029	0.768	—	—
polar metabolites	0.00016	0.155	1.218	—	0.010

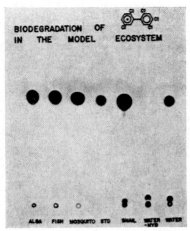

Fig. 1d

phthalate (MEHP), and phthalic acid (Table 5). A compound Rf 0.65 cochromatographed with phthalic anhydride and may be that compound formed from phthalic acid. The unknowns are probably ring hydroxylated products. DEHP is slowly degraded in living organisms by two mechanisms, hydrolysis by esterases and oxygenation by mixed function oxidases. Nevertheless it is a surprisingly stable micropollutant and is highly biomagnified.

D. Studies of Principles of Biodegradability

With exponentially rising organic chemical production, there is increasing urgency for the production of degradable substitutes for the wide array of poorly degradable pollutants such as the organochlorine pesticides, PCB's, alkyl phthalates, detergents, organomercurials, etc. The development of substitute compounds depends on systematic and fundamental knowledge of the sorts of biochemical degradations that occur in a variety of food web organisms and the effects of these on

bioconcentration and lipid storage. The laboratory model ecosystem is well suited for investigations of this nature and has been used in our laboratory to study the principles of biodegradability of the DDT-type compounds which is ideally constituted for examination of the affects of changes in aromatic and aliphatic substituents.[9, 12,13,14]

Studies with ^{14}C-labeled DDT, DDD, and DDE in the model system [9] showed that DDT was readily converted to DDE which comprised 52% of the radioactivity in snails, 58% in mosquito larvae, and 54% in fish. In the *Gambusia* fish at the top of the food chain, DDE was present at 110,000 times the concentration in water, and DDT 84,500 times. DDE proved highly persistent and was stored largely as the intact molecule with only slight evidence of the production of polar degradation products. Investigations of DDT analogues where the aryl chlorine atoms were replaced with degradophores such as CH_3O, C_2H_5O, CH_3, CH_3S which are substrates for the multifunction oxidase enzymes revealed substantially different behavior.[12, 13, 14] The action of the oxygenases on these substituents converted the lipid-soluble insecticides into a variety of water-partitioning derivatives which could be eliminated from the organisms rather than stored in body fats. Thus CH_3O- and C_2H_5O- are O-dealkylated to HO-,CH_3-oxidized to HOOC-, and CH_3S-oxidized to CH_3SO- and CH_3SO_2-. Examples of the degradation products produced in the model ecosystem and of the effects of these changes on the parameters of E.M. and B.I. are shown in Table 6. Clearly it is possible to tailor insecticidal molecules of the DDT-type so that they are substantially degraded in food chain or-

Table 6   Behavior of DDT Analogues in the Model Ecosystem [12, 13, 14]

	water	conc. ppm (equivalents)		
		snail	mosquito	fish
A. DDT				
total ^{14}C	0.004	22.9	8.9	54.2
$ClC_6H_4HCCCl_3C_6H_4Cl$	0.00022	7.6	1.8	18.6
$ClC_6H_4CCCl_2C_6H_4Cl$	0.00026	12.0	5.2	29.2
$ClC_6H_4HCCHCl_2C_6H_4Cl$	0.00012	1.6	0.4	5.3
polar metabolites	0.0032	0.98	1.5	0.8
	E.M. 84,500, B.I. 0.015 (fish)			
B. methoxychlor				
total ^3H	0.0016	15.7	0.48	0.33
$CH_3OC_6H_4HCCCl_3C_6H_4OCH_3$	0.00011	13.2	—	—
$CH_3OC_6H_4CCCl_2C_6H_4OCH_3$	—	0.7	—	—
$HOC_6H_4HCCCl_3C_6H_4OCH_3$	0.00013	1.0	—	—
$HOC_6H_4HCCCl_3C_6H_4OH$	0.00003	—	—	—
$HOC_6H_4CCCl_2C_6H_4OH$	0.00003	—	—	—
Unknown	0.00009	—	—	—
polar metabolites	0.00125	0.8	—	0.16
	E.M. 1545, B.I. 0.94 (fish)			
C. methylchlor				
total ^{14}C	0.216	101.00	1.00	0.68
$CH_3C_6H_4HCCCl_3C_6H_4CH_3$	0.0006	72.16	0.83	0.084
$CH_3C_6H_4HCCCl_3C_6H_4CH_2OH$	—	6.64	+	—
$CH_3C_6H_4HCCCl_3C_6H_4COOH$	0.034	—	+	—
$HOOCC_6H_4HCCCl_3C_6H_4COOH$	0.018	—	—	—
Unknowns	0.055	21.34	—	—
polar metabolites	0.118	0.851	0.167	0.60
	E.M. 140, B.I. 7.14 (fish)			
D. methyl ethoxychlor				
total ^3H	0.0184	10.83	0.74	0.24
$CH_3C_6H_4HCCCl_3C_6H_4OC_2H_5$	0.00012	5.03	0.11	0.04
$CH_3C_6H_4CCCl_2C_6H_4OC_2H_5$	0.00020	2.99	—	0.03
$CH_3C_6H_4HCCCl_3C_6H_4OH$	0.00039	—	—	0.02
$CH_3C_6H_4CCCl_2C_6H_4OH$	0.00015	—	0.11	0.02
$HOOCC_6H_4HCCCl_3C_6H_4OC_2H_5$	0.00271	—	0.06	0.03
Unknowns	0.00119	1.25	0.19	0.06
polar metabolites	0.0128	0.21	0.09	0.04
	E.M. 400, B.I. 1.20 (fish)			

ganisms. Studies are presently being made upon the effects of changes in the aliphatic portion of the DDT-type molecule, i.e. to $-C(CH_3)_3$ and $CH(CH_3)NO_2$ upon biodegradability. From these experiments we hope to obtain not only a better theoretical understanding of degradability but also potentially useful biodegradable insecticides.(12, 23)

Acknowledgements

This report is a contribution from the WHO International Reference Centre for Insecticides, University of Illinois, Urbana-Champaign. The research surveyed here was supported in part by grants from the World Health Organization; the Herman Frasch Foundation, American Chemical Society; The

Rockefeller Foundation; the Federal Water Quality Administration, Water Resources Center Project B-050 Illinois; and U.S. Public Health Service, National Institute of Health, Biomedical Sciences Grant FR 07030.

Especial thanks are due to Dr. I. P. Kapoor, Dr. A. S. Hirwe, Dr. J. Sanborn, Dr. G. K. Sangha, Mr. Po-Yung Lu, Mr. Joel Coats, Mrs. Carter Schuth, Miss Patricia Sherman, and Mr. Rick Furman, who participated in various phases of the work reviewed.

References

1 U. S. Tarif Commission, Chem. Eng. News May 7 (1973) 10.

2 Chem. Eng. News June 4 (1973) 11.

3 Chem. Eng. News April 16 (1973) 52.

4 Merck Index of Chemicals and Drugs 8th ed. Merck & Co. Rahway, N. J. (1968).

5 Lee, D. H. K.: Environmental health and human ecology. Suppl. Amer. J. Pub. Health 54 (1964) 7.

6 Chem. Eng. News Jan. 10 (1972) 6.

7 Warner, R. E.: Bioassays for microchemical environmental contaminants with special reference to water supplies. Bull. World Health Organ. 36 (1967) 181.

8 Winteringham, F. P. W.: Foreign chemicals and radioactive substances in food and environment; a comparative and integrated approach to the problem. Kemian Teollisuus 29 (1972) 561.

9 Metcalf, R. L., G. K. Sangha, and I. P. Kapoor: Model ecosystem for evaluation of pesticide biodegradability and ecological magnification. Envir. Sci. Technol. 5 (1971) 709.

10 Metcalf, R. L.: A laboratory model ecosystem to evaluate compounds producing biological magnification. Essays in Toxicology, 5 (1973) 17.

11 Freeman, L.: A standardized method for determining the toxicity of pure compounds

to fish. Sewage Ind. Waste 25 (1953) 845, 1331.

12 Kapoor, I. P., R. L. Metcalf, A. S. Hirwe, J. R. Coats, and M. S. Khalsa: Structure activity correlations of biodegradability of DDT analogs. J. Agr. Food Chem. 21 (1973) 310.

13 Kapoor, I. P., R. L. Metcalf, R. F. Nystrom, and G. K. Sangha: Comparative metabolism of DDT, methoxychlor, methiochlor and DDT in mouse, insects, and in a model ecosystem. J. Agr. Food Chem. 18 (1970) 1145.

14 Kapoor, I. P., R. L. Metcalf, A. S. Hirwe, Po-Yung Lu, J. R. Coats, and R. F. Nystrom: Comparative metabolism of DDT, methoxychlor, and ethoxychlor in mouse, insects, and a model ecosystem. J. Agr. Food Chem. 20 (1972) 1.

15 Metcalf, R. L., I. P. Kapoor, Po-Yung Lu, C. K. Schuth, and P. Sherman: Model ecosystem studies of the environmental fate of six organochlorine pesticides. Envir. Health Perspectives, June (1973) 35.

16 World Health Organization, Vector Biology and Control Evaluation of Insecticides for Vector Control, Pt. I. WHO/VBC/68.66 Geneva, Switzerland (1968).

17 Nisbet, I. C. T., and A. F. Sarafim: Rates and routes of transport of PCB's in the environment. Envir. Health Perspectives April (1972) 21.

18 Metcalf, R. L., J. Sanborn, and Po-Yung Lu: ms. in preparation.

19 Faith, W. L., D. B. Keyes, and R. L. Clark: "Industrial Chemicals" 3rd ed. John Wiley & Sons, N. Y. (1965).

20 Jaeger, R. J., and R. J. Rubin: Plasticizers from plastic devices: extraction, metabolism, and accumulation by biological systems. Science 170 (1970) 460.

21 Singh, A. R., W. H. Laurence, and J. Autian: Teratogenicity of phthalate esters in rats. J. Pharm. Sci. 61 (1972) 51.

22 Metcalf, R. L., G. M. Booth, C. K. Schuth, D. J. Hansen, and Po-Yung Lu: Envir. Health Perspectives. June (1973) 27.

23 Metcalf, R. L., I. P. Kapoor, and A. S. Hirwe: Biodegradable analogues of DDT. Bull. World Health Org. 41 (1971) 363.

Multielement Characterization of Atmospheric Aerosols by Instrumental Neutron Activation Analysis and X-Ray Fluorescence Analysis *)

By **L. A. Rancitelli, J. A. Cooper,** and **R. W. Perkins**

Battelle, Pacific Northwest Laboratories, Richland, Washington 99352 U.S.A.

Abstract

A large variety of natural and anthropogenic sources contribute to the total inventory of atmospheric aerosols. Many of these sources have unique chemical compositions which can be employed to assess the impact of a specific source to the total inventory on a local and worldwide basis. The simultaneous measurement of a wide spectrum of elements collected on air filters and in rain water can yield a great deal of information on the origin, transport, and removal of atmospheric pollutants. A pair of highly sensitive, precise and complementary instrumental techniques, X-ray fluorescence and neutron activation analysis have been developed and employed to measure over 40 elements on air filters and residues from fossil fuel plants.

The X-ray fluorescence analysis is based on the direct excitation of characteristic X-rays of the elements with 100 μCi ^{55}Fe and ^{109}Cd sources and their measurement with Si(Li) X-ray energy analyzers. Quantitative results are obtained both by comparing the air filter results to thin film standards and by using those elements measured by neutron activation as internal standards. The instrumental neutron activation procedure involves irradiation of the filter samples together with the appropriate standards followed by a γ-ray analysis of the activation products, after appropriate decay intervals, on either conventional or anti-coincidence shielded Ge(Li) gamma-ray spectrometers.

Several important observations concerning the origin of elements in urban aerosols can be inferred from the ratio of the trace element concentrations to the Sc concentrations. From a comparison of these ratios in granite and diabase with those of the filter, it is evident that Zn, Se, Sb, Hg and Pb levels have been increased by as much as several orders of magnitude. Al, Co, La, Fe, Eu, Sm, Tb, Ta, Hf, and Th appear to exist at levels compatible with an earth's crust origin.

*) This paper is based on work performed for the United States Atomic Energy Commission Contract AT (45–1)–1830.

**) Published in Comparative Studies of Food and Environmental Contamination (Proc. FAO/IAEA/WHO Symp. Otaniemi, 1973), IAEA, Vienna (1974) 3–22.

Introduction

Man is becoming increasingly aware of the delicate balance which permits the existence of life on this planet and of his own ability to inadvertently produce devastating effects upon this balance. While some of the serious effects of air pollution from automotive exhausts and industrial stack effluents are obvious in the immediate environs, the more subtle long-term and distant effects may require years for their full environmental impact to be recognized. The gaseous air pollutants SO_2, No_x, CO, CO_2 and organic degradation products, together with PCB's, the herbicides and pesticides, make up the bulk of pollutants entering the atmosphere. However, there are also large amounts of the toxic trace elements Hg, Cd, Pb, Se, As, Sb, Ag, Cu, Be, Cr, Zn, V and F which are associated with air pollutants from a wide spectrum of anthropogenic sources.[1–7] While serious concern regarding this latter group is only now being manifested, its long-range effect may be of far greater consequence. The toxic trace elements are permanent pollutants and their continuous release into the atmosphere could result in the irreversible poisoning of our terrestrial agricultural areas, together with our freshwater and ocean basins. Indeed, many of the rivers and lakes of the world have become polluted to the point where their edible aquatic organisms contain "dangerous levels" of toxic trace elements. The poisoning of vegetation and cattle in the environs of smelting operations and other industrial plants is well known.

In fallout studies sponsored by the Atomic Energy Commission over the past two decades, it has been shown that radioactive gases and particulates released at northern latitudes become distributed over the entire northern hemisphere and are eventually distributed throughout the southern hemisphere. This is true even for ground level releases. It is thus clear that while much of today's nonradioactive air pollutants are generated in localized areas and may produce serious regional effects, some of these pollutants are also carried great distances and thus contribute to the hemispheric and worldwide air pollution problems.[7, 8] The seriousness of global pollution will certainly increase with the decentralization of pollution sources, which is occurring as new giant fossil fuel plants and industrial operations are located over wider regions of industrialized countries of the world.

If we are to properly describe the long-range effect of pollutants on our environment, it is essential that their total environmental behavior be understood. Once injected into the atmosphere, pollutants are carried by prevailing winds. They undergo both physical and chemical reactions in the atmosphere and are eventually deposited on the earth's surface by either dry deposition or by precipitation scavenging. Where they reach terrestrial areas, they may enter the food chain by direct foliate adsorption or plant root uptake from the soil and thus become available for eventual uptake by man. Where they reach fresh or salt water surfaces, they may be incorporated directly into the base of the aquatic food chain and subsequently move to higher trophic levels. Once in the terrestrial or aquatic environment, the pollutants will undergo some recycling and with time become less available for continued uptake in the biosphere.

In our research program conducted for the USAEC over the past several years, rapid analytical techniques have been developed based on instrumental neutron activation analysis, and X-ray fluorescence, allowing the measurement of some 40 trace elements, including all those designated as pollutants, in air filters, water samples, biota, and other environmental materials.[7–9] The importance of a multi-element analytical approach, which includes essentially all of the elements of anthropogenic origin, plus some of natural

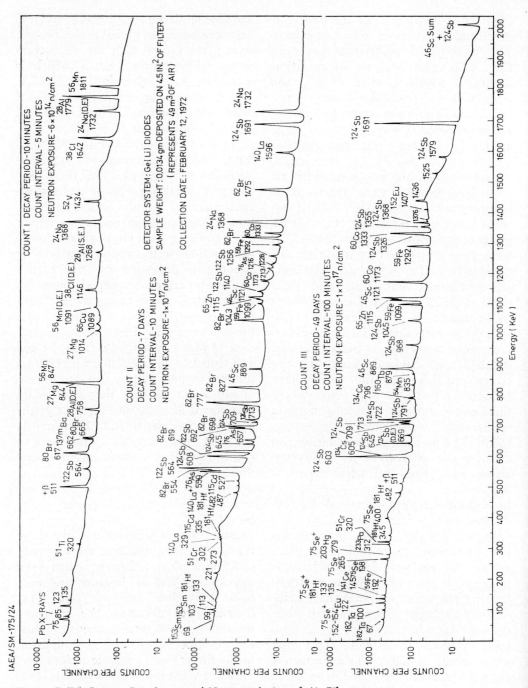

Fig. 1   Ge(Li) Gamma Ray Spectra of Neutron Activated Air Filter.

earth's crust origin, becomes obvious when one considers that, first, any trace element will present a hazard if its concentration exceeds "critical limits" in a given environment, and second, the environmental behavior of the toxic trace elements is dependent on the other elements and compounds present.

Experimental

The collection and preparation of aerosol samples for trace element analysis, particularly in remote areas, requires extreme care to avoid contaminating the sample during handling and pumping operations. Duce *et al.* (10) have given practical recommendations for avoiding contamination errors in air sampling.

Atmospheric particulate material is collected by pumping air through suitable filter media or cascade impactors. Air volumes are measured with recording or integrating flow meters. The most popular filter media presently in use are Whatman 41® cellulose filters and Millipore® membrane filters. Dams *et al.* (11) have published an excellent paper evaluating 10 various filter media for suitability for atmospheric sampling and elemental analysis by nondestructive INAA using GE(Li) gamma-ray spectrometry. We concur with them that the commercially available filter, Whatman 41® cellulose, is optimum from the standpoint of low blanks, particle retentivity and ease of handling. However, these filters do have a tendency to clog during prolonged sampling of dirty air. Stafford and Ettinger (12) have carefully evaluated the retention efficiencies of Whatman 41® filters as a function of particle size and air sampling velocity,

and these data are very useful for quantifying air sampling with these filters.

Duplicate samples from each exposed filter are removed for trace element analyses while the remaining filter is retained for the analysis of fallout radionuclides. Each sample consists of a 29 cm² section of filter. One piece is rolled tightly and placed in a clean polyethylene vial 2.5 cm long by 0.95 cm ID (2/5 dram), while the other is retained for X-ray fluorescence analysis. The vials, containing samples from each filter, are then placed in another clean vial 5.5 cm long by 1.43 cm ID (2 dram) and shipped to the reactor for neutron activation analysis.

Neutron Activation Analysis

The method of analysis consisted of submitting the samples to two irradiation periods and three counting periods.

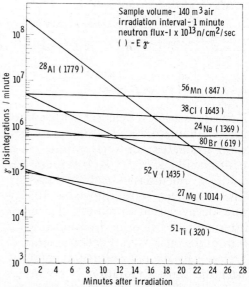

Fig. 2 *Typical Relative Activities of Short Lived Radionuclides in a Neutron Activated Air Filter.*

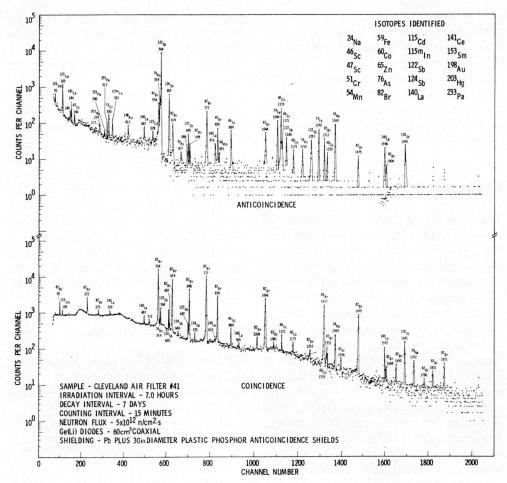

Fig. 3   Anticoincidence-Shielded *Gamma-Ray Spectra* of Neutron Activated Air Filter — Intermediate Lived Radionuclides.

First, the samples, together with appropriate standards and flux monitors, were irradiated for one minute at a neutron flux of $\sim 1.3 \times 10^{13}$ n/cm²-sec using a pneumatic "rabbit" system. These samples were allowed a 10 minute decay period to remove the major interference ^{28}Al prior to counting for 5 minutes on a Ge(Li) spectrometer. A typical gamma ray spectrum obtained after the short irradiation is presented in Fig. 1. The major activity ^{28}Al also has the highest energy photon which produced an interference to the meas-

urement of the other radionuclides. Fig. 2 contains the major activities found in an activated sample after irradiation. It is evident from this plot that a decay interval of 10 minutes represents the optimum compromise to permit the simultaneous measurement of these radionuclides. If one counts the sample immediately after irradiation the ^{28}Al masks many of the other radionuclides while permitting the sample to decay 20 minutes makes it difficult to determine minor constituents such as ^{52}V, ^{27}Mg and ^{51}Ti.

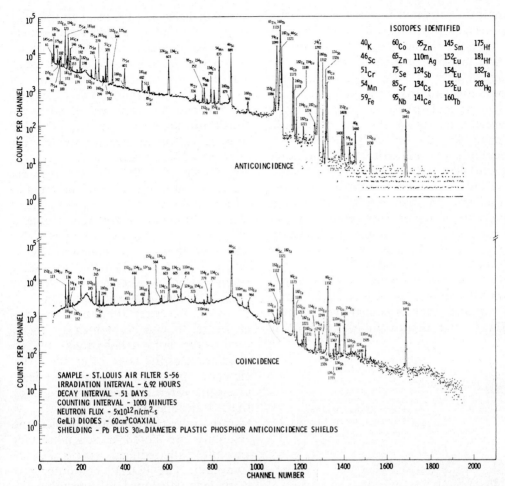

Fig. 4   Anticoincidence-Shielded Gamma-Ray Spectra of Neutron Activated Air Filter — Long Lived Radionuclides.

Direct comparison to standards containing known amounts of the elements of interest permits the simultaneous measurement of Al, Br, Cl, Mg, Mn, Na, Ti and V concentrations on the filters. Due to the short half life of the (n, γ) product of some of these elements, only one sample could be irradiated at any one time.

The second step in the analysis was to re-irradiate the samples and standards to an integral neutron flux of $\sim 10^{17}$ n/cm², allow 7 to 30 day decay periods, and count for 30 and

100 minutes, respectively on an anticoincidence shielded Ge(Li) spectrometer.[13] The major advantage of this detection system over a normal diode becomes readily apparent by comparing Fig. 1 with Fig. 2 and 3. The anticoincidence shielded spectrometer significantly improves the measurement of radionuclides such as ^{51}Cr, ^{198}Au or ^{65}Zn which emit a single gamma ray without altering the detection efficiency for radionuclides which emit coincidence gamma-rays. Up to 48 samples could be simultaneously

neutron activated in a rotating irradiation facility to ensure a uniform irradiation. Typical gamma-ray spectra of a neutron activated air filter obtained after a 5-day decay interval is presented in Fig. 3 and after a 50-day interval in Fig. 4. In addition we present in Fig. 5 the activity of the major gamma ray radionuclides as a function of time after the irradiation. From Fig. 3 and 5 it is evident that initially the major activity ^{24}Na which also emits a high energy gamma-ray, presents a major interference to the measurement of other intermediate lived radionuclides. Permitting an initial decay interval of 7 days reduces the ^{24}Na interference and improves the detections of ^{82}Br, ^{76}As, ^{140}La, ^{153}Sm and ^{177}Lu. It is also evident from Fig. 5 that permitting the sample to

decay at least 25 days prior to the long count on a Ge(Li) spectrometer reduces the interference from ^{82}Br without seriously affecting our ability to measure the remaining radionuclides of interest. The concentrations of As, Br, La, Na and Sm were calculated from the 30 minute count intervals. The 100 minute count intervals provided the concentration of Co, Cr, Eu, Fe, Hf, Hg, Sb, Sc, Se, Ta, Tb, Th and Zn. Table 1 summarizes the parameters which affect the measurement of each element; neutron activation products, half-lives, gamma-ray energies, possible interfering radionuclides, interfering reactions producing the same isotopes from another element, lengths of irradiation, decay and counting.

It should be noted that there are two elements, Br and Na, which are measured after both irradiations. By re-irradiation of the same filter, these two elements provided internal standards to ensure internal consistency between the two irradiations.

Gamma-ray spectra from samples and standards are stored on magnetic tape and the data reduced with a PDP-15 computer. The PDP-15 computer program was developed to determine the radionuclides present from their characteristic gamma-ray energies, calculate net peak areas, convert net peak areas to weight of elements by direct comparison to known standards and to calculate the concentration of the elements in the original air mass.

X-ray Fluorescence Analysis

The X-ray fluorescence analysis system consists of a high intensity (100 mCi) X-ray emitting radioisotopic source (^{55}Fe or ^{109}Cd) which excites the characteristic X-rays of the ele-

vi

IAEA / SM- 175 / 24

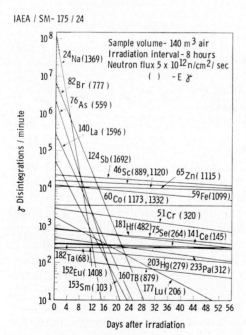

Fig. 5 Typical Relative Activities of Intermediate and Long Lived Radionuclides in a Neutron Activated Air Filter.

ments of interest in the sample (14—16) and a Si(Li) X-ray energy analyzer.(17) The air filter samples are mounted in 35 mm slide frames and positioned in an automatic sample changer. The automatic sample changer rotates the sample into position above the radio-isotopic source and Si(Li) detector at the proper time for analysis and an X-ray fluorescence spectrum is accumulated for a predetermined counting interval.

Fig. 6 and 7 show typical ^{55}Fe and ^{109}Cd X-ray excited spectra of urban aerosol trace elements collected on Whatman 41® filter paper. Also shown in these figures are the location of the X-ray lines of other elements which could be measured if they exceed the detection limit of the system. A total of 17 elements were detected in the analysis of this aerosol sample. However,

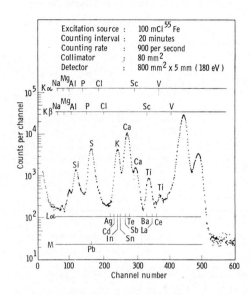

IAEA / SM- 175 / 24

Fig. 6 ^{55}Fe − X-ray Fluorescence Spectrum of Urban Air Filter.

Table 1   Nuclear Data Relating to the Measurement of the Elements in Air Filter Samples by Instrumental Neutron Activation Analysis

Irradiation parameters { Irradiation interval — 0.5-1.0 min
Neutron flux — 1x10¹³ neutrons/cm²/sec

Element	Target isotope	Isotopic abundance (%)	Product nuclide	Half-life	Thermal neutron cross section	Best γ for Measurement (KEV)	Number of γ's per 1000 decays	Associated γ-rays KEV (γ's/1000 disinteg)	Possible interfering nuclide in diode Measurement	Possible interfering nuclear reactions producing nuclides of interest
Al	^{27}Al	100	^{28}Al	2.31m	0.24b	1779	1000	NONE	^{151}Nd(1776)	^{28}Si(n,p)^{28}Al ^{31}P(n,a)^{28}Al
Br	^{79}Br	50.5	^{80}Br	17.6m	8.5b	617	70	666(10)	^{108}Ag(614)	^{80}Kr(n,p)^{80}Br
Cl	^{37}Cl	24.5	^{38}Cl	37.3m	0.4b	1643	380	2170(470)	NONE	^{38}Ar(n,p)^{38}Cl ^{41}K(n,a)^{38}Cl
Mg	^{26}Mg	11.2	^{27}Mg	9.45m	0.027b	844	700	180(7)	^{101}Mo(840) ^{151}Nd(841) ^{87}Kr(846) ^{56}Mn(847)	^{72}Al(n,p)^{27}Mg
						1014	300		^{101}Mo(1012) ^{151}Nd(1016) ^{125}Sn(1017) ^{188}Re(1018)	^{03}Si(n,a)^{27}Mg
Mn	^{55}Mn	100	^{56}Mn	2.58h	13.3b	847	990	2110(150)	^{27}Mg(844)	^{56}Fe(n,p)^{56}Mn
						1811	290		NONE	^{59}Co(n,a)^{56}Mn
Na	^{23}Na	100	^{24}Na	15.0h	0.53b	1369	1000	2754(1000)	^{125m}Sn(1369) ^{117}Cd(1373) ^{188}Re(1368)	^{24}Mg(n,p)^{24}Na ^{27}Al(n,a)^{24}Na
Ti	^{50}Ti	5.3	^{51}Ti	5.79m	0.14b	320	950	605(15) 928(50)	^{199}Pt(317) ^{151}Nd(320) ^{105}Ru(317) ^{182m}Ta(318)	^{51}V(n,p)^{51}Ti ^{54}Cr(n,a)^{51}Ti
V	^{51}V	99.8	^{52}V	3.76m	4.9b	1434	1000	NONE	^{139}Ba(1430)	^{52}Cr(n,p)^{52}V ^{55}Mn(n,a)^{52}V

Table 1 (contd) Nuclear Data Relating to the Measurement of the Elements in Air Filter Samples by Instrumental Neutron Activation Analysis

Irradiation parameters { Irradiation interval — 6-8 HRS
Neutron flux — 50x10^{12} neutrons/cm²/sec

Element	Target isotope	Isotopic abundance (%)	Product nuclide	Half-life	Thermal neutron cross section	Best γ for Measurement (KEV)	Number of γ's per 1000 decays	Associated γ-rays KEV (γ's/1000 disinteg)	Possible interfering nuclide in diode Measurement	Possible interfering nuclear reactions producing nuclides of interest
As	^{75}As	100	^{76}As	26.5h	4.5b	559	430	1220(50) 1440(7)	^{193}Os(558) ^{114m}In(558) ^{82}Br(554) ^{76}As(562) ^{122}Sb(564)	^{76}Se(n,p)^{76}As
						657	60	1789(3) 2100(3)	^{110m}Ag(657)	^{79}Br(n,a)^{76}As
Br	^{81}Br	49.48	^{82}Br	35.9h	3b	777	830	554(660) 619(410) 698(270) 828(250)1044(290) 1317(260) 1475(170)	^{187}W(772) ^{76}As(775) ^{152}Eu(779) ^{97}Mo(777) ^{131m}Te(774)	^{82}Kr(n,p)^{82}Br ^{85}Rb(n,a)^{82}Br
Co	^{59}Co	100	^{60}Co	5.24y	37b	1173	1000	NONE	^{188}Re(1176) ^{160}Tb(1178) ^{79}Kr(1332)	^{60}Ni(n,p)^{60}Co ^{63}Cu(n,a)^{60}Co
						1332	1000			
Cr	^{50}Cr	4.31	^{51}Cr	27.8d	17b	320	90	NONE	^{192}Ir(317)^{177m}Lu(319) ^{147}Nd(319) ^{193}Os(321) ^{177m}Lu(321) ^{177}Lu(321)	^{54}Fe(n,a)^{51}Cr
Eu	^{151}Eu	47.77	^{152}Eu	12.7y	5900b	1408	220	122(370) 245(80) 344(270) 779(140) 965(150) 1087(120) 1113(140)	NONE	^{152}Gd(n,p)^{152}E
Fe	^{58}Fe	0.31	^{59}Fe	45d	1.1b	1099	560	143(8) 192(28)	^{160}Tb(1103) ^{182}Ta(1289) ^{115m}Cd(1290) ^{152}Eu(1292)	^{59}Co(n,p)^{59}Fe ^{62}Ni(n,a)^{59}Fe
						1292	440			
Hf	^{180}Hf	35.22	^{181}Hf	43d	10b	482	810	133(480) 346(130)	^{192}Ir(485) ^{140}La(487)	^{181}Ta(n,p)^{181}H ^{184}W(n,a)^{181}Hf
Hg	^{202}Hg	29.8	^{203}Hg	46.9d	4b	279	770	NONE	^{197m}Hg(279) ^{75}Se(280) ^{193}Os(280) ^{175}Yb(283) ^{147}Nd(275) ^{133}Ba(276)	^{203}Tl(n,p)^{203}H ^{206}Pb(n,a)^{203}H
La	^{139}La	99.91	^{140}La	40.3h	8.9b	1596	960	329(200) 487(400) 815(190) 923(100) ^{154}Eu(1596) 2530(90)	^{72}Ga(1596)	^{140}Ce(n,p)^{140}L
Na	^{23}Na	100	^{24}Na	15h	0.53b	1369	1000	2754(1000)	^{134}Cs(1365)	^{24}Mg(n,p)^{24}N ^{27}Al(n,a)^{24}Na
Sb	^{123}Sb	42.75	^{124}Sb	60.2d	3.3b	1691	500	602(980) 723(98) 2091(69)	NONE	^{124}Te(n,p)^{124}S ^{127}I(n,a)^{124}Sb
Sc	^{45}Sc	100	^{46}Sc	83.8d	13b	889	1000	NONE	^{110m}Ag(884) ^{192}Ir(884) ^{65}Zn(1115) ^{182}Ta(1121) ^{131m}Te(1125)	^{46}Ti(n,p)^{46}Sc
						1120	1000			
Se	^{74}Se	0.87	^{75}Sc	120d	30b	265	600	97(33) 121(170) 136(570) 401(120)	^{169}Yb(261) ^{115}Cd(263) ^{182}Ta(264) ^{203}Hg(279) ^{133}Ba(296) ^{192}Ir(283)	^{78}Kr(n,a)^{75}Se
						280	250			
Sm	^{152}Sm	26.6	^{153}Sm	47.1h	210b	103	280	70(54)	^{182}Ta(100)	^{153}Eu(n,p)^{153}S ^{156}Gd(n,a)^{153}S
Ta	^{181}Ta	99.99	^{182}Ta	115d	21b	68	420	100(140) 152(70) 222(80) 1122(340) 1189(160)	^{169}Yb(63) ^{121}Te(65) ^{182}Ta(65) ^{75}Sc(66)	^{182}W(n,p)^{182}T ^{185}Re(n,a)^{182}T
						1221	270	1231(130)	NONE	
Tb	^{159}Tb	100	^{160}Tb	72d	46b	879	310	87(120) 299(300) 966(310) 1178(150)	^{185}Os(879)	^{160}Dy(n,p)^{160}T
Th	^{232}Th	100	^{233}Pa	27.0d	7.4b	312	270	300(50)	^{192}Ir(317)	NONE
Zn	^{64}Zn	48.89	^{65}Zn	243d	0.46b	1116	506	511(500β+)	^{129m}Te(1112) ^{152}Eu(1212) ^{182}Ta(1113) ^{160}Tb(1115) ^{46}Sc(1126)	NONE

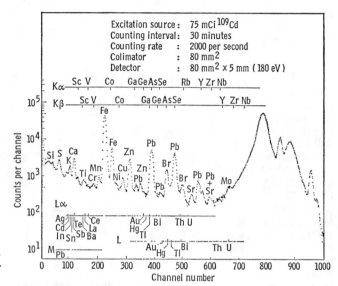

Excitation source : 75 mCi ^{109}Cd
Counting interval: 30 minutes
Counting rate   : 2000 per second
Colimator   : 80 mm^2
Detector   : 80 mm^2 x 5 mm (180 eV)

Fig. 7 ^{109}Cd — X-ray Fluores-cence Spectrum of Urban Air Filter.

at the present time not all of these elements can be quantitatively measured. Those elements that are typically detected by X-ray fluorescence but are presently only semiquantitatively measured because of absorption uncertainties include Na, Mg, Al, Si, S, Cl, K and Ca. The elements Ti, V, Cr, Mn, Fe, Ni, Cu, Zn, As, Br, Sr, Zr, Mo, Cd, Sn, Ba and Pb are usually measured quantitatively. The Pb peaks shown in Fig. 7 represent 1800 ng/m^3 of air while the Sr peaks represent 60 ng/m^3 or about 20 ppm of Sr on the filter as analyzed. The elemental concentrations are determined by a direct comparison of the sample characteristic X-ray counting rate with the counting rate obtained from a vacuum-deposited thin film standard of known concentration in ng/cm^2.(18) This is based on the thin film assumption that enhancement and absorption effects are negligible. This assumption has been demonstrated valid for elements above Mn in the periodic table when collected on Whatman 41® filter paper and for elements above Ca when collected on

membrane filters such as Nuclepore® in which there is minimal particle penetration.(16) An additional independent quantitative check on the accuracy of the X-ray analysis is obtained by using the instrumental neutron activation analysis values obtained for elements detected by both techniques. These elements (Fe, Mn, and Br) are used as standards together with theoretical X-ray production efficiencies to determine those elements which are not readily measured by instrumental neutron activation analysis.

Results and Discussion

The accuracy and precision for a single determination is better than 10% for most elements in air filters. Our technique of instrumental neutron activation analysis has been shown to be capable of an accuracy of better than 10% by the replicate analyses of well characterized standards such as the USGS Basalt Standard BCR-1, and NBS-Orchard Leaves.(19)

Table 2   Worldwide Trace Element Air Concentrations at Ground Level (ng/m³)

Sampling location	Al	Br	Cl	Cu	Mg	Mn	Na	Pb	V
Nord, Greenland (81° N, 15° W)	240	14	300	—	160	2.8	200	22	0.8
Thule, Greenland (77° N, 69° W)	380	20	200	—	130	4.5	440	15	1.4
New York, New York (41° N, 74° W)	1790	50	2300	—	150	60.0	2300	997	174.0
Salt Lake City, Utah (41° N, 112° W)	2330	620	1160	50	800	36.0	810	1567	3.9
Sterling, Virginia (39° N, 77° W)	1130	6	120	—	260	25.0	290	127	13.6
Miami, Florida (26° N, 80° W)	600	820	6280	—	550	25.0	4130	1956	15.8
Mauna Loa, Hawaii (20° N, 156° W)	410	<2	350	—	360	8.2	220	1.5	0.8
San Juan, Puerto Rico (18° N, 66° W)	760	460	8660	70	950	14.0	4100	0.21	6.8
Balboa, P.C.Z. (9° N, 80° W)	880	65	7200	100	1070	16.0	5200	317	200.0
Guyaquil, Ecuador (2° S, 81° W)	1210	130	200	—	340	24.0	250	228	15.1
Lima, Peru (12° S, 77° W)	3440	200	8930	30	2060	82.0	5600	344	91.1
Chaltaya, Bolivia (20° S, 68° W)	460	2	150	—	<30	4.3	120	11	0.7
Antofagasta, Chile (23° S, 71° W)	870	14	4170	70	1380	18.0	2600	41	14.7
Santiago, Chile (33° S, 71° W)	15,000	50	2100	180	3610	330.0	3300	—	40.9

A set of air filters collected from the AEC's 80th meridian sampling network in March 1972 was analyzed by instrumental neutron activation analysis and X-ray fluorescence analysis for a total of 25 major, minor and trace elements. These sampling stations operate for the purpose of monitoring airborne radioactivity and are not properly sited to avoid local contamination. Results of these analyses are presented in Table 2. The 14 stations ranged in latitude from Nord, Greenland (81° N–15° W) to Santiago, Chile (33°–71° W). Concentrations of a series of elements (Al, Cr, Fe, Hg, La, Mn, Pb, Sb, Se, V and Zn) were normalized to scandium, an element known to originate from geological sources (6) to determine their relationship to natural terrestrial sources. In Table 3 the concentration of these 12 normalized elements in the air filters are compared with similar values in Basalt, Granite and earth's crustal average.(20) The normalized concentrations in air filters can then be compared with the normalized concentrations in the earth's crust to provide an indication of the excess of an element in the air over the natural levels one would expect if the aerosol were derived from weathered terrestrial material. These excess ratios are presented in Table 4. The normalized concentrations in Table 3 and excess ratios

Table 2 (contd)   Worldwide Trace Element Air Concentrations at Ground Level (ng/m₃)

Sampling location	La	Sm	Eu	Co	Fe	Sc	Hf	Cr	Th	Tb	Ta	Sample volume (m³)
Nord, Greenland (81° N, 15° W)	0.046	0.012	—	0.15	166	0.040	0.059	0.8	0.02	0.005	0.027	646
Thule, Greenland (41° N, 112° W)	0.11	0.010	—	0.071	171	0.037	0.044	0.6	0.04	0.001	0.010	2012
New York, New York (41° N, 74° W)	9.1	0.84	0.021	2.2	4820	0.44	0.29	153	0.08	0.018	0.14	402
Salt Lake City, Utah (41° N, 112° W)	4.3	0.34	0.042	1.1	2124	0.39	0.23	6.7	0.44	0.034	0.28	583
Sterling, Virginia (39° N, 77° W)	0.51	0.066	0.013	0.39	829	0.21	0.091	2.5	0.16	0.002	0.041	640
Miami, Florida (26° N, 80° W)	0.49	0.065	—	0.52	2004	0.084	0.17	4.4	0.12	0.019	0.049	570
Mauna Loa, Hawaii (20° N, 156° W)	0.22	0.034	0.014	0.12	430	0.070	0.064	0.32	0.075	0.011	0.026	486
San Juan, Puerto Rico (18° N, 66° W)	—	0.030	0.002	0.25	554	0.15	0.067	1.3	0.022	0.008	0.018	617
Balboa, P.C.Z. (9°N, 80° W)	0.44	0.078	—	0.65	1175	0.22	0.065	2.0	0.018	0.027	0.045	715
Guyaquil, Ecuador (2° S, 81° W)	0.29	0.058	0.079	0.42	1126	0.35	0.043	2.3	0.028	0.020	0.020	1144
Lima, Peru (12° S, 77° W)	0.82	0.13	0.005	1.3	4020	0.57	0.26	4.2	0.059	0.029	0.068	665
Chaltaya, Bolivia (20° S, 68° W)	0.24	0.032	0.007	0.12	312	0.062	0.063	0.68	0.062	—	0.021	501
Antofagasta, Chile (23° S, 71° W)	0.30	0.058	0.006	0.28	826	0.20	0.074	1.5	0.12	0.022	0.037	707
Santiago, Chile (33° S, 71° W)	3.4	0.63	0.14	3.6	9225	3.0	0.76	8.3	1.05	0.12	0.15	488

in Table 4 indicate that at most sites Al, Co, Fe, La and Mn have a natural geochemical origin. A similar analysis of the air concentrations of Mg, Sm, Eu, Hf, Th, Tb and Ta in Table 2 indicates these elements also have a geochemical origin. Since basalt represents the major natural earth's crust component it is not surprising that in most cases the local air concentrations of the above elements can be represented at most sites by a mixture of approximately 90% basalt and 10% granite. However, if one compares the La/Sc ratios in Table 3, it is evident that aerosols over New York City; Miami, Florida; and Chataya, Bolivia have a high granite component. This is readily apparent in the La/Sc ratios which are about 40 in granite, 0.9 in basalts, and 1.5 in the earth crust average. The concentration of terrestrial derived elements such as Al, Fe and Sc indicate that variations of more than an order of magnitude (in the soil derived component of the local aerosol) exist between sites at any given time.

Nord and Thule, Greenland had the lowest Sc, Al and Fe concentrations and presumably the lowest dust burden, while the highest concentrations were observed in Santiago, Chile; Li-

Table 2 (contd)   Worldwide Trace Element Air Concentrations at Ground Level (ng/m³)

Sampling location	Sb	Hg	Se	Zn
Nord, Greenland (81° N, 15° W)	0.9±0.31	0.04±0.05	0.36±0.12	41±2
Thule, Greenland (77° N, 69° W)	<45	0.078±0.01	0.17±0.02	18±1
New York, New York (41° N, 74° W)	12.0±0.65	1.26±0.22	5.67±0.35	741±73
Salt Lake City, Utah (41° N, 112° W)	<15.7	0.12±0.09	3.10±0.18	151±4
Sterling, Virginia (39° N, 77° W)	<14.2	0.12±0.03	1.40±0.07	91±2
Miami, Florida (26° N, 80° W)	0.8±0.20	0.12±0.04	0.56±0.07	88±2
Mauna Loa, Hawaii (20° N, 156° W)	<18.8	0.22±0.04	0.18±0.07	85±2
San Juan, Puerto Rico (18° N, 66° W)	<14.81	0.12±0.01	0.28±0.03	60±1
Balboa, P.C.Z. (9° N, 80° W)	0.8±0.3	0.07±0.05	1.21±0.14	182±4
Guyaquil, Ecuador (2° S, 81° W)	1.1±0.18	0.21±0.05	0.71±0.09	25.4±0.4
Lima, Peru (12° S, 77° W)	21.4±0.32	0.69±0.06	1.53±0.12	632±5
Chaltaya, Bolivia (20° S, 68° W)	<18.2	0.07±0.06	0.07±0.03	—
Antofagasta, Chile (23° S, 71° W)	12.3±0.44	0.26±0.08	0.05±0.19	60±2
Santiago, Chile (33° S, 71° W)	24.0±0.40	0.60±0.09	1.35±0.19	1358±11

ma, Peru; and New York City. The normalization of the air concentrations of elements from known or suspected anthropogenic sources such as V, Mn, Co, Zn, As, Pb and Se to elements of terrestrial origin such as Al, Sc, or Fe provides an index of the local population by these elements (see Tables 3 and 4). This approach showed that Pb is increased over natural levels by several orders of magnitude at all stations except Mauna Loa, Hawaii; Thule, Greenland; and San Juan, Puerto Rico. Abnormally high concentrations of vanadium were noted in New York City; Miami, Florida; and Balboa, Panama Canal Zone. The sampling location at Balboa was close

to the shipping lanes and presumably the high vanadium concentration can be attributed to the emissions from ship stacks. The natural level of Se was increased by two or four orders of magnitude at all stations, while the particulate Hg and Zn levels were enhanced by one or two orders of magnitude. The possibility that the elements in Tables 3 and 4 which show excess ratios could have originated from the marine atmosphere, another major natural aerosol source, has been dismissed in an earlier paper.[7] While none of the sites can be considered truly pristine, the San Juan, Mauna Loa and Chaltayo, Bolivia stations most closely approximate natural conditions.

Table 3 Comparison of Trace Element Concentrations Normalized to Scandium in Ground Level Air Samples and Geological Materials

	Al	Co	Cr	Fe	Hg	La	Mn	Pb	Sb	Se	V	Zn
Nord, Greenland	6000	3.75	20	4150	1.0	1.15	70	550	22	9.0	20	1025
Thule, Greenland	10,270	1.92	16	4620	2.1	2.97	120	410	—	5.5	38	500
New York, New York	4070	5.00	350	10,950	2.9	20.7	140	2270	27	12.9	400	1680
Salt Lake City, Utah	5970	2.82	17	5450	0.3	11.0	92	4020	—	8.0	10	390
Sterling, Virginia	5380	1.86	12	3950	0.6	2.43	120	600	—	6.7	65	430
Miami, Florida	7140	6.18	52	23,860	1.4	5.83	300	23,290	9.5	6.7	190	1050
Mauna Loa, Hawaii	5860	1.71	4.6	6140	3.1	3.14	120	21.4	—	2.6	11	1210
San Juan, Puerto Rico	5070	1.67	12	3690	0.8	—	93	1.4	—	1.9	45	400
Balboa, P.C.Z.	4000	2.95	9.1	5340	0.3	2.0	73	1440	3.6	5.5	910	830
Guyaquil, Ecuador	3460	1.20	6.6	3220	0.6	0.83	69	650	3.1	2.0	43	73
Lima, Peru	6040	2.28	7.4	7050	1.2	1.44	140	600	38	2.7	160	1110
Chaltaya, Bolivia	7420	1.94	11.0	5030	1.1	3.87	69	180	—	1.1	11	—
Antofagasta, Chile	4350	1.40	7.5	4130	1.3	1.50	90	205	140	0.25	74	300
Santiago, Chile	5000	1.20	2.8	3080	0.2	1.13	110	—	8.0	0.5	14	450
Crustal average	3700	1.14	4.55	2270	0.00364	1.36	43	0.59	0.009	0.0027	6.14	3.182
Granite	24,770	0.80	7.3	4570	0.0667	40	77	16.3	0.133	—	5.3	15
Diabase	2310	1.47	3.5	2280	0.0059	0.88	39	0.24	0.032	—	7.1	2.41

Table 4 Trace Element Excesses in Worldwide Ground Level Air Samples

	Al	Co	Cr	Fe	Hg	La	Mn	Pb	Sb	Se	V	Zn
Nord, Greenland	1.6	3.3	4.4	1.8	270	0.8	1.6	930	2440	3960	3.3	320
Thule, Greenland	2.8	1.7	3.5	2.0	577	2.2	2.8	9690	—	2030	6.2	156
New York, New York	1.1	4.4	77	4.8	800	15	3.3	3850	3000	5680	65	530
Salt Lake City, Utah	1.6	2.5	3.7	2.4	85	8.1	2.1	6813	—	3520	1.6	120
Sterling, Virginia	1.5	1.6	2.6	1.4	160	1.8	2.8	1017	—	2950	10.5	140
Miami, Florida	1.9	5.4	11.4	10.5	380	4.3	7.0	39,475	1060	2950	30.9	330
Mauna Loa, Hawaii	1.6	1.5	1.0	2.7	850	2.3	2.8	36	—	1150	1.8	380
San Juan, Puerto Rico	1.4	1.5	2.6	1.6	220	—	2.2	2.4	—	840	7.3	130
Balboa, P.C.Z.	1.1	2.6	2.0	2.4	80	1.5	1.7	244	400	2420	148	260
Guyaquil, Ecuador	0.9	1.1	1.5	1.4	160	0.6	1.6	1100	340	880	7.0	25
Lima, Peru	1.6	2.0	1.6	3.1	330	1.1	3.3	1017	4220	1190	26.1	350
Chaltaya, Bolivia	2.0	1.7	2.4	2.2	302	2.8	1.6	305	—	480	1.8	—
Antofagasta, Chile	1.2	1.2	1.6	1.8	360	1.1	2.0	347	15,560	110	12.1	95
Santiago, Chile	1.4	1.1	0.6	1.4	55	0.8	2.6	—	890	220	2.3	140

$$\text{Excess} = \frac{\text{(Element: Sc)}\quad\text{Filter}}{\text{(Element: Sc)}\quad\text{Crustal Average}}$$

Acknowledgements

We are indebted to Dr. H. L. Volchok of the USAEC Health and Safety Laboratory for providing the air filter samples for this study. We also thank W. L. Butcher, R. W. Sanders and C. L. Wilkerson of this Laboratory for providing invaluable assistance in the trace element analyses. This paper is based on work performed under United States Atomic Energy Commission Contract AT (45–1)-1830.

References

1 Brar, S. S., D. M. Nelson, J. R. Kline, and P. F. Gustafson (1970): Instrumental analysis for trace elements present in Chicago surface air. J. Geophys. Res. 75, 2939–2945.

2 Dams, R., J. A. Robbins, K. A. Rahn, and J. W. Winchester (1970): Nondestructive neutron activation analysis of air pollution particulates. Analyt. Chem. 42, 861–867.

3 Dudey, N. D., L. E. Ross, and V. E. Noskin: Application of activation analysis and Ge (Li) detection techniques for the determination of stable elements in marine aerosols, in Modern Trends in Activation Analysis, National Bureau of Standards Special Publication 312, Vol. I, National Bureau of Stds., Washington, D. C., 1969, pp. 55–61.

4 Harrison, P. R., and J. W. Winchester (1971): Area-wide distribution of lead, copper, and cadmium in air particulates from Chicago and Northwest Indiana. Atmos. Environ. 5, 863–880.

5 Peirson, D. H., P. A. Cawse, L. Salmon, and R. S. Combray (1973): Trace elements in the atmospheric environment, Nature 241, pp. 252–256.

6 Rancitelli, L. A., and R. W. Perkins (1970): Trace element concentration in the troposphere and lower stratosphere. J. geophys. Res. 75, 3055–3064.

7 Rancitelli, L. A., R. W. Perkins, T. M. Tanner, and C. W. Thomas: Precipitation Scavenging (1970), Stable elements of the atmosphere as tracers of precipitation scavenging (Proc. Conf., Richland, Washington, 1970).

8 Rancitelli, L. A., J. A. Cooper, and R. W. Perkins: The multielement analysis of biological material by neutron activation and direct instrumental techniques, in Modern Trends in Activation Analysis, National Bureau of Standards Special Publication 312, Vol. I, National Bureau of Stds., Washington, D. C., 1969, pp. 101–109.

9 Perkins, R. W., and L. A. Rancitelli: Nuclear techniques for trace element and radionuclide measurements in natural waters, Battelle-Northwest Report BNWL-SA-3993, Richland, Washington, August 1971.

10 Duce, R. A.: Atmospheric sampling; to be published in a National Oceanic and Atmospheric Administration Manual for Measurement of Pollutants in the Marine Environment, E. D. Goldberg, Ed. 1973.

11 Dams, R., J. A. Robbins, K. A. Rahn, and J. W. Winchester: Evaluation of filter materials and impaction surfaces for nondestructive neutron activation analysis of aerosols Environ, Sci, Tech. 6, pp. 441–448 (1972).

12 Stafford, R. G., and H. J. Ettinger: Filter efficiency vs. particle size and velocity, LA-4650, Los Alamos Scientific Laboratory, Los Alamos, New Mexico, March 1971.

13 Cooper, J. A., and R. W. Perkins: A versatile Ge(Li)-NaI(Tl) Coincidence-Anticoincidence Gamma-Ray Spectrometer for Environmental and Biological Problems, Nucl. Inst. and Methods, 99: 125–146 (1972).

14 Cooper, J. A.: Comparison of particle and photon excited X-ray fluorescence applied to trace element measurements of environmental samples, Nucl. Instr. & Meth. 106 (1973) 525.

15 Cooper, J. A., and J. C. Langford: "Rapid trace element analysis of nail clippings by X-ray fluorescence", Trace Substances in Environmental Health IV, 1973. Ed. D. D. Hemphill and published by Univ. of Missouri, Columbia, Missouri, 1973.

16 Cooper, J. A.: Review of a Workshop on X-ray fluorescence analysis of aerosols, Pacific Northwest Laboratory Report No. BNWL-SA-4690 (June 1973).

17 Kevex, Inc., Burlingame, California.

18 Micro-Matter, Inc., Seattle, Washington.

19 Robertson, D. E., L. A. Rancitelli, J. C. Langford, and R. W. Perkings: Battelle-Northwest contribution to the IDOE Baseline studies of pollutants in the marine environment and research recommendations, 1972.

20 Mason, B.: Principles of Geochemistry, 3rd ed., pp. 45–46, John Wiley & Sons, Inc., New York, 1966.

Radioisotope Techniques in Delineation of the Environmental Behavior of Cadmium [*, **]

R. I. Van Hook, Jr., B. G. Blaylock, E. A. Bondietti, C. W. Francis, J. W. Huckabee, D. E. Reichle, F. H. Sweeton, J. P. Witherspoon

Environmental Sciences Division, Oak Ridge National Laboratory, Oak Ridge, Tennessee 37830

Abstract

Radioisotope techniques are being developed and utilized at Oak Ridge National Laboratory (ORNL) for evaluating the environmental behavior of toxic elements such as cadmium in aquatic and terrestrial ecosystems. Tracer techniques using ^{109}Cd in microcosm, field plot, and stream systems are providing information on biogeochemical cycling and distribution of cadmium in the environment. Parameters being measured include adsorption capacity for cadmium in mineral soils and sediments; uptake rates of cadmium in various plant species from both soils and nutrient solutions as affected by pH, competing cations, and chemical form of cadmium; and distribution of cadmium in various components of both aquatic and terrestrial ecosystems following application of ^{109}Cd to soil, vegetation, or directly to streams. Food chain parameters being estimated with ^{109}Cd include uptake, assimilation, and turnover by both aquatic and terrestrial organisms. Information obtained in these radiotracer studies is providing insight into the behavior of cadmium in aquatic and terrestrial ecosystems, especially transport rates of cadmium and potential biomagnification or dilution in food chains. The factors which influence the incorporation of cadmium into vegetative material as well as those affecting residence time in ecosystems have been identified. Use of ^{109}Cd also has permitted evaluation of a cadmium specific electrode as a tool for rapid assay of free cadmium ions in soil solutions.

Introduction

Cadmium is recognized as an important trace contaminant in both aquatic and terrestrial environments.[1, 2] This element has been identified in general environmental contamination around zinc smelters [3, 4] and coal-fired steam plants.[5] In Japan, cadmium has been directly linked to human disease (i.e., Itai-Itai) in the Jintsu River basin which received cadmium in effluent from zinc-lead mining and smelting operations.[6] Cadmium also is released to the environment through its dissipative uses such as cadmium fungicides, cadmium-plated hardware, certain paints, and by the large-scale use of materials in which cadmium is an impurity (e.g., zinc products, phosphate fertilizer, and coal). The esti-

* Research sponsored by the National Science Foundation's RANN Program and the U.S. Atomic Energy Commission under contract with the Union Carbide Corporation

** Published in Comparative Studies of Food and Environmental Contamination (Proc. FAO/IAEA/WHO Symp. Otaniemi, 1973), IAEA, Vienna (1974) 3–22

mated release of cadmium into the environment in 1968 was 5–8 million pounds.[7] This estimate excludes cadmium incorporated into products. The annual demand for cadmium is between 8 and 14 million pounds. Cadmium is toxic to animals in quite small concentrations (e.g., the acute lethal dose by ingestion is estimated to be between 0.35 and 3.5 g for man). Thus, stable element methodology is difficult to work with in studies involving animal systems such as food chains. This is one of the main reasons that at Oak Ridge National Laboratory (ORNL), we have been developing techniques for examining the behavior of cadmium in the environment with radioactive cadmium tracers (^{109}Cd and ^{115m}Cd).

Prior research at ORNL on the environmental behavior of radioisotopes (8, 9, 10) has provided our researchers with an insight into those radioisotope techniques that would be useful in answering questions concerning cadmium behavior. Use of radiocadmium in answering these questions allows the investigators to use very low, nontoxic concentrations of cadmium (< 0.01 ppm). These small additions of radiocadmium to biological and physical systems have little or no effect on the total cadmium chemistry and thus provide an excellent tracer for use in estimating rates of cadmium transfer from one environmental component to another. This feature permits accurate evaluation of the dynamics of cadmium in sediments, soils, water, and in biological systems such as food chains.

In sediment and soil systems, adsorption phenomena may be evaluated with radiocadmium through selected addition of cadmium tracers to systems of reference minerals, natural soil and sediment components, or to fractions of these systems that have been separated by zonal centrifugation. Following fractionation into different mineral components, cadmium sorption can be estimated for each component for identifying those most important in determining biological availability through plant uptake. Following the estimation of adsorption of cadmium in soil and sediment systems, radiocadmium may then be used in estimating the degree of plant uptake of cadmium from tagged nutrient solutions as well as from intact soils. Additionally, the distribution of cadmium in plant tissues is readily determined with radioisotopes of cadmium. After introduction of cadmium into the food chain, turn-over times in various tissues can be estimated for evaluating exposure of these organs to cadmium. Other uses of radiocadmium in environmental studies of cadmium include: microcosm studies for evaluating the movement of cadmium from terrestrial to aquatic components of natural ecosystems and stream tags for estimating transfers among aquatic ecosystem components.

The objectives of this presentation are (1) to outline the development of radioisotope techniques for studying the environmental behavior of cadmium, and (2) to illustrate the application of these radiotracer techniques to describe the behavior of cadmium in aquatic and terrestrial environments.

Characterization and Distribution of Cadmium in Sediments and Soils

Clay minerals, sesquioxides, and humic acids are the major components involved in adsorption reactions in sediments and soils. The importance of these adsorbers is related to their relative abundances as well as to the physicochemical conditions existing during the adsorption reactions. In river systems, the concentrations of toxic elements are very low relative to other ions, such as calcium or magnesium. Adsorption of the trace toxic elements thus must be measured in the presence of competing ions.

Cadmium Adsorption by Clay Minerals: In sediments, the importance of clay minerals in adsorption processes is due to (1) a high surface area to weight ratio; (2) exchange capacity based on permanent internal charge; and (3) large exchange capacity as a result of the extensive surface available between the silicate layers of the crystals in certain clays (i.e., vermiculite or montmorillonite).

Conditions in our study were selected to approximate the natural conditions existing when moderate levels of cadmium are released to streams. Since [109]Cd was used as a tracer, it was possible to work with concentrations of Cd^{2+} of 0.2 μM (22 pp10[9]) or less, which is in the range of the recommended upper limit for drinking water supplies (10 pp10[9]).

The solutions used in these tests all contained calcium acetate at a concentration of about 0.03 M. Calcium was present to simulate the large excess of this ion normally occurring in river water. The acetate anion was chosen because of its buffering action. The pH was adjusted by adding acetic acid or $Ca(OH)_2$. In each test, 30 ml of solution was equilibrated in a polypropylene tube with 20 to 60 mg of the clay mineral at 26° C. During equilibration, the samples were rotated in order to keep the solid suspended. At the end of several days of equilibration, the concentration of Cd^{2+} was determined from a radioassay of an aliquot of supernatant after centrifuging the sample. A radioassay was also made of the solid after filtering to remove most of the remaining solution. A small correction was made for the activity in the solution still remaining with the solid.

Adsorption was characterized by dividing the observed adsorption of cadmium on the solid (in milliequivalents per gram) by the ratio of the concentrations of the Cd^{2+} and Ca^{2+} ions in the equilibrated solution. This function, which is illustrated in Fig. 1, is expected to remain relatively stable over a range of Cd^{2+} and Ca^{2+} ion concentrations. The pH shown was that measured after equilibration.

The clay minerals studied were chosen to represent several different types. The Birch Pit kaolinite is a non-substituted clay with a 1:1 ratio of tetrahedral and octahedral layers. The illities, which have a 2:1 tetrahedral-octahedral ratio, have an internal charge as a result of some isomorphic substitution of aluminum for tetrahedral silicon. The African vermiculite is also a 2:1 clay and has some substitution of aluminum for tetrahedral silicon. However, potassium is not present between layers and the lattice can expand. The Clay Spur montmorillonite is similar to the vermiculite except that its isomorphic substitution is magnesium for the octahedral aluminum. As a result, the counter ions associated with this internal charge cannot approach as closely.

All the clay minerals adsorb Cd^{2+} with increasing strength as the pH is increased; with the illites, vermiculite, and kaolinite, this effect is stronger than with montmorillonite. The strong

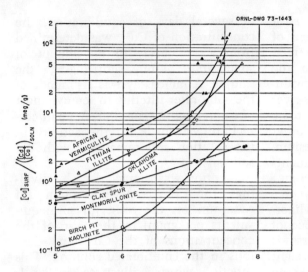

ORNL-DWG 73-1443

Fig. 1 Relative adsorptive capacities of clay minerals for low levels of cadmium (22 pp10⁹) in elevated calcium solutions at 20° C.

adsorption by vermiculite presumably is due to its large exchange capacity, much of this resulting from tetrahedral substitution, which appears much more effective than that from the octahedral substitution in montmorillonite.

These data indicate that, under some conditions, suspended clays might be used as an abatement technique by adsorbing a significant part of soluble cadmium released into a river. If there were a need to immobilize soluble cadmium, vermiculite and illite would be likely adsorbents. Use of limestone to raise the pH would help reduce the amount of clay required.

Evaluation of the Cadmium Selective-ion Electrode: Uptake of trace elements by plants is highly dependent on the equilibrium chemical activity of cations in the soil solution. While the total quantity of an element present may be in the ppm range, the concentration in solution may be in the $pp10^9$ range. Furthermore, the ions in solution may not be free but can be complexed with water-soluble organics or inorganics (e.g., SO_4^{-2}, HCO_3^{-}). It is important, therefore, to establish a method for rapidly measuring the free ion in

soil solutions or in some extractant used to estimate "availability". To this end, attention is being given to a cadmium selective-ion electrode (Orion Model 94-48A) which is portable and could be used for field evaluation of contaminated soils.

In order to test the electrode for this purpose, an experiment was designed using two soils, to which $^{109}Cd^{2+}$-tagged $Cd(NO_3)_2$ was added at the rate of 100 ppm. The soils were incubated in aluminum cores, 31 cm long and 8.5 cm in diameter. After incubation for 4 weeks at 70% of field capacity, the intact soil solution containing that cadmium in equilibrium with the soil solid phase was displaced by metering 0.01 M $CaCl_2$ at the top of the cores. The displaced soil solution was collected at the core bottoms in 50-ml increments. The initial 100 ml of displaced solution was relatively free of Cl^- and therefore was used for the Cd^{2+} determinations. In addition to gamma-ray spectrometry, atomic absorption analysis was used to determine stable cadmium. Since the solid-state cadmium electrode responds to cadmium ion activity and not directly to free ion concentration, the ionic strength of the soil solution must either be estimated or fixed by swamping the sample with a salt such as KNO_3. To avoid adding anything to the displaced soil solutions or upsetting any water-soluble cadmium-organic complexes, the ionic strength was estimated from the electrical conductivity of the solution with the assumption that Ca^{+2}, HCO_3^-, and NO_3^- were the principal ions present. From this estimate, the activity coefficient of Cd^{2+} was determined and thus the total cadmium present estimated.

The reasonable agreement between the three methods (Table 1) suggests that the electrode may be a useful tool for estimating solution concentrations of Cd^{2+}. By developing extracting solutions which reflect the "available" Cd^{2+}, the electrode may be used for rapid estimate of Cd^{2+} in soils and sediments. This is of particular interest in monitoring studies where first estimates of solution phase Cd^{2+} could be carried out in the field. This would allow the investigator to determine which' samples required more refined cadmium analyses in the laboratory. The use of ^{109}Cd enabled a rapid evaluation of the electrode by providing an accurate method of cadmium determination. The slightly higher cadmium values obtained by the electrode are probably due to interferences from other ions such as Fe^{3+}. Other sources of interference are also being investigated.

Earthworms as Factors Affecting the Distribution of Cadmium in Soils: An additional parameter which is important in describing the environmental behavior of a trace contaminant in soils is the physical re-distribution of the material. Distribution of elements in soils is a function of ion-exchange, water content, and mechanical mixing. Earthworms, a major component of the soil fauna, are probably the most important of the mechanical mixers and are responsible for maintaining soil characteristics and processes such as aeration, water permeability, and nutrient turnover.[11] With the major source of cadmium to terrestrial environments being atmospheric fallout (e.g., from coal-fired steam plants and Zn-Pb smelters), earthworms could play an important role in vertical re-distribution of cadmium from the litter into the soil profile.

In order to test this hypothesis, an experiment was initiated to examine the effects of the presence or absence of earthworms, different species of earthworms, and the presence or absence of water on vertical migration of ^{109}Cd added to the litter component of the system. In addition to providing information on the distribution of cadmium, this experiment illustrates the use of a radioisotope technique that allows the investigator to process a large number of samples which would be otherwise impractical using expensive conventional stable cadmium analyses. The experimental design consisted of 6.25-cm-diam. by 20-cm-long soil cores obtained from an Emory Silt Loam soil. These cores were covered with paraffin wax and sealed in heat-shrinkable tubing. Two earthworm genera were used: *Octolasium*, representative of

Table 1 Cadmium (as ppm Cd^{2+}) in Solution Obtained from a Captina Silt Loam with 1% Organic Matter at a pH of 5.5.

In-crement	^{109}Cd	Technique	
		Atomic absorption	Cadmium electrode
50 ml	0.88	0.89	1.4
50 ml	0.91	0.95	1.2

earthworms that are nonsurface soil feeders, and *Lumbricus*, representing a surface feeding species. Ten grams of *Festuca* litter containing one microcurie (μCi) of $^{109}Cd(NO_3)_2$ per gram dry weight was placed on the surface of each of the soil cores. The soil cores were then placed in constant temperature-humidity chambers and incubated for 6 weeks. All soil cores were brought to approximately 50% of field capacity before initiating the experiments. The cores receiving water amendments were watered at 4-day intervals with 45 ml of demineralized water. This amount of water represented the average rainfall for east Tennessee during summer months. At 3-week intervals, the soil cores were destructively sampled by removing litter and worms, and cutting the core horizontally into 1-cm sections. All samples of litter, worms, and soil were radioassayed for ^{109}Cd with a Nuclear Data Gamma

Spectrometer at a counting efficiency of 3%.

Vertical distribution of ^{109}Cd in the soil cores is illustrated in Fig. 2 as a function of earthworms, water, and time. It is obvious that in cores containing *Lumbricus*, penetration of ^{109}Cd was greatest. The influence of time on ^{109}Cd distribution with depth is more apparent in those cores receiving water than in those without water. In addition to the greater penetration of ^{109}Cd in those cores containing both *Lumbricus* and additional water, the distribution of cadmium throughout the core became more uniform with time. The depth of ^{109}Cd penetration in these cores is due to the mode of feeding by *Lumbricus*, i.e., these worms dig a burrow and carry surface litter into the burrow for feeding.

In soil cores containing *Octolasium*, ^{109}Cd had penetrated to about 7 cm by the end of 6 weeks, whereas the maximum penetration was about 4 cm in those cores receiving no water. The occurrence of ^{109}Cd at 10–15 cm depth after weeks 3 and 6 in those cores containing *Octolasium* and no water was due to worms channeling down the sides of the cores between the soil and the wax coating. Presumably, the penetration to 4 cm in these cores at the end of week 3 was also due to similar channelization.

Comparison of ^{109}Cd penetration between cores containing *Lumbricus* and those containing no earthworms shows a pronounced effect of these worms in distributing cadmium applied to the core surface in the form of soaked litter. If a similar comparison is made between cores containing *Octolasium* and those containing no earthworms, there are no apparent effects of earthworms. In these cores, the rate of pene-

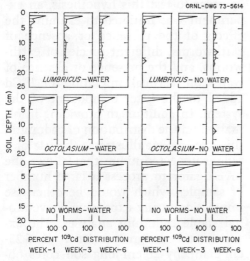

Fig. 2 *Vertical distribution of ^{109}Cd in soil expressed as a function of earthworm species and water following application of 10 grams of Festuca litter containing 1 μCi $^{109}Cd(NO_3)_2$ per grams dry weight to the surface of each core.*

tration and the distribution patterns of [109]Cd are almost identical. Apparently, the behavior of *Octolasium* in soils, i.e., burrowing and feeding in the soil and not at the surface, does not promote vertical migration of cadmium ions. There are, however, suggestions that these earthworms may act as bio-accumulators of cadmium and may serve to remove this element from the soil system by concentrating it in their tissues.[12, 13] This has broader implications in food chain analyses of cadmium behavior.

Cadmium Uptake and Distribution in Vegetation

Currently, there is a paucity of information on the uptake of cadmium from soils or the various soil properties which control its availability to plants. It is generally recognized that cadmium uptake by plants is greater from acid soils than from soils neutral in pH or slightly alkaline.[14] Soils containing high amounts of organic matter often contain greater quantities of cadmium than mineral soils, however, plant uptake from organic soils is often lower than uptake from mineral soils. Recent work has shown that even low levels of cadmium in nutrient solutions and soils can be toxic to plants.[15, 16, 17, 18] Due to the importance of chemical properties of soil solutions that may influence cadmium uptake and distribution in plants, at ORNL we have been developing radioisotope techniques using [109]Cd to determine the effects of chemical form, pH, and nitrogen, zinc, and selenium amendments on plant uptake.

Plant Uptake of Cadmium: Bush beans (*Phaseolus vulgaris* L., variety "Kentucky Wonder") used as a test crop in these experiments, were germinated in wet paper toweling and then transferred to 300 ml of nutrient solution made up of 6.6 mM KNO_3, 2.0 mM $NH_4H_2PO_4$, 3.75 mM $Ca(NO_3)_2 \cdot 4\ H_2O$, 1.0 mM $MgSO_4$, 0.021 mM FeDTPA, 0.0025 mM $ZnSO_4$, 0.004 mM H_3BO_3, and 0.003 mM $NaMoO_4$. Plants were maintained in a growth chamber at 65–70°F for 12 hr (day) and 50 to 55°F for 12 hr (night). After ten to 14 days, nutrient solutions were exchanged for the appropriate treatment solution containing ~ 3 pp10^9 cadmium as [109]Cd. Plants remained in the treatment solutions for 5 to 10 days and were subsequently harvested. Immediately after harvesting, roots of each plant were washed in 1 N $MgCl_2$ and distilled water to remove excess [109]Cd. Each plant was separated into root, leaf (exclusive of petiole), and stem (petioles included) for counting purposes.

The pH (4.6 to 6.5 range) of the nutrient solution affected plant uptake only when the iron chelate, diethylene-triaminepentaacetate (FeDTPA) was used as a source of iron in the nutrient solution. Under this condition, cadmium uptake was tenfold lower at pH 6.5 than at pH 4.6 after 24 hr. When the same experiment was carried out in the absence of iron DTPA, pH did not affect cadmium uptake. At a low pH, hydrogen ions compete with Cd^{2+} for the chelate thus causing cadmium ionic activity in the nutrient solution and consequently that available for plant uptake to be considerably greater than that at pH 6.5.

Amendments of nitrogen in the form of NH_4NO_3 significantly ($P \leq 0.05$) lowered foliar [109]Cd concentrations in Bush bean plants (Table 2). Significantly higher [109]Cd concentrations ($P \leq 0.05$) were found in the upper

Table 2 Cadmium-109 Concentrations and Dry Matter Production of Bush Beans Grown at Different Levels of Nitrogen in Nutrient Solution.*

Nitrogen concentration in nutrient solution (ppm)	Roots	Stems	Lower leaves	Top leaves	Leaves and stems	Total plant
			^{109}Cd concentration (pCi/mg)			
225	3707a	116.2a	48.4a	68.3a	72.9a	
338	4553b	93.1a	29.1b	40.5b	49.0b	
1463	4325a	72.6a	10.0c	25.5c	29.8c	
			Dry weight (mg)			
225	85.1a	135.1a	105.9a	145.0a		577a
338	65.1b	103.6b	98.2a	79.5b		444b
1463	62.0b	113.2b	110.1a	81.9b		468b

* Seven days growth, four replicates, NH₄NO₃ form of nitrogen. Numbers having an identical superscript within the same column and category are not statistically different at the 5% level as tested by Duncan's multiple range test.

leaves than in the lower leaves at all nitrogen levels. This implies that cadmium translocation in plants occurs early in leaf development and may be analogous to translocation of calcium which has an ionic radius very similar in size to that of cadmium (0.97 and 0.99 Å, respectively, for calcium and cadmium).

Because of the evidence that reasonably large quantities of selenium can be emitted from coal-fired steam plants (19) and its protective effects against cadmium toxicities in biological systems (20), experiments in nutrient solution were carried out to assess the importance of selenium in plant uptake of cadmium. After 7 days of growth, bean plants contained significantly higher (P ≤ 0.05) concentrations of ^{109}Cd in low selenium treatments (0.001 ppm Se as SeO₂) than those grown in high selenium treatments (0.1 ppm Se). Amendments of selenium did not appear to affect the growth of plants since there was no significant difference (P ≤ 0.05) in dry matter production between the two selenium levels.

Lagerwerff and Biersdorff (21) observed that zinc enhanced cadmium uptake in plants. They, concluded however, that the enhancement was an artifact caused by root damage at high cadmium and zinc concentrations. Our work with carrier-free ^{109}Cd indicates the enhancement to be real; significantly greater (P ≤ 0.05) concentrations of ^{109}Cd were found in stems and leaves of plants grown for 5 days in

Table 3 The Influence of Zinc in Nutrient Solutions on the Uptake of ^{109}Cd in Bush Beans.*

| Zinc concentration (ppm) | ^{109}Cd uptake | | |
	Roots (μCi/mg)	Leaves (pCi/mg)	Stems (pCi/mg)
0	16.3a	43.7b	68.0b
1	13.6a	64.4b	57.2b
5	13.2a	156.3a	195.9a
10	8.3b	180.6a	215.7a
100	0.86c	50.0b	168.5a

* Average of three replicates, five days growth. Numbers having an identical superscript within the same column are not statistically different at the 5% level as tested by Duncan's multiple range test.

5 ppm zinc than in 1 ppm zinc cultures (Table 3). The exact physiological process is difficult to evaluate, but it appears that in cases where zinc concentrations in the soil solution are at a high level, the cadmium concentrations in plants will be at a higher level than for plants growing in soil deficient in zinc. The major reason is that in addition to the apparent enhancement of cadmium uptake by plants by zinc concentrations, there also exists a close geochemical relationship between zinc and cadmium in soil.

The availability of two chemical forms of cadmium (water-soluble $CdCl_2$, and water-insoluble CdO) from a Captina silt loam at three levels of lime was evaluated in greenhouse studies. The first harvest (26 days growth) of Japanese millet (Echinochloa frumentacia) revealed that ^{109}Cd concentrations from the soluble form were two to three times greater than those from the insoluble form (Table 4). Liming the

Table 4 Concentration of ^{109}Cd in Japanese Millet Derived from two Chemical Forms of Cadmium at three Levels of Calcium Carbonate.*

CaCO₃ level (tons/ha)	^{109}Cd concentration (pCi/mg)		
	First harvest	Second harvest	
	total plant	Leaves	Heads
	Derived from $CdCl_2$		
0	46.4a	14.9a	4.6a
11	45.5a	3.6bc	0.79c
22	29.5b	4.0bc	0.46c
	Derived from CdO		
0	26.3bc	6.8b	2.4b
11	15.7d	2.6c	0.62c
22	18.1cd	2.2c	0.56c

* Average of three replicates. Numbers with an identical superscript within the same column are not statistically different at the 5% level as tested by Duncan's multiple range test.

soil tended to reduce the availability of both forms; availability from CdO was more affected than from $CdCl_2$. The second harvest, 63 days later showed that the level of liming was more important in plant uptake of cadmium than the chemical form of cadmium. For example, no significant difference (P ≤ 0.05) in ^{109}Cd concentrations was observed in the leaves from either form of cadmium in the limed treatments, while in the absence of lime significantly higher leaf concentrations resulted from the $CdCl_2$ than from CdO. These experiments indicate that the solution chemistry of cadmium in soils is an important aspect of plant uptake of this element. The cadmium soil reaction products which ultimately control plant availability from soils are definitely influenced by such variables as pH, chemical form of elements, and the antagonism and synergisms of other ions present in the system. Use of radiocadmium allowed very sophisticated experimental approaches at a level of resolution not obtainable at a reasonable cost through stable cadmium chemistry analyses.

Cadmium Distribution in Plants: Following uptake, the distribution of trace elements throughout the plant as well as subsequent losses from the plant are important considerations in defining source terms for food chains based upon particular plant parts. To provide information on the mobility of cadmium in an old field community, an experiment was initiated in which thirty 1-m² plots were each sprayed with a 1-liter solution containing 200 μCi of ^{109}Cd as ^{109}CdCl_2. Prior to application, half (15) of these plots were clipped bare of vegetation and litter to determine uptake by plants from soil. The remaining 15 plots were left vegetated.   Plots   were   harvested

Table 5 Average Percent of Applied ^{109}Cd in Components of Clipped and Unclipped Old-Field 1—m² Plots for 5 Months after a Spray Application.

	Time following application in April 1972 (months)				
	1	2	3	4	5
Clipped plots					
New vegetation	2.0±0.9*	0.7±0.3	0.5±0.1	0.3±0.02	0.3± 0.05
Soil	98.0±0.9	99.3±0.4	99.5±0.1	99.7±0.1	99.7± 0.1
Unclipped plots					
Vegetation	6.9±1.9	3.7±1.4	2.7±1.1	2.1± 0.8	1.7± 0.2
Litter	54.9±8.8	46.7±4.2	58.7±9.3	29.9±12.4	42.8±13.7
Soil	38.2±9.9	49.6±5.4	38.6±9.0	68.0± 8.3	55.5±11.3
Cummulated rainfall (cm)	6.8	17.2	26.1	39.7	44.1

* Average percent ± 1 standard error for three plots.

monthly for 5 months, and the distribution of ^{109}Cd was determined for plants (by species), litter and soil (by depth).

Table 5 gives the average per cent of total ^{109}Cd in major components of these plots for 5 months following application. This distribution is analogous to the movement of cadmium that may enter such a system in rain or dry fallout. Relatively small amounts (<7%) of this element were incorporated in living vegetation — either via direct uptake from revegetated clipped plots or where the radiocadmium was sprayed on unclipped plots (direct foliar application). Over 90% of the ^{109}Cd in soil remained in the top 2.5-cm layer where organic matter content was high. There was a net loss of ^{109}Cd from the litter layer over the 5-month period, but intense rainfall apparently moved some of the cadmium-contaminated surface soil back into the litter layer prior to sampling in the third and fifth months. At 5 months almost one half of the applied ^{109}Cd still remained in the litter. Thus, in ecosystems subjected to a chronic input of this element via rain or dry fallout, it may be most available to animals whose food base is litter or detritus.

Cadmium Behavior in Food Chains

After defining the amount and form of cadmium in the environment utilizing radiocadmium techniques for obtaining the required information, a most important question to be answered is how this element behaves in food chains. Food chain behavior of cadmium in either food chains leading to man or food chains involving wild species is largely unknown.(2) At ORNL, we have a continuous program devoted to development of new radioisotope techniques and modification of existing ones for measuring assimilation, equilibrium, and turnover in food chain components.(22—25) Additional effort is now going into establishing methods for estimating rates of cadmium as well as other trace contaminant transfers among food chain components in stream and microcosm experiments.(26, 27)

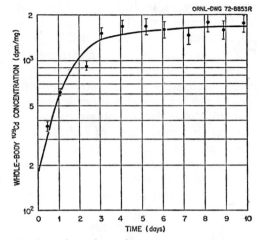

Fig. 3 Biological uptake of cadmium by crickets, Acheta domesticus, during chronic ingestion of old field litter tagged with $^{109}Cd(NO_3)_2$. Each point represents a mean of 40 observations with one standard error of the mean.

zero, the intercept represents the percentage of the initial dose which is incorporated into body tissues.[27] In this case, the assimilation rate was 17%. Analysis of the data in Fig. 4 indicates a biological half-life of 7 days for cadmium in Acheta.

Biological half-lives for cadmium also have been estimated for other animal species using ^{109}Cd (Table 6). Values for cadmium half-lives ranged from 7 days in insects to almost 100 days in birds and predacious arthropods. In each of these species, except mosquito fish (Gambusia), the half-lives were determined by feeding the animal on ^{109}Cd-tagged food. For Gambusia, the half-life was determined from direct uptake of cadmium from water. This value may differ from a half-life deter-

Cadmium Uptake and Turnover: Radioisotopes of cadmium have been utilized in determining rates of cadmium uptake by animals from different food sources, distribution in different tissues following uptake, and the biological elimination rates for cadmium incorporated into body tissues. Figures 3 and 4 illustrate the uptake and elimination of ^{109}Cd by crickets (Acheta) feeding on $^{109}Cd(NO_3)_2$-tagged old field litter. There was a rapid accumulation of cadmium (Fig. 3) followed by an equilibration as input of ^{109}Cd balanced output. Elimination of ^{109}Cd (Fig. 4) involved two components: a rapid early component representing gut clearance and a slower late component representing biological elimination of cadmium from body tissues. The elimination curve illustrates a method of estimating an assimilation rate of cadmium from an ingested dose. If the curve is based on a single feeding, with extrapolation of the late component back to time

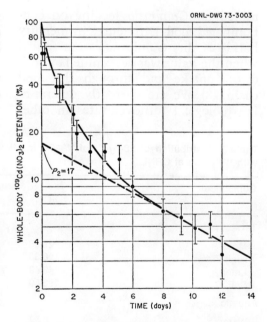

Fig. 4 Biological elimination of cadmium by crickets, Acheta domesticus, following a single feeding on old field litter tagged with $^{109}Cd(NO_3)_2$. Each point represents a mean of 40 observations with one standard error of the mean.

Table 6   Biological Half-Lives and Food Sources of ^{109}Cd in Selected Aquatic and Terrestrial Animals.

Animal	T_b (days)	Food source of cadmium
Aquatic		
minnow (Gambusia affinis)	61	*
fish (Micropterus salmoides)	21	Gambusia
Terrestrial		
insect (Acheta domesticus)	7	Grassland vegetation
spider (Lycosa spp.)	99	Acheta
mammal (Sidmodon hispidus)	18	Grassland vegetation
bird (Spizella passerina)	99	Mixed seeds

* Determined from direct uptake from water rather than from the food chain.

mined for *Gambusia* feeding on contaminated food. Differences in mercury half-lives in *Gambusia* were noted when the value was determined by both direct uptake and through the food chain.[28] We are currently determining biological turnover rates in other animal species including bloodworms (*Chironomus*), quail (*Colinus*), owls (*Otus*), and mice (*Peromyscus*).

Cadmium Microcosm Experiments: There is evidence that indicates that cadmium enters the aquatic environment from terrestrial sources at a slow rate.[26] In order to obtain information on this movement of cadmium, a microcosm experiment was designed to quantify the cadmium ion transfer among system components, and to analyze the sites and quantities of accumulation of cadmium within the terrestrial and aquatic systems.

The experimental design consisted of two replicate microcosms which were established and maintained at a natural photoperiod, rainfall, and temperature for 3 weeks in May 1971. Both microcosms consisted of soil, litter, and vegetation removed intact from stream side and sediments, water, snails (*Goniobasis*), and mosquito fish (*Gambusia*) from the stream. After ^{115m}CdCl$_2$ (284 μCi) was applied to the terrestrial portion of each microcosm in a simulated rainfall, sampling was conducted on a weekly basis with

Table 7   Distribution of 284 μCi of ^{115m}Cd in Microcosm Experiments at 27 Days After Tagging with ^{115m}CdCl$_2$ at ORNL.

Component	Microcosm A		Microcosm B	
	μCi	% Activity	μCi	% Activity
Terrestrial		93.8		95.5
Moss	29.90	10.2	23.34	7.9
High Plants	0.38	0.1	0.78	0.3
Litter	43.40	15.3	30.20	10.7
Soil	193.50	68.2	218.00	76.8
Aquatic		3.4		4.6
Water	0.52	0.2	0.56	0.2
Sediment	8.93	3.1	8.63	3.3
Fish	<0.01	<0.1	<0.09	<0.1
Snails	0.22	0.1	0.14	<0.09
Watercress	0.02	<0.09	0.04	<0.09
Plastic Liner	7.89	2.8	3.26	1.1

radiocadmium determinations being made on each sample.

A total cadmium budget (Table 7) was determined after 27 days by harvesting all components of each system. A leaching treatment of ammonium acetate was used to determine the cadmium in the soil that was biologically available. These tests indicated that only 2–3% of the total cadmium present was leachable or easily available for uptake by plants or soil fauna. Most of the cadmium (93.8 and 95.9%) remained in the terrestrial portion with about 70% of the material tied-up in the soil. Approximately 4% of the initial cadmium amendment was transported to the aquatic portion of the microcosm, and of this, the majority (3%) was in the sediments. Thus, it appears that cadmium movement from terrestrial to aquatic environments is in fact, quite slow.

Cadmium Stream Tag Experiment: Once cadmium enters the aquatic environment, it is necessary to understand its distribution in order to identify those components important in accumulation and retention of cadmium. To this end, an experiment was initiated in which ^{109}Cd was added to a small stream in east Tennessee.

Cadmium-109 concentrations in the water column during isotope flow-through at each sampling station (10 m apart) are illustrated in Fig. 5. The discharge rate of this stream was 360 liters/min. Cadmium proved to be relatively immobile when introduced into the stream, since 95% of the ^{109}Cd added was retained in the first 100 m below the input point. The per cent of the ^{109}Cd activity associated with particulate matter suspended in the water did not change systematically with stream position, but remained at about 20% activity on $>0.45\ \mu$ particles,

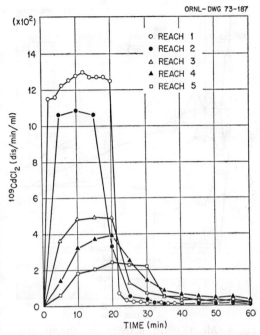

ORNL–DWG 73–187

Fig. 5 *Activity of ^{109}Cd in 5-ml aliquots of stream water at successive downstream sampling stations during a $^{109}CdCl_2$ stream tagging experiment in east Tennessee.*

with about 80% of the activity in solution.

Transport of ^{109}Cd between the major compartments of the stream ecosystem in reach 3 is illustrated as a function of time in Fig. 6. The concentration of ^{109}Cd in the stream sediments was variable, probably due to the turbulence of the water and the non-uniform distribution of the sediments. Periphyton had a very rapid loss of ^{109}Cd during the first 5 days following the tag. The loss rate for ^{109}Cd in this compartment then slowed considerably during the remainder of the study. Both snails and fish exhibited peak concentrations of ^{109}Cd at about one week. After this peak, the snails showed an apparent exponential loss of ^{109}Cd, but the fish showed what seems to be a bimodal elimination

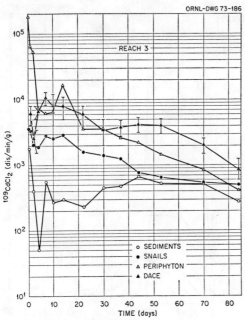

ORNL–DWG 73–186

Fig. 6 Mean ¹⁰⁹Cd activity in stream eco-system compartments at one sampling station (reach 3) during a ¹⁰⁹CdCl₂ stream tagging experiment. Values for snails represent composite samples; those for dace four to six observations; and those for sediments and periphyton, three observations.

curve; the elimination rate increasing 3-months after the tag.

In general, the fish had one and one-half times higher ¹⁰⁹Cd concentrations than periphyton after about one week, and twice the ¹⁰⁹Cd concentrations in periphyton after 2–¹/₂ months. These results indicate that cadmium accumulates in animals, but not nearly to the extent as does methylmercury.(26) Cadmium also appears to be generally less mobile in the aquatic environment than mercury.

Conclusions

The previous sections have summarized experimental procedures utilizing radioisotopes of cadmium for delineating the environmental behavior of cadmium. Our aim in these discussions has been twofold: (1) to highlight those radioisotope techniques utilizing ¹⁰⁹Cd and ¹¹⁵ᵐCd that have proved useful in answering questions relating to characterization of cadmium behavior in soils and sediments, distribution of cadmium in soils, uptake and distribution of cadmium in plants, and food chain transport of cadmium; and (2) to present data and analyses obtained from these radioisotope studies illustrating the type of insight into trace contaminant behavior that can be gained from such investigations.

There appear to be three major advantages to the use of radioisotopes in studying the environmental behavior of trace elements. The first is the very low levels of the element that can be used because of the high specific activities available. The second is the relative ease of detection of the gamma-ray-emitting radioisotopes of these trace elements. A third advantage in using radioisotopes in these studies concerns the cost of trace element analysis. The cost of stable-element analyses by any of the current techniques (e.g., atomic absorption, neutron activation, or spark source mass spectrometry) would prohibit studies of rates of trace element transfer such as determining biological half-lives because of the large number of samples involved. In addition, radioisotope techniques allow the investigator to analyze living plants and animals for determining concentrations, whereas any stable-element analysis is destructive in nature. The primary difficulty of using radioisotopes in short-term environmental studies is that true equilibrium between some added radioisotope and the naturally occurring stable element is usually not achieved in biological systems. This can influence the interpretation of data from studies dealing with amounts of elements moving from one component to another. However, if this limitation is understood, the advantage of the data obtained from rate studies far overshadows this disadvantage.

The results of these radiocadmium studies indicate that in lake and river sediments the clay minerals adsorb Cd^{2+} with increasing strength as pH is increased. While this effect is greatest with illites, both illites and vermiculites could be used to remove dissolved Cd^{2+} from aqueous systems. The cadmium selective electrode appears to be quite useful as a field tool for rapidly estimating solution concentrations of Cd^{2+} in soils and sediments. This would reduce the number of samples requiring laboratory analyses in exploratory monitoring. Cadmium-109 experiments also indicated that deep burrowing earthworms had a greater effect on the distribution of cadmium in soils than did the shallow burrowing species. In both cases, water amendments produced a more pronounced effect than when water was withheld. Plant uptake studies with ^{109}Cd indicate that the cadmium-soil reaction products which ultimately control plant availability from soils are definitely influenced by such variables as pH, chemical form of elements, and the antagonisms and synergisms of other ions present in the system. Results from plant distribution studies revealed that cadmium would be most available to animals whose food base was litter or detritus when the ecosystem was subjected to a chronic input from rain or dry fall. In food chain studies, biological half-lives of cadmium ranged from 7 to 99 days for a variety of animals. Stream tagging experiments demonstrated cadmium to be accumulated by animals, but not to the extent of methylmercury. Cadmium appeared to be generally less mobile in the aquatic environment than mercury. In land-water (microcosm) studies, the majority of cadmium added via atmospheric input was tied-up in the soil while of the small amount going to the aquatic system, the majority accumulated in the sediments.

Acknowledgments

We are indebted to the technical personnel and students who participated in this research. We especially thank Drs. W. Fulkerson and T. Tamura for timely remarks during planning, execution, and writing stages of these research projects.

References

1 Friberg, L., M. Piscator, and G. Nordberg: *Cadmium in the Environment*, Chemical Rubber Company, Cleveland, Ohio (1971).

2 Fulkerson, W., and H. E. Goeller: (Eds.), "Cadmium the dissipated element", ORNL-NSF-EP-21 (1973).

3 Munshower, F.: "Cadmium compartmentation and cycling in a grassland ecosystem in Deer Lodge Valley, Montana", Ph. D. Dissertation, University of Montana, Missoula, Montana (1972).

4 Kobayashi, J.: "Air and water pollution by cadmium, lead, and zinc attributed to the largest zinc refinery in Japan", In Trace Substances in Environmental Health — V. (Hemphill, D. D., Ed.), University of Missouri, Columbia, Missouri (June 1972) 515.

5 Klein, D. H., and P. Russell: "Heavy Metals: Fallout around a power plant", Environ. Sci. and Technol. 7 (1973) 357.

6 Yamagata, N., and I. Shigematsu: "Cadmium pollution in perspective", Bull. Inst. Public Health 19 (1970) 1.

7 Goeller, H. E., E. C. Hise, and H. B. Flora: "Societal flow of zinc and cadmium", In Cadmium the Dissipated Element (Fulkerson, W., and Goeller, H. E., Eds.), ORNL-NSF-EP-21 (1973) 61.

8 Witherspoon, J. P.: "Cycling of cesium-134 in white oak trees", Ecological Monographs 34 (1962) 403.

9 Auerbach, S. I.: "Radionuclide cycling: current status and future needs", Health Phys. 11 (1965) 1355.

10 Olson, J. S.: "Use of tracer techniques for the study of biogeochemical cycles", In Proceedings of the Copenhagen Symposium on Ecosystems (1967).

11 Satchell, J. E.: "Lumbricidae", In Soil Biology (Burges, A., and F. Raw, Eds.), London (1967).

12 Andren, A. W., et al.: "Environmental monitoring of toxic materials in ecosystems", In Ecology and Analysis of Trace Contaminants Progress Report, ORNL-NSF-EATC-1 (January 1973) 61.

13 Earthworms High in Metals, Chemistry and Engineering News 50 (1972) 1.

14 Lagerwerff, J. W.: "Uptake of cadmium, lead and zinc by radish from soil and air", Soil Sci. 111 (1971) 129.

15 Bingham, F. T., and A. L. Page: "Relationship of substrate Cd to accumulation of Cd and growth of economic plants", In Agronomy Abstracts, 1972 Annual Meeting, American Society of Agronomy (October 1972) 176.

16 Haghiri, F.: "Cadmium uptake by plants", J. Environ. Quality 2 (1973) 93.

17 John, M. K., H. H. Chuah, and C. J. van Laerhoven: "Cadmium contamination of soil and its uptake by oats", Environ. Sci. and Technol. 6 (1972) 555.

18 Root, R. A., and D. E. Koeppe: "Uptake of cadmium and its toxicity on hydroponically growth corn", In Agronomy Abstracts, 1972 Annual Meeting, American Society of Agronomy (October 1972) 113.

19 Bolton, N. E., et al.: "Trace element measurements at the coal-fired Allen Steam Plant", Progress Report, ORNL-NSF-EP-43 (January 1973) 83.

20 Parizek, J. I., et al.: "Metabolic interrelations of trace elements — the effects of some inorganic and organic compounds of selenium on the metabolism of cadmium and mercury in the rat", Physiologia Bohemoslovaca 18 (1969) 95.

21 Lagerwerff, J. V., and G. T. Biersdorf: "Interaction of zinc with uptake and translocation of cadmium in radish", In Trace Substances in Environmental Health — V (Hemphill, D. D., Ed.), University of Missouri, Columbia, Missouri (June 1972) 515.

22 Kevern, N. R.: "Feeding rate of carp estimated by radioisotope method", Transactions of the American Fisheries Society 95 (1966) 363.

23 Reichle, D. E.: "Measurement of elemental assimilation by animals from radioisotope retention patterns", Ecology 50 (1969) 1402.

24 Reichle, D. E., and R. I. Van Hook: "Radionuclide dynamics in insect food chains", Manitoba Entomologist 4 (1970) 22.

25 Kolehmainen, S. E.: "The balances of cesium-137, stable cesium, and potassium of bluegill (Lepomis macrochirus Raf.) and other fish in White Oak Lake", Health Physics 23 (1972) 301.

26 Huckabee, J. W., and B. G. Blaylock: "Transfer of mercury and cadmium from terrestrial to aquatic ecosystems", In Proc. Sym. on the Role of Metal Ions in Biological Systems, CSUI-ANL, Argonne National Laboratory, Chicago (Nov. 1972).

27 Van Hook, R. I., and A. J. Yates: "Cadmium behavior in a grassland arthropod food chain", Submitted to Ecology.

28 Blaylock, B. G., et al.: "Ecology of toxic metals", in Ecology and Analysis of Trace Contaminants Progress Report, ORNL-NSF-EATC-1 (January 1973).

Global Inputs and Trends of Chemical Residues in the Biosphere *)

By **Prof. Dr. F. Korte**

FAO/IAEA Symposium on Nuclear Techniques in Comparative Studies of Food and Environmental Contamination

Mr. Chairman, Ladies and Gentlemen, First of all I would like to express my thanks for the opportunity to discuss here the topic "Global inputs and trends of chemical residues in the biosphere".

The use of the term Environmental Quality becomes more and more popular and it is even used today within politics as a part of quality of life. It might be appropriate to try to define Environmental Quality. The part of the environment which is noticeable and easy to describe could be called primary environmental quality. Criteria being part of the material environmental quality then are factors like colors, forests, waters, but also garbage, bad odor, noise, etc. Additionally, those factors which cannot be directly noticed but have been in the focal point for some years are of importance. These latter are affected by presence of substances in molecular dispersion, no matter whether they are radioactive or not. Within this approach to environmental quality, foreign compounds play an important role. I would like to point out that allelo-chemicals, i.e., compounds occurring naturally in the living environment and having controlling functions in ecological systems, have to be included. Considering

this, primary environmental quality appears to be quite complex.

But there may be secondary changes in environmental quality, too. In this context, I consider, for example, the fact that the chemistry of the environment may remain unchanged, but that man consciously or unconsciously may change the individual intake of chemicals due to changing his way of life and eating habits.

Tertiary changes of environmental quality could be summarized under the effects noticed from the increasing use of psychotropic products, be it caffein, nicotine, and alcohol, or hashish, opium, etc. These tertiary changes also include the increasing use of any kind of drugs.

In this paper, I would like to concentrate on the discussion of problems based on direct human industrial activity, the influence of changing eating habits and, to a lesser extent, the changes in way of life. This paper will not deal with the importance of these aspects as related to nature conservation and town planning.

Looking to today's environmental quality, we notice that undesired effects for man, flora and faune, respectively, occur only locally under special situations. In saying this, I refer to the conditions resulting in the occurrence of the Minamata disease.

The release of high amounts of mercury into Minamata Bay, the slow water flow in the bay resulting in an accumulation, and high conversion to

*) Published in Comparative Studies of Food and Environmental Contamination (Proc. FAO/IAEA/WHO Symp. Otaniemi, 1973), IAEA, Vienna (1974) 3-22

methyl mercury in fish had to occur simultaneously. These together with the high fish consumption of the population caused the Minamata disease. Correspondingly, a fish slimming-diet popular in North America caused a change in the intake of foreign compounds — in this case, increased amounts of methyl mercury. According to our definition, a secondary change of environmental quality had effects similar to Minamata disease.

The Itai-Itai disease which occurred near Kunazawa is of similar local importance. This disease was caused by the slow and continuous leaching of cadmium salts out of the waste salts of a closed mine into a river, the water of which was used by the people there without purification as drinking water.

A further example for a local effect is the London fog of 1952. The synergistic action of sulphur dioxide and dust, which was unknown at that time, and, simultaneously, the special meteorological situation was harmful to man. During this fog incident, 4000 people died more than in average. Such changes of air quality due to a synergistic effect seem already to be of importance today in larger areas for unhealthy people and especially for sensitive groups like children, old people, and pregnant women.

Another problem occurring in some areas of the world is the accumulation of nitrate from fertilizers in ground and drinking water. This accumulation is of increased importance if there is a frequent recycling of the water.

This might be necessary under special climatic and geological situations, e.g., in some areas of Israel, the San Joaquin Valley and some other coastal valleys of California. Nitrate concentrations there occasionally exceed WHO-tolerances.

Another example for local effects are the consequences of photochemical reactions by atmospheric sunlight, resulting in situations hazardous to man under special meteorological conditions and high air humidity. The described situation is known as Los Angeles Smog. Local vegetation damage by high sulphur dioxide emission from special industries and fossil fueled power stations is a corrresponding example.

However, I think it is right to say that up to now no effects being harmful for the health of the whole population (comparable to the Minamata disease) have been demonstrated globally. Obviously, this results from the very small global concentrations of the respective compounds as compared to local and regional concentrations. Whilst sulphur dioxide, for example, may reach local concentrations of several ppm and regional concentrations of 200—300 ppb, it will hardly exceed a few ppb globally.

The differences between local and global concentrations become even more distinct when one looks at synthetic compounds and, for example, compares the concentrations in a river of an industrial area with a river in a virgin area or even in the ocean. This comparison shows differences in concentration frequently greater than by the factor 1000. This applies also for ubiquitous foreign compounds as for example DDT and PCB. Different conditions apply in the atmosphere, where transport and dispersion can easily take place.

According to our present knowledge, the low global levels of these substances have no or at least insignificant effects on animals and plants. Effects of compounds present in ppb-concentrations are known so far only in a few examples: for example, photosynthetic activity of phytoplankton is reduced by concentrations of about 100 ppb DDT. This has been shown only for a few plankton species and, on the other hand, this is not the DDT-concentration in the world ocean. The potential synergistic action of small concentrations of foreign compounds nevertheless remains a problem to be considered.

When saying that today environmental quality in its chemical part has changed for the worse under special local conditions but shows no signs of deterioration globally, we must admit on the other hand that upon increasing use of synthetic products, the trend must be in the direction of a further and large scale deterioration. Therefore, we could expect the known local effects to occur in larger areas in the future. This prediction of the trend would be changed

a) if the consumed industrial products would break down quickly under environmental conditions, resulting

in compounds which could be catabolized by organisms. That means then, that relative to total production, the manufacture of persistent compounds should decrease. Considering the production-pattern of the chemical industry, the condition for this, namely the development of favorable substitutes according to an ecological point of view, would be an effort which can hardly be underestimated.

b) In the case that man would re-collect used products and handle them locally. They then should be re-cycled or should be returned into natural compound cycles by means of complete and controlled combustion. As regards waste, there is a principle difference between radioactive and other foreign compounds. Radioactive wastes with long half-lifes should be disposed of exclusively under controlled conditions.

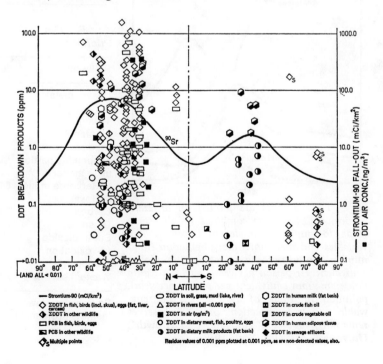

Fig. 1

According to my opinion, it would be quite sufficient if, in spite of increasing world population and production due to the forced development of new technologies, environmental contamination by non-reusable waste could be kept at today's level. However, even under the presumption of this favorable prognosis, there is still the uncontrollable phenomenon of dispersion tendency. Regionally or locally used substances with high water solubility are washed out of waste deposits by rain and are carried into the ocean by rivers, if they are not metabolized or adsorbed during transport. Many of the persistent organic environmental chemicals have a low water solubility and, frequently, a low vapour pressure as well. At lower concentrations, however, they may be quickly vaporized

from soil, waters, and plants and thus may be transported over large areas in the atmosphere. This dispersion has been detected with substances like DDT and PCB, and to a smaller extent with hexachlorobenzene within the last 10 years. This dispersion, together with the enrichment of these substances in

Pesticides and Metals Top Priorities in Environmental Stress . . .

Pollutant	Index	Pollutant	Index
Pesticides	140	Oxides of nitrogen	24
Heavy metals	90	Radioactive waste	20
Carbon dioxide	75	(for storage)	
Sulfur dioxide	72	Litter	16
(+ oxidation products)		Tritium, krypton-85	16
Suspended particulate matter	72	(nuclear power)	
		Photochemical	12
Oil spills	48	oxidants	
Waterborne industrial wastes	48	Hydrocarbons in air	10
		Carbon monoxide	9
Solid waste	35	Waste heat	5
Chemical fertilizer	30	Community noise	4
Organic sewage	24	(+ sonic boom)	

. . . But Metals Predominate in Projections of Priorities

Pollutant	Index	Pollutant	Index
Heavy metals	135	Chemical fertilizer	63
Solid waste	120	Organic sewage	48
Tritium, krypton-85	120	Oxides of nitrogen	42
(nuclear power)		Litter	40
Suspended particulate matter	90	Radioactive waste (for storage)	40
		Pesticides	30
Waterborne industrial wastes	84	Hydrocarbons in air	18
		Photochemical	18
Carbon dioxide	75	oxidants	
Oil spills	72	Community noise	15
Sulfur dioxide	72	(+ sonic boom)	
(+ oxidation products)		Carbon monoxide	12
Waste heat	72		

Source: Dr. Howard Reiquam, Battelle Memorial Institute (Chem. Eng. News, Jan. 10, 1972)

Fig. 3

Exhaustion of resources cuts industrial output

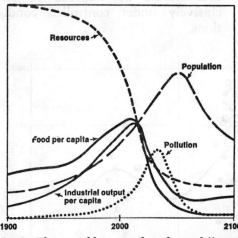

Resources

Population

Food per capita

Pollution

Industrial output per capita

1900   2000   2100

Fig. 2 The variables are plotted on different vertical scales but are combined on the same graph to emphasize modes of behavior. The variables and their units are: population (total number), industrial output per capita (dollar equivalent per person per year), food per capita (kilogram-grain equivalent per person per year), pollution (multiple of 1970 level), nonrenewable resources (fraction of 1900 reserves remaining). Source: "The Limits to Growth", Universe Books, New York, 1972.

food chains, has led to supraregional effects on wildlife animals, for example to the reduction of egg shell thickness of some kinds of birds of prey.

Even the distribution of Sr^{90} resulting from tests with the atomic bomb is in principle analogous to the dispersion of DDT, where higher concentrations have been measured on the Northern hemisphere. The left part of the figure given here shows the fallout of Strontium 90 and of the sum of DDT plus conversion products in the Northern hemisphere, the right part, the corresponding data for the Southern hemisphere. The scale for Strontium 90 (right) is mCi/km^2, that for DDT (left) is ppm. The full black curve represents the fallout for Strontium 90, the signs represent DDT in various samples, e.g., the horizontal ellipse in soil, grass, and sediments, the horizontal rhombus in fish, birds, and eggs, the circle in meat, fish, and poultry.

Perhaps I should mention the MIT-Club of Rome-study of Meadows as an example of an unfavorable prognosis. Certainly, you know the book and the discussion on the applicability of such simplified computer models. Many ex-

perts are of the opinion that the methods used to obtain conclusions of this kind are not without weaknesses. The curves on depletion of resources, food per capita industrial output and on air pollution show critical situations between the year 2000 and 2050 (Fig. 2).

The Battelle Memorial Institute tried to rank chemicals according to a hazard-index. According to this list, the hazard-index or the contamination index for all substances, with the exception of pesticides, will increase in the future (Fig. 3).

Given the fact that these more or less pessimistic pictures of environmental quality are not universally accepted, I do not want to discuss them in more detail, but I prefer to concentrate on the analytically detectable alteration of environmental quality due to chemicals. The primary parameter, which determines whether a measurable change occurs, is given by the production height and use pattern. I would like to give you here two mathematical examples:

If we assume that the organic chemicals produced at present — 100 million tons per year — were not decomposed or altered but released into the environment quantitatively, there would be, with an even distribution over the total land surface of the earth, a load of 700 mg/m^2 or, in case of penetration into a soil layer of 10 cm depth, a concentration of 2.5 ppm (Fig. 4).

Although this mathematical example contains some perhaps inadmissible simplifications, it shows that, due to the industrial activity of man, the material environment can be changed within a short period and can be globally measured. If you would take as an example a naturally occurring substance like mercury, the natural zero level should be added to the calculated numbers.

Maximum Global Concentrations of Org. Chemicals

Land surface of the earth	149×10^6 km²
Volume of oceans	1.3×10^9 km³
Weight of the atmosphere	5.1×10^{15} to

Org. chemicals production
1973, total	app. 100×10^6 to

Basis of calculation: no breakdown, dispersion of total amount in one medium only

Dispersion on total	
land surface	700 mg/m²
or in 10 cm soil layer	2.5 ppm
Dispersion in *oceans*	0.8×10^{-4} ppm
or in 1 m layer	0.3 ppm
Dispersion in *atmosphere*	0.02 ppm

Fig. 4

| Land Surface of the Earth | 13.392 x 10⁹ ha | | | |

Agriculture	1.424 x 10⁹ ha
Pastures	3.001 x 10⁹ ha
Forests	4.091 x 10⁹ ha

Soil Input per Year   (no conversion, no evaporation or leaching)

	World production Mio. tons/year	kg/ha in area of use locally	globally	kg/ha when distributed on total land surface
Pesticides	1	2—4	0.12 (a/p/f)	0.07
Nitrogen	15	70	3.38 (a/p)	1.12
fertilizers			1.76 (a/p/f)	
org.	100			7.47
Chemicals total				

Fig. 5   Maximum Concentrations of Organic Chemicals in Soil.

If we consider under corresponding conditions — no conversion, evaporation or washing out — the possible environmental contamination from pesticides, nitrogen fertilizers, or organic chemicals, the result would be the shown contamination in kg/ha, where especially the last column not containing the use area, is of global importance (Fig. 5). If all used pesticides were distributed on the whole land surface of the earth, there would result a contamination of 7 mg/m²; the used nitrogen fertilizer would be present in concentrations of about 100 mg/m²; all organic chemicals distributed on the surface would represent annually more than 700 mg/m². Similar quantities could be obtained with other fertilizers

	1950	1970	1985	Release in Env. 1970
grand total 10⁶ t	7	63	250	20

Organic Chemicals — World Production

Manufactured	10⁶ t	Natural sources	10⁶ t
Solvents	10	Methane	1600
Detergents	1,5	Terpene type	
Pesticides	1	hydrocarbons	170
Gaseous base chemicals	1	Lubricating and industrial oils	2—5
Miscellaneous	7		

Release of Organic Compounds in the Environment, 1970 (Ilff. 1971)

Fig. 6

Organic chemicals will remain the leader

Key chemical product classes	Shipments, billions of dollars 1970	1975	1980
Chemical and allied products, total	54.4	78.8	114.4
Alkalies and chlorine	0.86	1.25	1.86
Industrial gases	0.71	1.05	1.55
Industrial organic chemicals	8.29	13.26	20.95
Industrial inorganic chemicals	5.13	7.64	11.46
Plastics and resin materials	4.81	7.03	10.41
Synthetic rubber	1.25	1.67	2.29
Man-made fibers	3.76	5.71	8.64

Dec. 15, 1969 C & EN

Fig. 7

(phosphate, potash). Apart from the fact that such a distribution does not take place under natural conditions, most of the substances are metabolized and/or altered to natural products, as e.g. fertilizers, so that the results of this mathematical example represent a maximum contamination.

As regards the world production of chemicals, there are only estimations which I would like to mention here because of their significance. In 1950, the world production of totally synthetic organic chemicals amounted to 7 million tons, in 1970 it was already 63 million tons; for 1985, the author cited here expects an increase up to 250 million tons. This would be an increase by a factor of 2.5 for the period of ten years. Published production numbers from the USA and the FRG show the factor estimated here for organic chemicals (Fig. 6).

In Fig. 7, you can see that industrial organic chemicals have the highest production increase in this decade, at least in the USA.

The data given in Fig. 8 for the use of chemicals in the FRG show an average trend of increase by a factor of 2 for the past ten years.

What further knowledge would be necessary for the global evaluation of environmental chemicals; are there models, which we could use for the evaluation of environmental chemicals? There are international agreements on the evaluation of radioactivity by

Product	Consumption in 1969 (1,000 t)	Factor	Compared to
Chlorine	1,749	2.5	1960
Sodium hydroxide	1,511	2.0	1960
Hydrochloric acid	600	2.4	1960
Sulphur dioxide	38	1.5	1963
Carbon disulfide	84	1.35	1960
Sulphuric acid	3,619	1.3	1960
Nitric acid	718	1.5	1962
Lead oxides	37.6	1.5	1960
Lead carbonates	0.9	0.4	1960
Liquid ammonia	2,106	1.8	1960
Ethylene glycol	171	2.0	1966
Formaldehyde	388	2.8	1960
Tri- and Tetrachloroethylene	200		
Derivatives of phthalic acid (without plasticizers)	165	2.0	1960
Plasticizers on the basis of phthalic acid	166	2.7	1960
Synthetic rubber	315	2.8	1960
Plasticizers, total	190	1.5	1966
Detergents	176	1.4	1966
Fertilizers:			
Nitrogen	1,011	1.5	1959/60
Phosphorus pentoxide	802	1.1	1959/60
Insecticides, active materials	1.0		
Herbicides, active materials	8.8	1.8	1960
Fungicides, active materials	4.7		
Mercury	0.76		
Motor benzine	14,084	2.8	1960
Mineral oil products, total	102,083	3.7	1960

Fig. 8   Consumption and Trend of Chemicals in FRG.

Evaluation of the Importance of Environmental Chemicals

Production and industrial waste	Use pattern	Persistence

Dispersion tendency

Conversion under biotic and abiotic conditions	Biological consequences (structure-activity-relationship)

Fig. 9

means of evaluation parameters. The WHO has provided evaluation parameters for medicines (drugs) and there are good scales, too, for the evaluation of pesticides, which we could consider today as models of environmental chemicals.

Within national and international legislation, these evaluation parameters vary for certain groups of substances — up to now, there are only national evaluations for solvents, detergents, cosmetics, food additives, etc. From the internationally treated groups of substances, pesticides certainly represent the best model for the evaluation of environmental chemicals, as they correspond to almost all organic chemicals due to their use and dispersion tendency.

The parameters shown in Fig. 9, which I have already discussed partially, could represent a scale for environmental contamination by chemicals, an indica-

tion whether there are local or long-term global contaminations, and which consequences the use of the corresponding chemicals would have for man and his living environment. A useful focussing of the 6 parameters against one another could make possible the setting up of a priority list for environmental hazards due to chemicals.

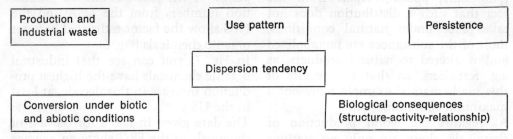

$C_8H_8ClNO_2$

$C_{13}H_{11}Cl_2NO_2S$

$C_8H_8BrCl_2O_3PS$

Fig. 11   Halogen-containing Pesticides not classified as halogenated hydrocarbon Pesticides.

Chlorinated paraffins	25,793
Chloroform	86,662
Carbon tetrachloride	323,974
Vinyl chloride (monomer)	1,100,302
Trichloroethylene	222,444
Tetrachloroethylene	241,973
Dichlorodifluoromethane	82,726
Other halogenated hydrocarbons	3,232,505

US-production of non-pesticidal halogenated chemicals, 1967 (tons)

Fig. 10

What knowledge do we have for environmental chemicals in general, and for the model environmental chemicals: "Pesticides", respectively?

First of all, we should consider those environmental chemicals which, due to their structure, are persistent, namely

Common trade name	DDT

Formula:	

Purity:	75—80%
	main by-product o,p'-DDT
Production level:	not exactly known
Use pattern:	use against more than 100 different insects on large variety of crops, in public health, as mothproofing agent
	quantities unknown

Metabolites:	mainly DDE, DDD, DDA, DCB
Occurrence residues:	up to 7 ppm including analogues
Outside area of use:	concentration up to more than 10 ppm in wildlife, fish including analogues; atmosphere: up to 200 pp 10^{12}

LD_{50}:	DDT rat — oral 250 mg/kg bodyweight
	DDD rat — oral 3400 mg/kg bodyweight
	DDE rat — oral 1000 mg/kg bodyweight
	DDT monkey oral >200 mg/kg bodyweight
Chronic toxicity:	great number of data on DDT and analogues in animals and men-e.g. neoplastic disorders in mice (multigeneration)
	no abnormalities in men after 11—19 years occupational exposure to around 18 mg/man/day
Wildlife effects:	accumulation in food chains, reduction eggshell thickness certain bird species

(FAO/WHO, 1965—1969)

Fig. 12   Present Knowledge on DDT.

Common trade name	Toxaphene

Formula:	average emp. $C_{10}H_{10}Cl_8$
	appr. 25 camphene related 8Cl containing compounds, structures *unknown*
Purity:	*unknown*
Production level:	*unknown*, however widely used in quantity and number
Use pattern:	agronomic, vegetable, fruit crops, ectoparasites on cattle, sheep, swine
	quantities *unknown*

Metabolites:	*unknown*
Occurrence residues:	0—<0.5 ppm
Outside area of use:	*unknown*

LD_{50}:	guinea pig oral 365 mg/kg bodyweight
	rat oral 60—120 mg/kg bodyweight
	dog i. v. 5—10 mg/kg bodyweight
Chronic toxicity:	rats − 25 ppm long term − no effects
	monkey − two years-0.64-0.78 ppm in diet − no effects
	volunteers ten days during 30 min./day, 500 mg/m³ air − no effects
Wildlife effects:	*unknown*

Fig. 13   Present Knowledge on Toxaphene.   (FAO/WHO, 1969)

Common trade name	Parathion
Formula:	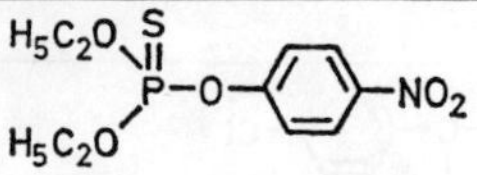 H_5C_2O and H_5C_2O with $P-O$—⟨benzene ring⟩—NO_2, S double bond on P
Purity:	98.76%
Production level:	1966: 15,000 tons
Use pattern:	wide spectrum insecticide, use on food crops, fruits, vegetables, tea quantities *unknown*
Metabolites:	paraoxon, amino-parathion, amino-paraoxon, p-amino-phenol, p-nitro-phenol
Occurrence residues:	0.02-7 ppm
Outside area of use:	*unknown*
LD$_{50}$:	male rat — oral — 5.0—30.0 mg/kg bodyweight guinea pig — oral 9.3—32.0 mg/kg bodyweight
Chronic toxicity:	no effect level on cholinesterase activity: rat: 0.05 mg/kg bodyweight man: 0.05 mg/kg bodyweight/day
Wildlife effects:	*unknown*

(FAO/WHO 1965, 1968)

Fig. 14   Present Knowledge on Parathion.

Common trade name	Zineb
Formula:	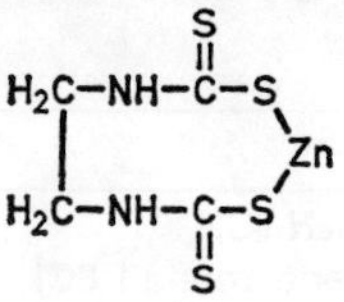 $H_2C-NH-C-S$ (C with S double bond) and $H_2C-NH-C-S$ (C with S double bond), bonded to Zn
Purity:	*unknown*
Production level:	*unknown*
Use pattern:	fungicide on crop and fruit plants quantities *unknown*
Metabolites:	ethylene-thiuram disulfide (ETD) ethylene-thiuram monosulfide (ETM) ethylene-thiourea (ETU)
Occurrence residues:	0—11 ppm, e.g. lettuce 3.6 ppm
Outside area of use:	*unknown*
LD$_{50}$:	rat — oral — >5200 mg/kg bodyweight
Chronic toxicity:	rat — 2 years — 500 ppm goitrogenic effect dog — 1 year — 2000 ppm no effects rat reproduction study — 100 mg/kg/day orally resulted in sterility, resorption of fetuses and anomalous tails
Wildlife effects:	*unknown*

(FAO/WHO)

Fig. 15   Present Knowledge on Zineb.

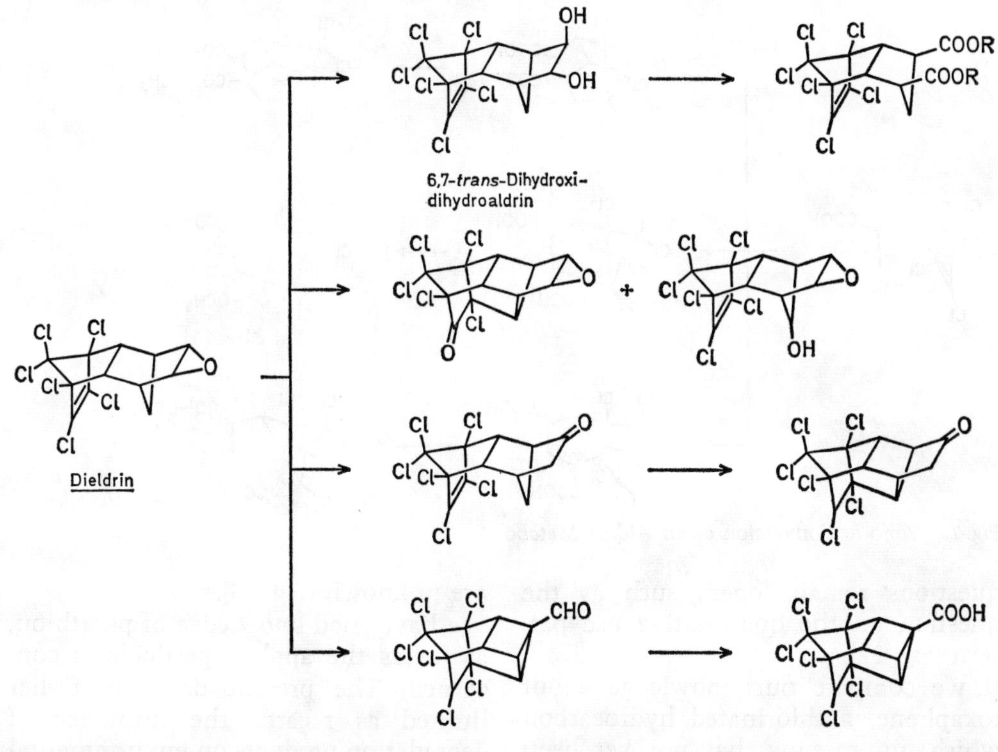

Fig. 16   Metabolites of Aldrin/Dieldrin.

chlorinated hydrocarbons. Today, only a part of those pesticides which are organic chlorine-compounds are included in discussions on environmental quality.

As you can see in Fig. 10 (USA), other chlorinated hydrocarbons are produced in larger quantities. These substances, especially chlorinated paraffins, can get into the environment and into contact with man in a similar way as do pesticides and PCBs.

Besides this situation as regards chlorinated environmental chemicals in general, a number of halogen-containing pesticides, e.g. the three active principles shown in Fig. 11, are not classified as halogenated hydrocarbon pesticides. Further examples would be 2,4-D, 2,4,5-T, pentachlorophenol, DDVP and ronnel. About 40% of all known pesticides contain halogens. The classification of pesticides as organophosphates, carbamates, etc., is based on the assumption that, after the first conversion, e.g. the saponification of the C-O-P-bonds in organophosphates, the fractions have no more influence on environmental quality. According to our present knowledge, this assumption is not valid. It would be more reasonable to classify environmental chemicals according to chemical structures, independent of the fact whether pesticides, solvents, or other industrial chemicals are the primary products.

About DDT — as an example for the so-called pesticidal chlorinated hydrocarbons — we have considerable information regarding the parameters which can be used as an evaluation base. Nevertheless, here too, some

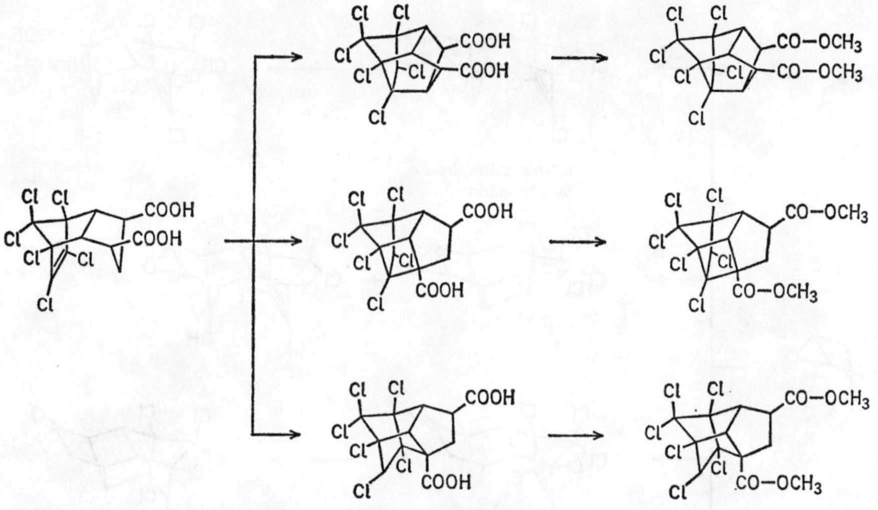

Fig. 17   Abiotic Conversion of an Aldrin Metabolite.

questions remain open, such as the question on the quantitative use pattern (Fig. 12).

If we compare our knowledge about toxaphene, a chlorinated hydrocarbon which, up to now, has not yet been classified in this group, we notice that even the structures of the components are unknown (Fig. 13).

We have good knowledge of parathion, as far as the applied pesticide is concerned. The present data are rather limited as regards the influence of degradation products on environmental quality (Fig. 14).

Finally, I would like to mention, as a

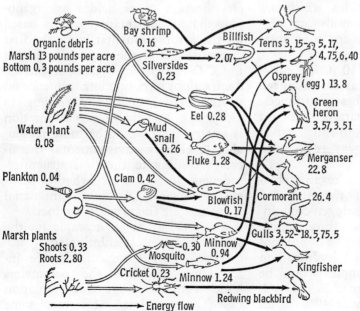

Fig. 18   Accumulation of DDT in Food Chains.

last example of this comparison, zineb which is a metal-organic compound for which there are only a few more data available than for toxaphene (Fig. 15).

It is surprising that just those substances about which we have the best knowledge are frequently discussed in public and that these are submitted to strong legal restrictions of use, whereas there is hardly any interest as regards less known substances. In this connection, besides DDT there should be mentioned the cyclodiene insecticides and, to some extent, PCB, as well known models of environmental chemicals.

For aldrin and dieldrin, for example, a large number of degradation products in various organisms is known, as well as their structures and concentrations of the total residues in the environment (Fig. 16).

Besides the information on the abiotic conversion of the applied pesticides, the substances are known which can result from some metabolites under abiotic conditions, in Fig. 17 the main metabolite in higher plants. For this class of substances, at least contributions to the estimation of the longterm alteration of the material environment are present today.

A further intensively discussed criterium used to classify pesticidal chlorinated hydrocarbons and PCBs as especially

Ash Compositions of Different Water Plants

| | Percentage | | | |
	K_2O	Na_2O	CaO	SiO_2
Stratiotes aloides	30.8	2.7	10.7	1.1
Nymphaea alba	14.4	29.7	18.9	0.5
Chara spec.	0.2	0.1	54.8	5.9
Phragmites communis	1.1	0.5	0.3	71.5

Fig. 20

hazardous to the environment is the known accumulation of residues in food chains, which is shown in Fig. 18 with the example of DDT. Let me point out that an enrichment in food chains of naturally occurring substances could take place as well.

In Fig. 19, I show you as an example the concentrations of some elements in sea water compared to a mine organism, namely Calanus finmarchicus. As compared to its environment, the copepode contains some elements in lower concentrations, most elements, however, in a higher concentration than the sea water, for instance phosphorus is 20 thousand fold of the environment concentration.

	Ocean water	Calanus fin-marchicus	Concen-tration factor
Oxygen	85.966	79.99	0.93
Hydrogen	10.726	10.21	0.95
Chlorine	1.935	1.05	0.54
Sodium	1.075	0.54	0.50
Magnesium	0.130	0.03	0.23
Sulfur	0.090	0.14	1.6
Calcium	0.042	0.04	1.0
Potassium	0.039	0.29	7.4
Carbon	0.003	6.10	2000
Nitrogen	0.001	1.52	1500
Phosphorus	<0.0001	0.13	20000
Iron	<0.0001	0.007	1500

Elements in ocean water as compared to Calanus finmarchicus (% weight, total 100)

Fig. 19

Different water plants growing in the same environment show broad variations in the ash composition, calculated for e.g. potassium oxide from 0.2 to 30.8%, for silicium dioxide from 0.5 to 71.5% etc. The last two figures shown demonstrate that organisms principally use the basic chemicals of their milieu or their food in different amounts. This is one of the causes why in the course of evolution the variety of species which we see today has developed. Whether organisms living today can adapt correspondingly to foreign compounds or whether they will develop this adaptation is still an open question.

Conclusion

With the examples given here I wanted first, to show which parameters are influencing the environment today,
second, to discuss the chemical relationships of the changes in the material environmental quality
and third, to point out the conditions necessary for a decrease in the evolution of environmental contamination.
The picture of constant environmental contamination despite increasing production appears at first to be an optimistic goal. A necessary condition for realization of this ideal would be given, if, on an international scale, it would be possible to bring all producers and users of industrial products to the point that they would publish their production levels, the use patterns of industrial products, and the quantities of the corresponding products used in various sectors. Then it would be possible for the first time to predict the alteration of the material environmental quality by means of the results of scientific work, in which radioisotopes play a central role, and we would be no longer in the unsatisfactory situation of detecting the presence of unintended and/or undesired products only after the appearance of the negative effects which have been luckily, up to now, only local. Unfortunately, the possibilities of acquiring the worldwide use patterns and the quantities used in the various sectors are not very favorable, because there arise apprehensions and fears of competition which, obviously, reduce the publications of the concerned industrial branches to a minimum. It seems to me that this problem requires a cooperative international effort, because the availability of these data would essentially facilitate the detection and evaluation of the real environmental contamination and would bypass the frequent polemic and emotional discussions.

Some Experiences Relating to Food Safety

By **Dr. Emil M. Mrak**

When Dr. Ikeda asked me to participate, I suggested he send me a letter outlining what he had in mind. This he did, and I was overwhelmed at the breadth of the suggestions he made, which included something that might be going on behind the scenes in the politics in Washington, insecticides, additives, getting scientific data to legislators, and impact of industry groups and food faddists. This could take a couple volumes, but I will try to cover some of these suggestions as best I can.

Perhaps the best place to start is at about the time I was to retire as Chancellor of U. C. Davis. Some months before I retired, I was wondering what I might do and how I might occupy myself. Most certainly this has been answered, and the question confronting me, has confronted me ever since, and even today, is how to put the brakes on with respect to activities. In any event, one of the invitations was to be Chairman of Secretary Finch's Commission on Pesticides and Their Relationship to Environmental Health. This appealed to me and most certainly I was happy to become involved. Not only did I become involved, but I had an opportunity to learn an enormous amount about what goes on with respect to legislation, understanding, judgments, and so on. Some of the impressions have been good, and many of them have been bad.

One might ask, how did the Finch Commission come into existence. As you probably know, the Food and Drug Administration is in HEW and, of course, under the Secretary of HEW.

The FDA had collected fish from Lake Michigan that, they indicated, contained DDT up to 15 ppm of DDT, whereas the tolerance was 5 ppm. The question was, what to to do about this, and one might say why a tolerance at all, when DDT has never really been shown to be harmful to humans. This did, indeed, raise some serious questions.

At any rate, an interim tolerance higher than 5 ppm was set up and then a Commission was appointed and asked to study the matter very carefully. This was indeed important, because not only was it a matter of continuing the tolerance at 5 ppm, raising it, or taking some other action, but it was important to consider the States surrounding the Great Lakes. The governors of these states got into the matter and raised the questions as to what to do. The Finch Commission, therefore, was charged with studying the matter of the relation of pesticides to health and coming up with a report in about six months. This seemed like an impossibility but as it actually developed, the 700-page document was prepared and issued in less than six months. The Commission, to a large extent, was picked by the Secretary and his assistants and, in fact, had been picked before I became Chairman. It contained ecologists, entomologists, industry representatives, health people, a consumer representative, and a cancer expert. I requested that three people be added to the Commission, namely Dr. William Darby, now President of the Nutrition Foundation; Dr. Leon Golberg, the distinguished toxicologist at

Albany Medical School; and a person who participated in Dr. Jenson's NRC Committee on DDT. I also asked that Dr. Dale Lindsay, former director of the Division of Research Grants of the NIG and subsequently my assistant chancellor at Davis, be a consultant to the Committee. I asked Dr. Lindsay's assistance because of his understanding of the workings of the bureaucracies in and about Washington, and believe me, I can give you my fullest assurance, this was and is important to any Committee or Commission. The Commission met and the question was, how was it to go about its work. This is when Dr. Lindsay first became very active. He pointed out that the areas for study must be concerned with the needs and uses, contamination of the environment, non target organisms other than man, effects on man, and finally, criteria and recommendations.

The Commission was divided into subcommittees to cover these areas. It was soon realized that the funds allotted for the operation of the committee — $ 100,000 — were completetly inadequate for the task imposed on it. Dr. Lindsay, therefore, suggested that we prepare a series of letters to be signed by the Surgeon General directing various agencies within HEW to assign certain personnel to work with us as staff members. Furthermore, it was indicated that their expenses, travel, typing and so forth was to be taken care of by their respective departments. This is something I would never had done — I wouldn't have thought of it — I did not know enough about the workings of Washington, so I would have tried to get along with the meager funds assigned to it. By this action alone, Dr. Lindsay contributed tremendously. In any event, we were able then to assign what we might call

working or executive secretaries, to each of the subcommittees. But this wasn't all.

We soon learned that when it came to the matter of health, there were four distinct areas that were so important, it was necessary to set up under the subcommittee concerned with health, panels concerned with carcinogenesis, teratogenesis, mutagenesis and interaction.

It was then necessary to find people to work with the individual panels. The magic letters prepared by Lindsay, of course, took care of this. So, we had full time people working with the subcommittees and panels, and this enabled us to move and move fast.

This was a great experience for me and should I ever be assigned to another such task in Washington, and if the funds are inadequate, I shall certainly follow the Lindsay procedure and prepare letters for the Secretary or Administrator concerned with the agency and ask for help. This is so very important that it cannot be over-emphasized. This, I would say is the main reason we were able to produce such an extensive report — over 700 pages in a period of about four months. But, we needed a few other pushers and I will come to these.

It was necessary to obtain, through Secretary Finch, a Presidential priority for publication, and this was done — another trick of the game I would never have thought of. It meant that the Government Printing Office would give this the highest priority possible, rather than letting the material lay for two years before getting it in print. This type of operation is so much more effective than is possible with National Research Council or National Academy committees. I once thought the sun rose and set on National Academy committees, and while I serve on three now, and while I think there is a place for them, I certainly think they have their limitations. They are slow, cumbersome, slow, and do not seem to dwell on a subject until all come to an agreement, which I do not believe is a desirable product. Their final results or recommendations are bound to be watered down.

One other point in connection with the panels, subcommittees and Commission. The panels, of course, report to the subcommittee, and the subcommittees to the Commission.

We made it very clear that the reports of the panels and the subcommittees would be published, but not necessarily approved or accepted by the Commission as a whole. As a matter of fact, the Commission published only 12 pages of recommendations and discussion. The rest was back-up material consisting of reports of the panels and subcommittees.

I made it very clear that we had to come to conclusions, and that the report would not be watered down. I made it very clear, too, that if necessary, we would have a minority, middle ground, and the majority report. I made it very clear that if this were the case, I would give my own interpretations of the report when presented to the Secretary.

Strange as it may seem, this firm and hard attitude compelled even the most recalcitrant, representing one extreme or the other, to go along with what the majority desired. The report most certainly was not watered down.

I attended some of the subcommittee and panel meetings. The people worked hard, they obtained data from various sources, and of course, eventually assembled it into the report that was sent to the Commission. I would like to comment on what went on during the deliberations of some of the panels concerned with health.

The panel concerned with cancer was the one that left me with a permanent and almost distastful impression. It was made up of people from various organizations, including the Cancer Institute, and industry. It contained people who call themselves oncologists, statisticians, toxicologists, pathologists, and some concerned with environmental health.

I must say they reviewed carefully every bit of literature relating to the effect of DDT and cancer. They judged the various works and treated them as either acceptable or unacceptable contributions. It was needless to say that there was plenty of discussion and arguments which were indeed intense. As a matter of fact, while attending one panel meeting as Chairman of the total Commission, I spoke up on a few ocassions emphatically, and pounded the table too, with the view of inducing the members to get down to business. Members from the Cancer Institute, in particular, were inclined to be extremists. Pathologists of the Cancer Institute made it known they did not think much of the pathologists representing the Food and Drug Administration. The statisticians of the Cancer Institute took one point of view and the industry member another point of view. Here is one case where we did get a majority and minority report — the statisticians on the Cancer Panel never did agree.

I was quite shocked to find that pathologists were even uncertain with respect to some of the observations. This, in particular, pertained to the so-called bionetics test for cancer. DDT was force fed to weanling mice to the extent of about thirty thousand times the amount we may take in per day on a weight-per-weight basis. After some time, the mice were sacrificed and the various organs examined. In some cases, abnormal structures were noted on the livers, though there were not a great many. Whether or not the number was statistically significant was never agreed on.

In any event, there was a great argument as to whether the structures were nodules or tumors. The argument on this very subject lasted almost a full day. I could not believe my ears, for I had always thought a pathologist was infallible. I always thought they were certain, but now I know — now I know that in spite of the fact their incomes can be tremendous in commercial practice, their judgment is not always certain.

In any event, it was finally concluded that perhaps the structures were tumors. Remember I pointed out that they perhaps were tumors.

The next question then was, if they were tumors, would they or would they not metastasize into cancer. This took a great deal of discussion and finally it was generally agreed that if they were tumors, of course they might metastasize, and since they were considered to be tumors — even though this was not entirely certain — the only thing the panel could do was to say they might metastasize. This is the type thing that went on behind the scenes in our panels, and it was a shocker to me.

Finally, came the argument of the statisticians and they never did agree as to whether the observations were statistically significant or not. Those in the Cancer Institute said yes. One individual in industry said no, and so it went. There was a discussion on threshhold values. Those in the Cancer Institute do not believe there is such a thing, yet to my knowledge, this question hats not been answered by research. Is there a level below which compounds have no effect? Members of the Cancer Institute say no. If this is the case, we can start right in by outlawing almost everything. Then again, what is the protocol to use in testing for cancer. This was never agreed on, and although the bionetics test seemed completely irrational, members of the Cancer Institute were inclined to accept it, others questioned it.

I might go on to discuss the panel on mutagenesis. It came up with the same type of arguments and uncertainties. Likewise, for the one on teratogenesis. It is quite apparent that in both these cases, methodology is something that is lacking. We just do not have reliable or generally accepted protocols for testing for safety.

As far as interaction is concerned, here it was agreed that this does occur, but getting down to specific details seems almost hopeless.

The panel reports did not show these arguments, but the confusion is there. We have had criticism on some of the reports, unfortunately, on the assumption that since they were included in the total final Commission report, the Commission agreed with everything that was said by the panels and subcommittees. As already pointed out, however, this is anything but the truth. The Commission actually made fourteen recommendations. The first was considered a very important one and suggested closer cooperation among departments of HEW, Agriculture and Interior on pesticide problems through the establishment of a new interagency agreement.

I must say that Secretary Finch wasted no time in bringing this about. It was rather astounding to me to find that our government agencies were not co-operating to the extent they should. A classical example related to the BHC vaporizer of which some of you may be aware.

The Department of Agriculture, which had responsibility for registering pesticides, accepted the vaporizer, but the FDA suggested that tests should be made on the effect on humans and on residues on exposed foods. It was a long time before the latter was done, and to my knowledge, tests to determine the effect on animals were not done until after restrictions were placed on this chemical.

There is now a so-called interagency committee. Ostensibly, this is made up of cabinet officers who consider the use of pesticides by their various departments. A cabinet officer, of course, is not going to bother about pesticides, so each one assigns a person to make up a group known as the working group. If you have not heard of this group, I strongly recommend you not only learn about it, but you get acquainted with Dr. Wymer, Mrs. Ruth West, Dr. Prescott, and others who represent the secretariat of this operation. It makes recommendations and actually policy relating to the use of pesticides by the federal government. The policies they establish in turn, of course, are bound to effect all of us.

The Commission also talked about improving elements within HEW. Here, again, we found that some departments were not working with each other — some had a great deal of data, and were not giving it to others, and some were severely criticizing others.

We suggested that DDT and DDD be eliminated in two years where possible. We also recommended that it be retained for health and welfare of man. The latter statement, of course, meant for agriculture where needed. This recommendation was made not because of health hazards, but rather because we were fully aware if such a recommendation had not been made, there very well could have been laws to make it illegal to manufacture or use these chemicals. This indeed would have been tragic.

We talked about minimizing human exposure to pesticides and, of course, this is now being considered by various departments, and unfortunately by the Department of Labor as a result of the authority granted in the Occupational Safety and Health Act (OSHA).

We suggested creating a committee to advise the Secretary, this was done, and it was called the SPAC Committee. Later SPAC was transferred to the Environmental Protection Agency and became known as "H-MAC" for Hazardous Materials Advisory Committee.

It was suggested that suitable standards for the pesticide content of food and water and other aspects of environmental quality be studied. This, of course, is now being done by the Environmental Protection Agency.

We strongly recommended that the Delaney Clause be modified, but this is something that does not appear to be possible. I might point out, however, that while we were developing our report, some of our people did talk with some of Delaney's assistants to explore the possibility of reconsidering his clause to modify it in light of new information — for example, our ability to now analyze in parts per trillion. We thought they might emphasize the words in the clause that refer to "ingestion" and also to "appropriate tests" — words that have been consistently avoided by the FDA attorneys. Unfortunately, about the time we were talking with some of Delaney's people, the cyclamate ban came along and, of course, it was based on the Delaney Clause. To me this was a tragic situation for there really was no basis for it. In any event, Delaney then found it was very newsworthy. In fact, he went on TV with Gloria Swanson, and said it was the best thing he ever did. When a person does this, it is not going to be very easy to induce him to back up. So, there it stands. However, it may be possible to modify the interpretation of the Clause and this, I think, is going to be the answer in the end. I will come back to this later.

We talked about establishing a clearinghouse for information, and I think EPA is doing this. We also indicated there should be incentives for industry, but thus far nothing much has been done about this, although I must say that there is an indemnity clause in the new pesticide bill which is a great concession in the proper direction.

We talked about working closer with the State governments, and this is something that I know Mr. Ruckelshaus wants to do.

Finally, we suggested better international efforts with respect to communicating information. This is something for the future.

These, very briefly, were the recommendations and I am pleased to say that I believe that EPA is considering them very seriously.

When the Secretary's new advisory committee was formed, it immediately started studying some very broad problems. It was apparent to this Committee that there was a need for a generally accepted and reliable protocols for testing for safety. People were and still are using all types of tests, some very superficial, some unbeliev-ably irresponsible, some almost based on opinion.

It was quite apparent to us that the cyclamate work was not really reliable and that the ban that took place was largely at the instigation of the company manufacturing it because of the fear of suits. The work done on this material was repeated by the British, and they were not able to confirm the observations made in this country. There are all kinds of questions with respect to what the results obtained in USA really meant. These questions related pretty much to the methods used and the interpretation.

But that was not all. There was the work on 2,4,5-T and its ostensible teratogenic effect. When the bionetics people ran tests with 2,4,5-T, they thought they observed teratogenic effects. This may have been true because the samples used were made by a company that had discontinued making 2,4,5-T. Apparently in the manufacture of substances such as 2,4,5-T, the control of heat during the manufacturing process is extremely important. If not handled properly, a group of chemicals known as diaoxins are formed, and these are indeed toxic. This was well known by the largest manufacturer of 2,4,5-T and this company eventually eliminated practically all of this contaminant. This came out in our Commission hearings, and it was decided at that time that it would be desirable to have more tests on 2,4,5-T. As a consequence, the Food and Drug Administration, in collaboration with the Dow Chemical Company, had tests underway in Indianapolis under the supervision of the Food and Drug Administration. Frequent observations were made, and it appeared that when ingested under controlled conditions, the animals were fairing very well. Unfortunately at this time, Senator Hart become interested in 2,4,5-T — perhaps because it was valuable from a publicity standpoint. Hart had hearings and called in people from the Public Health Service to testify. The Surgeon General spoke and indicated that in a few weeks there would be results from tests being conducted at the NIEHS. It appeared that he would not be able to deliver in time to satisfy Senator Hart. However, the NIEHS, under the direction of Dr. Paul Kotin, carried on tests whereby they injected 2,4,5-T into the intraperitoneal cavity of hamsters, in a carrying agent called DMSO. The latter, as you may know, is in itself a very active chemical and is said by some to be a teratogen. In any event, is it logical to test teratogenesis by injecting it in the carrying agent into the intraperitoneal cavity? What does it mean? Dr. Kotin took the results to Washington and gave them to the Surgeon General and they were discussed in secret by a number of people. I judge those who first saw the results felt they tended to verify the view that 2,4,5-T was a very toxic material. The results were later made public, but not before the Surgeon General had called in a group to discuss the demise of this chemical.

I happened to be in Washington at the time a young man by the name of Pauley was working for Under Secretary Veneman. He indicated to me that there was a meeting down the hall that I might find interesting and why didn't I join him in attending it. This I did, and when I entered the room, I was astounded to find they were discussing 2,4,5-T and its demise. The Surgeon General was giving assignments as to who would prepare various sections of the death knell report. This

was indeed a shocking situation for Ted Bierly of the Department of Agriculture was there. He wasn't certain of what 2,4,5-T was used for, but phoned his office to find out. In any event, the ban was to be placed on 2,4,5-T because of an imminent hazard. This bothered me for I wanted to know what was meant by imminent hazard. I also wanted to know if they had data, and the Surgeon General said yes, and I was shown a table that did not mean anything to me. The Surgeon General did point out, however, by now it might be all over town and I can't say I appreciated the statement.

It so happened that some time after I entered the room, where 2,4,5-T was being discussed, Bill Darby came in. He asked about the results and was given the table. He then proceeded to question Dr. Kotin on the results and, to my astonishment, Dr. Kotin could not answer him. He said he did not know.

In any event, Dr. Darby's analysis was that the data was really unreliable and should go back to the laboratory for more data. There were serious questions as to whether or not they had enough animals to do the work and to come up with reliable results. In any event, here was the material that went to the Hart Committee.

On the other side of town, about the same time, the Office of Science and Technology was studying 2,4,5-T and they came to the conclusion that it was a harmless substance and was even ready to recommend a tolerance in parts per billion. This, of course, was ignored, because at that time people were supposed to find something wrong with chemicals rather than something good about them.

When the EPA was formed, they established a committee to study 2,4,5-

T, and the total committee, except one statistician, recommended its continued use. The report was given to four or five other individuals, including myself, and we all, independently, recommended approval of the majority report. In spite of this, its use was cancelled, which means that there were restrictions placed on its use. This is still a mess and is in the courts for clarification.

I believe today, however, this type thing is far less apt to happen; because EPA is much better organized and has a better understanding for the need for scientific input. Nevertheless, it is this type thing that helped create a very chaotic situation.

But, those were not the only chaotic situations. There was the question of the safety of accent. It was injected in the base of the brain of young monkeys and adverse effects were noted. There was a situation of brominated oils which the Canadians said caused heart lesions and which were later disproved by the British. There was the matter of nitrosamines which is still unsettled.

There is some evidence that coffee is carcinogenic. In the meantime, Diethylstilbestrol has been greatly restricted and Diethylpyrocarbonate is no longer used in wines and juices. The methods involved are not only those of testing for safety, but for analyses. This is well manifested of what eventually came out of the DDT story. I would like, therefore, to go back to DDT.

Since our report was written, a great many studies have been made, and it is quite clear that the Food and Drug Administration confused DDT and Polychlorinated biphenyls in its analyses for DDT. It was clear too that the material that is causing the death of preborn mink are PCB's rather than DDT. This has been well demonstrated. It has also been demonstrated that what was thought to be DDT in milk, poultry, and fish, were actually PCB's. Why did this happen? It happened because we did not have satisfactory methods of analyses or perhaps those making analyses were sloppy.

It may surprise you to know that the Campbell Soup Company first called the presence of PCB in chicken to the attention of the USDA and the FDA. It is for this reason, therefore, we need better methods of analyses. So, in addition to protocols for testing for safety, we need procedures for analyses.

We can consider the mercury story and tell the same story. To make it brief, I might point out that the Dow Chemical Company, in running tests by the atomic absorption procedure pointed out to the Food and Drug that their wet methods of analyses was only picking up about half the mercury present. The FDA studied the matter and found it was true and changed to the new method of analysis, but retained the tolerance at a half part per million — which really meant they cut the tolerance in half.

The SPAC committee studied very carefully the possibility of the government's setting up a laboratory to develop protocols for methods of testing for safety and perhaps monitoring. Why would such a laboratory be necessary when the government already has so many laboratories, and especially the Cancer Institute and the laboratories at the Triangle in North Carolina. The facts of the case are, the people carrying on work in these laboratories are dedicated to fundamental work rather than mission-oriented work. You might say then, why doesn't the FDA and

EPA do this work? Here again, these are enforcement agencies and have relatively little money for research. Because of this, then, it was decided that a laboratory should be established and operated by both FDA and EPA. Dr. Norton Nelson of our Committee studied the matter very carefully and recommended that it be established at the Pine Bluff, Arkansas, Arsenal which was being decommissioned as a biological warfare facility. We looked at Fort Detrick, but it was too costly and too large.

The recommendation went to the Office of Science and Technology, which was enthusiastic about it. It went through HEW and here a number of the people, including the Surgeon General, and people in the Cancer Institute tried to kill it. The Commissioner of the FDA was in favor of it. In attending one HEW meeting at which it was being discussed, Bob Pauley really kept the thing from being killed. The people in the Cancer Institute and the Surgeon General did not want it to go ahead, because it conflicted, in their opinion, with the work of the Cancer Institute. I just do not believe this is true.

In any event, the OST went to the President, and he approved the conversion of the Pine Bluff facility into the National Center for Toxicological Research. At present there is great enthusiasm for this in the Food and Drug, but I am not sure of the EPA interest in this facility. I should point out that the senators and congressmen (especially Wilbur Mills) have our aces in the hole in this matter.

Eventually the Environmental Protection Agency was formed, and this was done by cannibalizing a great many departments. When this is done, it is a common procedure in Washington to cast off the undesirable people, let them go to the new agency. Because of this, EPA has been burdened with a lack of talent in some areas. I do believe, however, that in general its Administrators are excellent people. I have great confidence in Ruckelshaus and Dominick, in particular. When this reorganization was made, I could not help but thinking of a statement made by the Roman Petronius 2000 years ago. He said: "We trained hard — but it seemed that every time we were beginning to form into teams, we would be reorganized. I was to learn later in life that we tend to meet any new situation by reorganizing; and what a wonderful method it can be for creating the illusion of progress while producing confusion, inefficiency and demoralization." In a way this is almost what happened to the EPA to start with, but I must say that the able administration is pulling it out of this chaotic situation into a very fine working body.

I indicated that the SPAC committee was transferred to EPA and it was termed the HMAC, Hazardous Materials Advisory Committee. It was located under the Administrator for Science and Monitoring. It is strange, however, that since it has been active in EPA that most of its contact has been with Mr. Dominick who is concerned with pesticides, hazardous materials, solid waste and radiation.

We have also had contact with Mr. Marshall Miller in the Administrator's office and to some extent with the people concerned with water. It has very little contact with the Office of Research and Monitoring.

I have thought about this situation a great deal and wondered why. I think the nature of our committee — the scientists, the business people, in general the people of great distinction, are really more concerned with policy than molecules. In line with this, matters relating to policy have been referred to us. I believe, in general, we have handled them well. Furthermore, since EPA is an operational and enforcement organization, research is really a secondary activity. It does not, therefore, have the place in the total picture of operation that the other programs do. I think this is the reason our committee has been so concerned with the policy matters, such as what about the use of chemicals around food plants, what about aerosols, what about biological control replacing pesticides, what about pesticide production, cost, and so on, and now — what about herbicides? Are they used carelessly? Do we need them? What about the toxicity. This is a matter being studied quite extensively now.

No matter what one might say, when it comes to decisions in government, there must be good and considered decisions based on good judgment. Unfortunately, this isn't always the case. Very often emotions and pressure are unbelievable. Furthermore, the lay scientist comes into the picture, and how is the average person to know whether a scientist of great distinction speaking out of his field is reliable or not. In this connection, Bill Darby made the following very interesting statement in his AMA Symposium: "Evaluation of the desirability of use of various food additives, chemicals, or processing techniques demands a specialized background of knowledge and experience, without which scientific judgment falls short of what is desirable in the public interests. Public misinformation about food safety is an inevitable consequence of misplaced confidence of scientists in their individual ability and authority to pass opinions on the question of food safety. We, as scientists, must recognize that outside our own field, or specialized background and knowledge, we too are laymen. We must help in concert with others of diverse and different scientific, professional and public backgrounds to formulate policy. The individual scientist is unlikely to be so all wise that he should dictate the policy decisions and public positions." This to me is a very important statement.

The news media take up what these scientists say, and since some of them are almost unscrupulous in their statements and accusations, they are dramatic. A person of distinction once said that the news media not only want dramatic things, but they want to respond to the public and since the public seems to like lies better than the dull truths, they give them lies. There is indeed a need for sound information, and if we could only make it dramatic. There is a need to give the bureaucrats strength, and I believe support the sound views and those who may have them.

I believe Dr. Darby has found a way to do this. In the case of saccharin, he brought in scientists from all over the world who were doing research on the carcinogenesis of saccharin. These people sat down at a table to discuss their methods, their work, and so on for about three days. Food and Drug officials from Canada and the U. S. were present. Out of the fourteen researchers present, only one had obtained positive results. It was then agreed that there should be another meeting in about eight months and this will be held next March. In the interim, the FDA has decided to take no action on saccharin. This has given them support and strength to withstand the pressures and emotions to which they are exposed.

A meeting was held in Nashville on sugar and nutrition. I believe this Symposium will do much to allay the fears of those who would be inclined to follow people like Yuhdkin and others who would ban the use of sugars — the answer being, of course, moderation and a balanced diet. Just don't eliminate anything termed empty calories, whatever they might be. To me this is a ridiculous, catchy phrase — for what is an empty calorie? Is a vitamin alone an empty calorie? At any rate, the Commissioner of the FDA has talked to Darby about having a world conference sponsored by the Nutrition and other foundations, where the safety of foods will be discussed. I just hope that this will give the FDA Commissioner great strength in making future decisions — decisions that won't have to be like those relating to diethylstilbestrol.

Finally, I would like to end up by saying that we must broaden the views of the bureaucrats so we do not fall into the trap of reductionism. On the other hand, they must not divest themselves of some evidence of depth as a result of being victims of "reductio ad absurdum."

Medical and Toxicological Research on Dieldrin and Related Compounds

By **J. D. Jansen**

Shell Internationale Research Maatschappij, B.V., The Hague

Currently in the U.S.A. Public Hearings on dieldrin are underway which, according to inscrutable American custom, will try to settle in court a fundamental toxicological problem. However, whatever the outcome of these hearings, I sincerely believe that the problem is fundamental and that much of the development in industrial toxicological concepts and insights will be greatly influenced by its scientific solution.

Toxicology has for may years been exclusively concerned with poisons, which are those chemicals or mixtures which actually killed people. The study of the hazards was not of prime importance because the poisoned patient or dead body was there to prove the hazard.

In industrial toxicology the initial problem is different: we study the toxicology of a new chemical in order to be able to assess the magnitude of the future hazard of this chemical of which by definition we have no, or hardly any human medical experience. Therefore the methods used must be good enough to *forecast* dangers and it is not enough if they afterwards confirm in selected test animals what we already know from bitter medical experience.

Whether these methods are presently good enough may be illustrated by dieldrin, which may now be one of the toxicologically most fully investigated chemical compounds. As in all research each problem solved gives rise to further questions which are further investigated so that research is still going on. I do hope that the following, very general, description may give you some idea of the present status of knowledge in this field.

Two hundred and fifty years ago Alexander Pope (1688—1744) already said it:

"Know then thyself, presume not God to scan."

"The proper study of mankind is man."

From the very beginning in 1952 we have been involved in the health protection of our own — and some government-workers, handling dieldrin in its use in Public Health vector control programmes (i.e. spraying houses and premises in yellow-fever — and malaria eradication campaigns). Toxicological Research including long-term feeding studies on rats and dogs had been carried out in the U.S.(1) From 1954 onwards we have also been concerned with the health protection of our industrial workers exposed to aldrin and dieldrin in the manufacture and formulation of these compounds at the Shell Chemical plants at Pernis, near Rotterdam in the Netherlands.

Complete periodical clinical and laboratory examinations of the workers were carried out at least twice yearly or more frequently when indicated by complaints or intercurrent illness. Nevertheless, during the earlier years of production cases of intoxication, some with convulsions, occurred. In an attempt to be forwarned against these convulsive intoxications before they would occur, electro-encephalographic studies in co-operation with the neurology department of the Rotterdam University Teaching Hospital (Dr. M. de Vlieger) were included in the routine examinations of the insecticides workers. Thus the medical supervision and study of these (Pernis) industrial workers was expanded as new knowledge and clinical possibilities permitted, often with help from the medical faculty of the Dijkzicht Hospital.

Simultaneously toxicological and biochemical studies with animals were being carried out at our Tunstall Laboratory. Shell also sponsored such research at Utrecht University (the Netherlands) (Prof. H. van Genderen) (2) and sponsored research on the metabolism of dieldrin and allied compounds at the University of Bonn, Germany (Professor F. Korte).(3)

This latter study showed that, contrary to an — until that time — popular belief dieldrin is broken down and metabolized and that on continuous feeding a continuous accumulation in the mammalian body does not occur. On the contrary, after a certain time a plateau is reached in the concentration of dieldrin in the body after which the average amount daily absorbed is also excreted every day.

Assessment of the exposure or rather of the absorption in the Pernis workers by measuring dieldrin concentrations in the air or on the workclothes proved unpractical and unreliable. Therefore in our Tunstall Laboratory a sensitive analytical method was developed which enabled the determination of the very low concentrations of dieldrin in fat, blood and other tissues of man and animals.(4, 5, 6) Not only did this method enable precise monitoring of the general population but in addition a study on the pharmacokinetics in man could be initiated with our own scientists participating as volunteers.

From this study with human volunteers (7, 8) three important aspects became evident:

1. Intake of 232 μg of dieldrin per day over a period of two years had no effect on the health of man.
2. The plateau effect with respect to tissue accumulation as seen in laboratory animals was also confirmed in man.
3. A mathematical relationship was established between the average daily intake of dieldrin and its concentration in blood and fat.

The validity of this relationship or ratio was confirmed by calculating the daily intake of the population from the analytically determined concentrations of dieldrin in the bodies of people in the U.K. and to compare these with independently determined residue actually present in the representative daily diet in the U.K.

Once this ratio between intake and concentration in the body was known it could be applied to determine the total equivalent oral intake in the industrial workers and thus relationships between clinical and laboratory findings and total intake could be established. The routine supervision of the industrial workers at Pernis enabled us to correlate signs and symptoms as well as the absence of toxic manifestations with concentrations of dieldrin in the blood.

This continuing study of the Pernis workers, carried out with the help of Tunstall Laboratory, an outside laboratory (Searle) in the U.K. and the Rotterdam ("Erasmus") University teaching hospital enabled the establishment of a no-effect level in man.(9)

This no-effect level in man still stands and is in accord with the ever more exhaustive parameters introduced later. Since it had become evident from studies with laboratory animals that perhaps the earliest and reversible effects appear in the liver, about 10 years ago liver function tests were added to the routine periodic medical examinations of the workers. At about the same time determination of insecticide concentration in the blood became part of the routine. From 1969 onwards an auto-analyzer, measuring 12 bio-chemical parameters, linked to a computer, has been used.

At this moment in time about one thousand workers have been employed full-time for a shorter or longer period in the Pernis chemical plants where they have had exposure to the insecticides. Everyone of them had been under continuous medical supervision by periodical medical examinations during that period. In this connection it might be useful for a better understanding to realize that this group of workers did not consist of carefully selected perfectly healthy young men. Female workers, however, are not employed in chemical plants in our country. The age of the male workers ranged from 22–64 years and they were subject to the same infirmities, ailments, diseases and accidents to the same extent as occured in comparable groups of workers at the same site. All workers either with 4 or more years of exposure or those who had had an intoxication and who remained under this strict regimen of close medical supervision.

Over these 19 years four of these people have died:

One pensioned-off employee died in an automobile traffic accident. He was a passenger.

One operator died when his motorbike was involved in a traffic accident.

One day-foreman with comparatively little exposure died from carcinoma of the stomach.

One contractor-employee died from a myocardial infarction.

Inevitably over the course of these 19 years some (52) of the workers dropped out on account of changing employer, moved to other parts of the country or emigrated. Each and everyone of these "drop-outs" have recently been contacted again and it was ascertained that the "dropping-out" had not occurred on account of ill-health.(10) They are all alive and all in good health according to their own evaluation.

Bio-chemical- and electron-microscopical studies in rats, dogs, mice and monkeys revealed that the earliest, and readily reversible, effect of exposure to dieldrin is induction of liver microsomal enzyme systems. Therefore, to find at what level of intake this response can be demonstrated in man, tests for the activity of these enzymes were included in the medical examinations of members of the long-term, high-exposure group of workers.

In these people a no-effect level in man was established commensurate with a dieldrin concentration in the blood of 0.2 ppm. This dieldrin level in the blood corresponds with an equivalent oral intake of 33 μg/kg per day or about 2 mg/man/day.

This level is also a no-effect level with respect to the activity of alkaline phosphatase in the bloodserum, the increase of which may be an early indicator of enzyme induction.(11)

By the time other parameters for enzyme induction were included into the tests of medical supervision the exposure of the workers was further reduced by improved industrial hygiene and there were no workers with a dieldrin concentration in the blood exceeding 0.15 ppm and later 0.105 ppm. These levels correspond to an equivalent oral intake of 25 and 17.5 μg per kg per day respectively or 1.5 and approx. 1 mg/man/day.

This microsomal enzyme stimulation is an effect which becomes apparent in a short time both in man (12) and in the mouse where it occurs almost immediately.(13)

In the meantime Tunstall Laboratory had performed 2-year tests with dieldrin on rats and dogs (14) which showed no evidence of carcinogenic activity and Kettering Laboratory finished a 6-year study in monkeys, also with negative results.

However, Tunstall's first long-term study with mice confirmed the findings of Davis and Fitzhugh i.e. tumours were produced in the liver. It was felt we had to go to the scientific community to seek their opinion and their counsel. To this end a Round-Table discussion was organized at the Tunstall Laboratory to which the leading toxicologists in the world were invited and most actually participated. The Joint Meeting of the FAO Working Party of Experts and the WHO Expert Committee on Pesticide Residues in their 1966 meeting made a toxicological evaluation of dieldrin and having considered all the known facts estimated the Acceptable Daily Intake for man to be 0.0001 mg per kg body weight or about 6 μg/man/day (which may be compared with the level of about 2000 μg/man/day, established in our own workers). The following year, at the Joint Meeting 1967, the Commitee containing several other toxicologists, adhered to the toxicological evaluation published in the last report.

Our own studies with mice have been and are being continued.(15, 16) The results have been compared with those from studies in other species in which similar morphological electron-microscopical and bio-chemical studies were undertaken. In addition, in our and other laboratories, in mice and other species feeding studies were repeated with compounds other than dieldrin, notably phenobarbital. It is now established that the tumours produced by 1000 ppm phenobarbital in the feed of mice are the same as those produced by 10 ppm of dieldrin in the feed of the same mice. The same applies to 200 ppm β-BHC, 400 γ-BHC and 100 ppm DDT. Again the results were discussed at a second Round-Table in 1970. These results have been published, but part of this work is of course still continuing.

At that point in time the crucial question of the relevance of extrapolation of these laboratory data in mice to man became of course of the greatest importance.

The interpretation of these mouse tumours in relation to their significance to man depends not upon the results of any single study but upon a synthesis of several considerations. These are:

1. The response of the liver in various species to dieldrin.
2. The response of the liver in various species to some other compounds.
3. Experience from clinical medicine
4. The results from a continuing study in industrial workers with high exposure.

It has been shown that:
1. In the mouse the response of the liver to dieldrin is different from that in the rat, the dog and the monkey.

2.1. In some strains of mice DDT, alpha-, beta- and gamma-BHC and phenobarbital are associated with formation of primary livertumours just like dieldrin.

2.2. The livertumours in the mouse caused by phenobarbital are in every respect similar to those caused by dieldrin.

2.3. In the three species studied in this respect i.e. the dog, the rat and the mouse the early structural changes in the liver caused by dieldrin, as seen with the light- and electron-microscope are similar for each species to the early changes caused by phenobarbital.

2.4. These early and identical changes caused by dieldrin and phenobarbital are associated with increased microsomal enzyme activity in the liver in all species tested.

2.5. In the mouse the response to dieldrin and phenobarbital ultimately leading to tumour formation begins with liver enlargement. Other species react with liver enlargement but develop no tumours. Man and monkey appear to have neither response.

3.1. Human hepatomata develop mainly in cirrhotic livers where the process of cirrhosis is associated with clearly disturbed liver functions and precedes manifestations of the tumours by several years.

3.2. In the human primary liver cancer the presence of alpha-feto protein is associated with and might precede clinical signs of the tumour.

4.1. The continuing epidemiological and clinical study of the Pernis workers gives no indication whatsoever of a carcinogenic effect. In this carefully studied group of heavily exposed people, observed over a period of up to 19 years only one cancer of the stomach has occurred, after a relatively short period of comparatively mild exposure. This is an incidence less than expected in a comparable group of the general population.

4.2. In addition in this group of people there is no disturbance of liver-function and there is no detectable alpha-feto protein in the blood-serum.

4.3. In all species of laboratory animals studied (mouse, rat, dog, monkey) alterations in enzyme activity occur prior to the changes in liver size or morphology. In man at an equivalent oral intake of 25 μg per kg body weight (250 times the average daily intake in the general population), no changes in enzyme activity occur.

4.4. Druckrey found a dose/time response relationship in exposure to carcinogens (i.e. the higher the dose the earlier the tumour appears, the smaller the dose the later the tumour appears). According to this knowledge the above period of 19 years of high exposure in which no tumours developed would correspond to a longer period (in which no tumours would develop) at the current low exposure level of the general populations.

4.5. The results of this continuing clinical and epidemiological study in 230 workers over 19 plus years of course is not proof in itself of the non-carcinogenicity of dieldrin in man.

5.1. Such proof can only be given if a proper epidemiological study could be performed upon a large group of people, exposed for a life-time to dieldrin or any other compound which produces similar tumours in mice.

I assume that the next speaker may have some information on this question. Further work will establish whether the concepts above are confirmed and, if so, these will become part of the body of toxicological science.

If this is so, then Dr. Deichmann's view would be fully confirmed as he expressed it in a lecture in Kiel on DDT with the words: "I believe it is obvious that we need to review our testing procedures and the interpretation of data." It would also have a marked effect on our notion of what are human carcinogens, inevitably focusing our future research on the problems which remain established as real dangers to human health, which would be fewer in number but clearer defined and therefore more amenable to well-conducted research.

It might be that in the case of liver tumours in some strains of mice, these strains could be at fault and not the compounds.

References

1 Treon J. F., and F. P. Cleveland: Toxicity of certain chlorinated hydrocarbon insecticides for laboratory animals, with special reference to aldrin and dieldrin. Agric. and Food Chem., vol. 3 (5) May 1955, 402–408.

2 van Genderen, H.: The toxicology of the chlorinated hydrocarbon insecticides. A progress report with particular reference to the qualitative aspects of the action in warm-blooded animals. Mededelingen van de Landbouwhogeschool te Gent, 1965, deel XXX, No. 3, 1321–1335.

3 Ludwig, G., J. Weis, and F. Korte: Excretion and distribution of aldrin − 14C and its metabolites after oral administration for a long period of time. Life Sciences, vol. 3, 1964, 123–130.

4 Robinson, J.: The determination of dieldrin in the blood by gas liquid chromatography. Paper to be given at the 3rd International Meeting in Forensic Immunology, Medicide, Pathology and Toxicology, London, April 16–24, 1963.

5 Richardson, A., J. Robinson, B. Bush, and J. M. Davies: Determination of dieldrin (HEOD) in blood. Archiv. Envir. Health, vol. 14 (5) May 1967, 703–708.

6 Robinson, J., A. Richardson, and J. M. Davies: Comparison of analytical methods for determination of dieldrin (HEOD) in blood. Archiv. Envir. Hlth., Vol. 15 (1) July 1967, 67–69.

7 Hunter, C. G., and J. Robinson: Pharmacodynamics of dieldrin (HEOD). I. Ingestion by human subjects for 18 months. Archiv. Envir. Hlth., vol. 15 (5) Nov. 1967, 614–626.

8 Hunter, C. G., J. Robinson, and M. Roberts: Pharmacodynamics of dieldrin (HEOD). II. Ingestion by human subjects for 18 to 24 months, and postelposure for eight months. Archiv. Envir. Hlth., vol. 18 (1) Jan. 1969, 12–21.

9 Jager, K. W.: Aldrin, dieldrin, endrin and telodrin. An epidemiological and toxicological study of long-term occupational exposure. Amsterdam, 1970, Elsevier Publishing Company.

10 Versteeg, J. P. J., and K. W. Jager: Long-term occupational eposure to the insecticides aldrin, dieldrin, endrin, and telodrin. Brit. J. Ind. Med., vol. 30, 1973, 201–202.

11 Fujimoto, J. M., F. Eich, and H. R. Nichols: Enhanced sulfobromophthalein disappearance in mice pretreated with various drugs. Biochem. Pharmac., vol. 14, 1965, 515–524.

12 Kolmodin, B., D. L. Azarnoff, and F. Sjöqvist: Effect of environmental factors on drug metabolism: Decreased plasma half-life of antipyrine in workers exposed to chlorinated hydrocarbon insecticides. Clin. Pharm. Therap., vol. 10 (5) 1969, 638–642.

13 Wright, A. S., D. Potter, M. F. Wooder, and C. Donninger: The effects of dieldrin on the subcellular structure and function of mammalian liver cells. Food Cosm. Tox., vol. 10, 1972, 311–332.

14 Walker, A. I. T., D. E. Stevenson, J. Robinson, and E. Thorpe: The toxicology and pharmacodynamics of dieldrin (HEOD). Two-year oral exposures of rats and dogs. Tox. and Appl. Pharmac., vol. 15, 1969, 345–373.

15 Walker, A. I. T., E. Thorpe, and D. E. Stevenson: The toxicology of dieldrin

(HEOD). I. Long-term oral toxicity studies in mice. Food Cosm. Tox., vol. 11, 1973, 415–432.

16 Thorpe, E., and A. I. T. Walker: The toxicology of dieldrin (HEOD). II. Comparative long-term oral toxicity studies in mice with dieldrin, DDT, phenobarbitone, β-BHC and γ-BHC. Food Cosm. Tox., vol. 11, 1973, 433–442.

Ecological Chemical Evaluation of Waste Treatment Procedures

The Behavior of Xenobiotics in Waste Composting

By **Werner P. Müller, Friedhelm Korte**

Institut für Ökologische Chemie der Gesellschaft für Strahlen- und Umweltforschung, MBH München

Abstract

The object of this paper is to evaluate technical treatment of waste under ecological-chemical aspects. For this reason, the role of waste treatment as a critical factor in the environmental load by xenobiotics has been investigated, the concept "xenobiotics" is defined and illustrated by examples. Determining criteria for "environmental conformity" of waste treatment methods are compiled. The technology and microbiology of waste composting, which is regarded as a method "friendly" to the environment, are presented in short form.

In the experimental section the appearance and the behavior of organic xenobiotics in waste treatment through composting are investigated. Municipal waste in the Federal Republic of Germany contains polychlorinated biphenyls (PCB's) in concentrations between 0.4 and 9.7 $\mu g/g$. With the help of a simulation apparatus, whose construction and function are described, it was found that 2,2'-dichlorobiphenyl (PCB) remains unchanged in the waste composting process to at least 98%. Similar results were obtained from experiments with cyclodiene insecticides (aldrin, dieldrin, heptachlor), herbicides (buturon, monolinuron), fungicides (imugan), chlorinated anilines (p-chloroaniline) and fused-ring hydrocarbons (benzo-[a]-pyrene), which are either not converted at all or are converted only in negligible amounts. Some of the degradation products (e.g., dieldrin, p-chloroaniline) are even more persistent than their starting materials.

As a conclusion, these results suggest that composting, at least with respect to the investigated classes of substances, does not provide suitable treatment for decontamination of waste. The environmental conformity and thus the value of this waste treatment procedures are questioned under ecological-chemical criteria.

Ecological chemistry as a discipline of environmental research concerns itself with material balances in the ecosphere, the material exchange between the individual ecosystems and the material conversions within these systems. Emphasis is placed on investigation of the displacement of the ecological dynamic equilibrium through exploitation of resources on one hand and distribution of anthropogenic materials on the other. An essential aspect is the behavior of organic xenobiotics, i.e. substances which can only enter the environment by human activity. In this article the behavior of selected organic xenobiotics by treatment of solid wastes is investigated with the aim of

Translated by: Dr. Laeticia Kilzer, Mount Marty College, Yankton, South Dakota 57078, U.S.A.

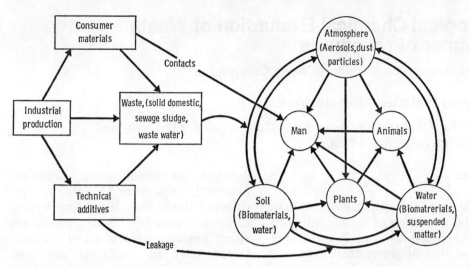

Fig. 1 *Sources of environmental contamination and significant transport routes of xenobiotics in the environment.*

evaluating the technique of waste composting under ecological-chemical criteria.

The natural dynamic equilibrium in present industrial countries is constantly being disturbed by two trends: first, the increasing depletion of raw material reserves by the inherent requirements of steadily expanding industrial production, second, the threatening increase of environmental pollution through increasing waste amounts. The discernible shortage of raw materials will lead inevitably to the opening of new sources of raw materials and to stronger regulations aiming at an extension of limited resources. As a consequence of these partially counteracting developments, it can be expected that at one time there will be increased contamination of the environment through xenobiotics especially in the form of waste products, at other times better recycling techniques will be developed for the re-utilization of waste materials in order to preserve limited raw material resources. Here waste treatment has a central role since

the treatment techniques determine extensively the nature and extent of the recycled and the discharged materials. Under these aspects there must be a comparison of various waste treatment procedures with a view toward their environmental conformity (compatibility) whereby hygienic, toxicological, and ecological-chemical criteria can be referred to for evaluation. Thus hypotheses may be developed in which ecological aspects (besides economic, geographic, and sociological implications) may be assessed in the choice of appropriate waste treatment methods.

The main concern in the intended ecological-chemical examination of various waste treatment methods, is the occurrence and behavior of organic xenobiotics which are found in varying concentrations in communal and industrial waste.

1. Xenobiotics in the Environment

Xenobiotics are materials of anthropogenic origin which can enter the environment only by the direct activity of man, intentionally or non-intentionally. Xenobiotics include inorganic and organic compounds; in the following, only organic xenobiotics will be examined.

Organic xenobiotics belong to distinct chemical classes and are used in various areas of application. The scale ranges from pesticides through synthetic products (polymers, adducts, and condensates), stabilizers, antioxidants, dyes, detergents, emulsifiers, fuels, fuel additives, synthetic resins, lacquers and gums, food preservatives, to drugs. Pollutants freed by technical processes, such as fused-ring aromatic hydrocarbons from burning processes, and also the numerous conversion products of xenobiotics which exist in the environment as the result of biotic and abiotic processes, belong to this class.

The total production of these xenobiotics is constantly increasing and as a consequence an increased burden of the environment where most xenobiotics finally arrive whether it is during production (e.g., side products from pharmaceutical production), through discharge at the area of use (e.g., pesticides), through physical processes during use (e.g., solvents), as waste after use (e.g., synthetic products), or by accidents independent of use (e.g., PCB's).

In the environment the pollutants are taken up and transported by soil, water, and air (Fig. 1). They are exchanged by evaporation, condensation, solution, adsorption and desorption. Global transport proceeds mainly through convection in the atmosphere and the major water currents. Thereby xenobiotics are either dissolved or ad-

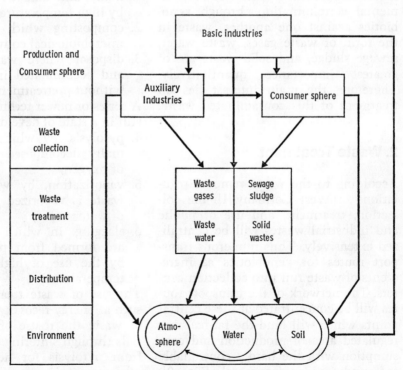

Fig. 2 *Localisation of waste treatment in the scheme of transport routes of xenobiotics.*

sorbed on support materials such as aerosols and suspended matter.

A considerable number of xenobiotics are not easily susceptible to biological degradation. After assimilation of the pollutants in the environment, an accumulation of these substances in organisms can occur. Via the food chain, the resorbed xenobiotics are transported further so that in carnivores, at the end of such a chain, acute concentrations are reached occasionally.

Already at very low concentrations (ppm range) a noticeable displacement of the dynamic equilibrium in an ecosystem (e.g., through freeing of ecological niches by insect control) becomes apparent, because of their extensive toxic behavior (growth inhibition, immune suppressive effect, teratogenic, mutagenic, and carcinogenic behavior, hormonal interference).

If one compares the sources of environmental contamination through xenobiotics against one another, waste in the form of waste gases, waste water, sewage sludge, and solid waste are of greatest importance quantitatively. Therefore, the main interest lies on treatment of this contaminated waste.

2. Waste Treatment

According to the waste removal regulations in West Germany (1), the collection, treatment, and use of public and industrial wastes will be centralized extensively. Thus numerous transport routes for xenobiotics as ingredients of waste run into collection centers. The network of the transport routes will be concentrated at these central points which will bind the far-reaching regulated area of production and consumption with the hardly controllable surroundings (environment) (Fig. 2).

Consequently, waste treatment is a critical factor in the control of this aspect of environmental pollution. At these central locations the actual treatment method determines both quantity and quality of xenobiotics being recovered or released in the environment. Therefore, regulated intervention in these locations allows an effective influence on environmental quality and safety.

In the treatment of solid wastes, just as in waste gas purification and waste water clarification, various techniques are available (Fig. 3). In chemical conversion treatment, recycling processes (2) can be introduced where raw materials such as paper, glass, synthetics, ferrometals, etc. can be removed. In further treatment of wastes, the FRG uses mainly three techniques (3, 4):

1. incineration where waste is thermally and oxidatively decomposed by high temperature,
2. composting which uses controlled microbiological conversion,
3. disposal where waste is removed and accumulated in suitable areas without pretreatment.

A series of newer techniques, which are still in a state of development, are:

4. pyrolysis, by which waste is thermally decomposed in the absence of oxygen,
5. vaporization, by which pulverized waste is vaporized in the presence of water vapor,
6. slagging, in which sintered stones are formed from pulverized waste by the use of high pressures and temperatures.

The use of waste treatment products, such as energy recovery by incineration of waste the re-use of building materials through sintering, the use of gases from pyrolysis for heating, etc., are other aspects in recycling.

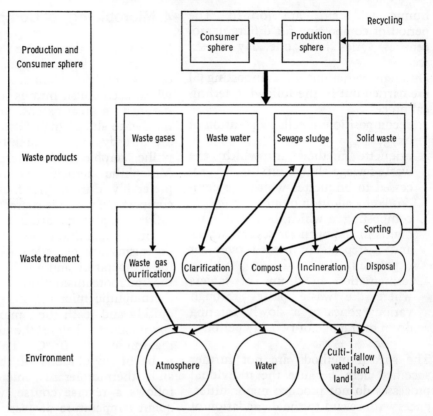

Fig. 3   Production and treatment of wastes.
Pathways of waste treatment products in the environment.

Procedures for special waste which are formed mainly in industrial processes and which require specific treatment (salt deposits, solvents, scrap metal, etc.), are not considered here.

Frequently, the mentioned procedures are combined with one another such as composting with combustion. Waste disposal has a special position among the mentioned procedures because all waste treatment techniques give solid residues which are finally released into the environment.

3. Technology of Composting

Composting methods utilize microbiological decomposition of dead organic materials by saprophytic bacteria and fungi. In the technical procedures, the waste is generally pulverized by grinding, freed from iron metal by a magnetic separator, and finally mixed vigorously. Sometimes up to 40% by weight sewage sludge can be added. The moisture content of the materials is adjusted to about 40%. In the subsequent decaying process, the waste-sewage sludge mixture is decomposed under aerobic conditions by saprophytic microorganisms. In the microbial metabolism processes large amounts of heat are liberated which leads to a temperature increase up to 70° C. In addition, metabolic products such as CO_2, H_2O, other mineral products, humic acid, etc., are formed. The period of composting for different procedures varies from one day to six weeks.

Different variations of composting (5) are carried out in the following technical processes:

1. decomposition in cells by addition of oxygen
2. briquette method in which the waste-sewage sludge mixture is processed to briquettes which are composted by addition of air
3. composting in which the material remains in the open air several weeks and is repeatedly turned over for exposure to air
4. fermentation in drums or towers where the waste passes through various zones of a slowly rotating drum over a one or two day period

The special methods are not further specified here, because the individual processes do not produce major differences with regard to the degradation of waste.

In a post treatment of the composting products the undecayed portion is separated and the resulting "fresh" compost is pulverized. Through "post-composting" one finally obtains the "ripe" compost.

The compost resulting from the biological decomposition process is used generally for the improvement of water retention and structure of soil. This finds use in agriculture, especially in monocultures such as fruit and wine growing, in forestry, in the reclaiming of wasteland (gravel pits, lignite mines, and waste heaps), in gardening, parks and grounds, formerly also as a feed additive in hog raising, etc. The frequently cited use as fertilizer, which the name "compost" suggests, is insignificant because of the high C:N ratio.

4. Microbiology of Composting

Industrially, the waste composting technique uses the degradative ability of saprophytic microorganisms (6) which accomplish humus formation in nature to a great extent. In waste and in sewage sludge high bacterial counts are already present, mainly in mesophilic aerobic and anaerobic forms. Mesophilic aerobic bateria multiply preferably during the first phase of composting, which is characterized by a rather steep temperature gradient. The optimum conditions for these microorganisms are exceeded with increasing temperature and a redistribution in the microorganism flora begins; now, thermophilic microorganisms multiply quickly and reach their maximum cell count shortly before the temperature maximum of 70° C. The development of populations of thermophilic and thermotolerant microorganisms follows a reverse course, with subsequent temperature decrease, that is, at the end of the process, the mesophilic aerobic species are abundant. In general, mesophilic, thermo-tolerant and thermophilic bacteria decompose the utilizable organic substances, particularly in the initial phase of composting (7); in the period after the temperature maximum, which includes the interval from about 30° to 60°, essentially fungi (e.g., Hyphomyces, yeasts or actinomycetes) are found.(8) The enzymes (e.g., glycosidases) of individual strains, isolated from waste compost, show a temperature optimum between 40° C and 70° C; the enzymes are still active even at 90° C.

Considering the multiplicity of microorganism species in composting and their capability for enzymatic conversion of organic materials, organic xenobiotics contained in wastes can also be regarded as potential substrates for microbiological metabolism.

5. Presence of Xenobiotics in Solid Wastes

The presence of organic pollutants in solid wastes has been reported quantitatively in only a few investigations.[9] A series of waste, sewage sludge, and waste compost samples were investigated for their content of polychlorinated biphenyls (PCB's) [10] which are known as global environmental contaminants.[11] PCB's are typical representatives of the persistent chlorinated hydrocarbons to which class certain pesticides like DDT belong.

The analytical method for PCB's in waste samples is described in detail elsewhere [12]; it is almost identical with the residue analysis of xenobiotics in waste compost described in section 8. In all of the analyzed samples, PCB's were found in concentrations between 0.4 and 9.7 $\mu g/g$.

A correlation between PCB concentrations and examined sample types cannot be made from the available data. However, it is noteworthy that the PCB concentration of compost and garden soil used for soil improvement and recultivation was found to be several $\mu g/g$.

Table 1. PCB-content of waste and compost samples (related to a mixture of 28.5% Clophen A 50 and 71.5% Clophen A 60). The test data are mean values of several tests. The relative standard deviations are related to respective single test.

No.	Sample	Origin	Date	pH[1]	Water[2] content%	PCB-Content[3, 4] ($\mu g/g$)
(1)	domestic waste	Geiselbullach	18. 6. 1973	6.2	47	0.5
(2)	dto.	dto.	14. 6. 1973	7.2	30.5	0.4
(3)	dto.	dto.	15. 6. 1973	7.1	31	1.0
(4)	dto.	dto.	18. 6. 1973	6.2	28.5	0.6
(5)	dto.	dto.	19. 6. 1973	7.3	24	0.7
(6)	dto.	dto.	20. 6. 1973	6.8	26.5	1.3
(7)	dto.	dto.	22. 6. 1973	7.3	28	2.2
(8)	dto.	dto.	25. 6. 1973	6.2	49.5	1.1
(9)	dto.	dto.	20. 2. 1973	5.9	43	5.9
(10)	waste	Blaubeuren	1. 8. 1972	–	37	9.7
(11)	waste/sewage sludge	dto.	1. 8. 1972	–	43.5	2.1
(12)	dto.	Schweinfurt	3. 10. 1972	–	45	2.5
(13)	sewage sludge	Blaubeuren	1. 8. 1972	–	–	2.0
(14)	dto.	Warendorf	1972	–	–	1.8
(15)	dto.	Geiselbullach	8. 8. 1973	7.9	–	
(16)	fresh compost	Blaubeuren	1. 8. 1972	–	37.5	1.9
(17)	ripe compost	Geiselbullach	20. 2. 1973	–	24	2.9
(18)	dto.	Blaubeuren	1. 8. 1972	–	31	3.8
(19)	dto.	dto.	1971	–	–	5.0
(20)	dto.	Schweinfurt	3. 10. 1972	–	29.5	3.7
(21)	garden soil	Blaubeuren	1. 8. 1972	–	26	2.5
(22)	dto.	dto.	1971	–	–	1.8
(23)	dto.	Amsterdam	1. 8. 1973	5.2	48	0.4

Relative standard deviation [1] 1.5%, [2] 5%, [3] 2.5%, [4] related to dry weight

A summary of the source and type, water content, and PCB concentration of several samples is given in Table 1. If the mean value of 2.3 μg/g is used as the average PCB concentration in domestic waste, and if one calculates on the basis of the total domestic waste of about 1.6×10^7 tons produced yearly in the FRG, one finds that about 37 tons of PCB's per year are present in solid domestic waste (with all statistical reservations). Besides the quantitative determination of PCB's, other chlorinated hydrocarbons such as DDT, DDE, aldrin, dieldrin, lindane, hexachlorobenzene, etc., were determined qualitatively in waste samples. Here questions arise as to whether and in which cases these xenobiotics are converted during the composting process or even whether they are decomposed.[13]

6. Simulation of Composting

Simulated practical conditions were used for the investigation of the behavior of xenobiotics during composting. This concers both the selection of suitable chemical testing objects (see Chapter 7) and the determination of their concentrations, which were chosen in accordance with the magnitude of PCB residues in domestic refuse, i.e. in the range of a few μg/g. In such dilutions, the use of radionuclides is essential. On the other hand, this requirement makes an investigation in an open system impossible. Therefore, the development of a simulation apparatus which enables the imitation of the composting process at the laboratory scale is necessary. To fulfill these requirements, an apparatus was constructed which could be adjusted and controlled through a series of regulators and control elements. Fig. 4 shows a scheme of the complete apparatus. The essential element is a Dewar flask in which the microbial decomposition of the waste material proceeds. In order to maintain the oxygen concentration essential for the microbial degradation process, a gas train system is attached to the Dewar flask; it removes gaseous metabolism products such as CO_2 and it is open for gases and closed for radioactive substances, so long as their condensation point is not below $-40°$ C. Gas transport is achieved by a vacuum pump; the air, at 100% relative humidity, is drawn through the compartments, two attached cold fingers (cooled to $-25°$ or $-40°$ by means of cryostats), and through two filters (activated charcoal, silicagel).

Regulation of the process is achieved by several control devices:

1. a thermostatic air humidifier which regulates the relative humidity and temperature
2. a metal spiral in the compartment by which the process temperature can be increased by heating
3. a water reservoir, so that the water lost from the compartment (evaporation, percolation) can be replaced
4. a timer which regulates the periodic air suction times

The following elements record the process:

1. a thermal element attached to a recorder which continually records the temperature of the compartment
2. a hygrometer to record the humidity
3. a gas gauge with digital recorder which registers the volume of the transported gas
4. an infrared absorption spectrometer with analog recorder to determine the CO_2 content of the withdrawn air (calibration with test gas)

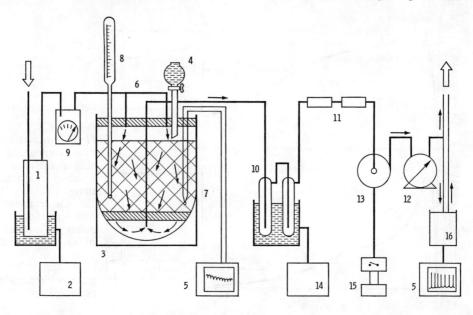

Fig. 4

Legend:
1 Air humidifier	5 Recorder	9 Hygrometer	13 Vacuum pump
2 Thermostat	6 Gas inlet	10 Cold reservoir	14 Cryostat
3 Dewar flask	7 Thermo element	11 Filter	15 Timer
4 Water reservoir	8 Thermometer	12 Gasmeter	16 CO_2-gauge

The parameters which can be registered continually during the process by these elements are:
1. temperature
2. CO_2 content of the removed air
3. gas flow
4. air humidity
5. water content of the waste sample
Before and after the process, the pH and moisture content of the samples are determined. The specific suitable parameters of temperature and CO_2 content of the withdrawn air, which indicate the acitivity of saprophytic microorganisms, are shown on the diagram as functions of time (Fig. 5). In general, the curves show the same characteristics as in the technical processes.
As is seen in Fig. 5, during the first 30 hours after beginning the process, the temperature increases rapidly; it flattens when reaching an intermediate stage, and then rises again to a maximum temperature. This course clearly mirrors the development and shifting in the population of saprophytic microorganisms: while in the initial phase, because of the optimal conditions (rich nutrient supply, sufficient oxygen), an exponential increase is reached quickly, the growth conditions for a portion of the microorganisms are worsened by the temperature increase resulting from generate metabolic energy as well as by the decrease of nutrients. For the mesophilic microorganisms the optimum temperature (between 20° and 45°) is exceeded quickly, while the growth conditions for the thermophilic bacteria and fungi becomes optimal. In these ways an shift occurs in the populations which is possibly observed in the gradations of the temperature-time curve.

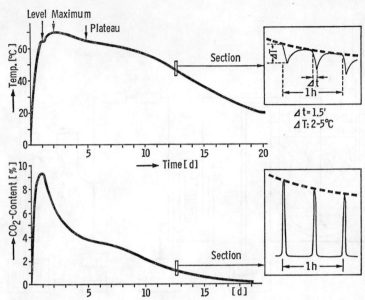

Fig. 5 Schematic graph of waste composting process. above: variation of temperature as a function of time; right: section of the diagram with time axis expanded. below: CO_2-concentration of the exhaust air as a function of time; right: section of the diagram with time axis expanded.

Finally the temperature reaches its maximum: at this temperature (about 70° C) the development of microorganisms may stop; however, the metabolic activity is not interrupted but simply reduced. An equilibrium is established between the amount of heat liberated by the microorganisms and the amount of heat removed through air circulation so that no further temperature increase occurs. Characteristically, it is observed that, during the development of the temperature curve, at each withdrawal interval, the temperature drops 3–5° C and in the following "rest phase" again slowly approaches its initial value (Fig. 5, right).

The conditions necessary for thermophilic (and thermotolerant) saprophytes have clearly reached an optimum at this temperature range (45° to 70° C). Therefore, with sufficient nutrient supply, aerobic conditions, and adequate humidity, microbial activity continues over a long period. Consequently, the temperature drops only slowly.

An additional corresponding function is the CO_2 time-curve (Fig. 5, bottom). Although a steeper rise is seen in the phase of the exponential increase, the curve flattens at the transition to the stationary section and finally drops before reaching the maximum temperature. The periodically measured CO_2 content of the removed air ranges, as seen in the expanded insertion, between 0.2 and 10% (removal period every 10 to 60 minutes).

With the help of the CO_2-time diagrams, the section with the steep temperature gradient in the initial process of the cycle can be associated with maximum microbial activity. The periodic flushing with gas is held constant; in general, it amounts to about 10 l/min; with exhaust times of 1.5 min and cycle times of an average of 30 minutes, about 20 m³ air passes through the apparatus during the total process.

The culture medium for the saprophytic microorganisms, a mixture of varying amounts of waste and sewage sludge, is taken directly from the waste treatment plant. The waste is ground previously and freed of ferromagnetic materials. For use in the simulation apparatus, noncompostable material (glass, stone, synthetics, etc., about 15—32% of the total waste) is removed. The remaining material is finely ground with a mixer.

In general, the samples are slightly acidic to neutral (pH between 5.5 and 8.7). During the composting process the pH-value varies mostly toward the neutral point. Variations up to $\pm$ 2 pH units are found. The moisture content of the samples ranges from 24 to 57%. If it is necessary, the samples are wetted to a moisture content of at least 40% before beginning the process in order to provide optimal growing conditions for the microorganisms. A definite amount of xenobiotic is then added (see Ch. 8) and the sample is placed in the Dewar vessel. This cell (cylinder, h = 30 cm, ϕ = 30 cm) is subdivided by means of a wire basket (10 $\times$ 10 cm) into a central (radioactive) zone and a peripheral (non-radioactive) zone in order to control the distribution of xenobiotics during the microbial decomposition process. During the entire process the simulation apparatus operates completely automatically; only the water content is adjusted manually. After about three weeks, the decomposed samples can be withdrawn; usually, they are worked up without further storage.

7. Selection of Xenobiotics

The selection of xenobiotics for investigation depends on several criteria
1. representative for a chemical class of compounds
2. evidence or expectation of appearance in waste
3. gradation in the persistency

4. consideration of primary and degradation products
5. amount of industrial production
Using these considerations, polychlorinated biphenyls, identified in waste samples, were used as model substances for chlorinated hydrocarbons of greater persistence (halogenated paraffins, terphenyls, etc.). As representative of numerous mixtures of lower chlorine content, a definite isomer (2,2'-dichlorobiphenyl) which is a major component of Clophen A30 or Aroclor 1242 (8%) and of Aroclor 1221 (9.2%) was selected. Although PCB isomers found in environmental samples range mostly in tetra- to hexachloro substitution, selection falls on dichlorobiphenyl, since lower chlorinated biphenyls are generally less persistent and might eventually replace the PCB's with higher chlorine content. Furthermore, the cyclodiene insecticides, aldrin, dieldrin, and heptachlor were selected as typical representatives of their class found in environmental samples and in waste and of continued importance in pest control. The persistence of these biocides is comparable to that of PCB's.

Finally buturon (active substance in Eptapur) and monolinuron (Aresin) are determined as representatives of phenyl urea herbicides. Indeed, these substances are brought into the environment in increasing amounts for weed control. However, little is known about the fate of these urea derivatives. The same is true for imugan (Milfaron), an aminal, as representative of fungicides. The compounds possibly have the least persistence of all the selected xenobiotics.

Benzo-[a]-pyrene, whose presence in waste compost was already established, was investigated as representative of fused-ring aromatic hydrocarbons

Table 2. Added amounts, specific activity, and recovery of xenobiotics subjected to simulated composting process.

xenobiotic[1] molecular weight (Dalton)	specific activity[2] (Ci/Mol)	applied radioactivity[3] (Ci)	applied quantity[3] (Mol)	conc.[3] (μg/g)
2,2'-dichlorobiphenyl	0.7	4.7	$6.7 \cdot 10^{-6}$	5
223.1		$1.04 \cdot 10^{7}$ dpm	1.5 mg	
aldrin	0.18	2.9	$1.6 \cdot 10^{-5}$	20
364.9		$6.44 \cdot 10^{6}$ dpm	6 mg	
dieldrin	0.44	6.89	$1.6 \cdot 10^{-5}$	20
380.9		$1.53 \cdot 10^{7}$ dpm	6 mg	
heptachlor	0.29	4.7	$1.6 \cdot 10^{-5}$	20
373.3		$1.05 \cdot 10^{6}$ dpm	6 mg	
imugan	0.42	3.7	$8.9 \cdot 10^{-6}$	10
336.5		$8.21 \cdot 10^{6}$ dpm	3 mg	
monolinuron	0.41	11.35	$2.8 \cdot 10^{-5}$	20
214.7		$2.52 \cdot 10^{7}$ dpm	6 mg	
buturon	0.24	3.1	$1.3 \cdot 10^{-5}$	10
236.7		$6.88 \cdot 10^{6}$ dpm	3 mg	
p-chloroaniline	1.7	4.1	$2.4 \cdot 10^{-6}$	1
127.6		$9.1 \cdot 10^{6}$ dpm	0.3 mg	
benzo-[a]-pyrene	0.98	11.7	$1.2 \cdot 10^{-5}$	10
252.3		$2.48 \cdot 10^{7}$ dpm	3 mg	

[1] radiochemical purity $\geq$ 99% (clean-up by thin layer chromatography)
[2] relative standard deviation of liquid scintillation counting 0.1 to 2.5 (related to single determination of the total counting rate)
[3] related to the central zone

whose carcinogenicity is already known. Since there are contradictory investigations regarding the synthesis or degradation of benzo-[a]-pyrene via plants and soil organisms (14), experiments on the behavior of these substances in composting should be carried out and reported for clarification of these questions.

Among the degradation products of xenobiotics, dieldrin was found as a conversion product of aldrin as well as p-chloroaniline, a hydrolysis product of buturon, monolinuron, and other xenobiotics, and 3,4-dichloroaniline, a degradation product of imugan.

8. Analysis of Xenobiotics in Waste Compost

The application of definite amounts of specific xenobiotics (not radioactively labelled) is achieved by a spraying process using minimal volumes of an easily vaporizable solvent (acetone, b. p. 56° C). Thus the development of microorganisms is scarcely affected by the applied amounts of solvent. In addition, the corresponding radioactively labelled sample is applied to the pulverized waste in the central zone of the cell using an acetone-water mixture. Table 2 summarizes the added amounts, specific activity, and radiochemical purity of the applied substances. The concentrations ranged between 1 and 20 μg/g (e.g., the PCB-content of the analyzed samples is between 0.4 and 9.7 μg/g; 2,2'-dichlorobiphenyl was applied in a concentration of 5 μg/g). Higher concentrations can have an increased effect on the development and composition of the populations.(15)

Table 2. Added amounts, specific activity, and recovery of xenobiotics subjected to simulated composting process (Continuation).

	percentage of applied radioactivity			percentage of recovered radioactivity		
yield of extraction	non-extractable residues	total balance	starting materials	degradation products		non-identified substances[4]
96.1	0.6	96.7	98.2	—		1.8
92.5	5.7	98.2	88.4	dieldrin 5.2 aldrin-trans-diol 0.7		5.7
87.9	7.9	95.8	97.3	—		2.7
91.2	6.3	97.5	64.1	heptachloroepoxide 21.9 hydroxychlordene 10.2		3.8
90.7	4.1	94.8	91.8	3,4-dichloroaniline 5.5		2.7
76.5	24.9	101.4	86.2	N-methoxy-N'-4-chloro-phenyl-urea 0.4		13.4
81.4	14.3	95.7	50.5	p-chloroaniline 44.8		4.7
78.7	16.6	95.3	86	—		14
93.1	5.6	98.7	99.5	—		0.5

4) see also section 9

Three weeks after incubation, the nutrient in the waste-sewage sludge mixture was used to such an extent that the growth and metabolic activity of the microorganisms were reduced greatly. At this point the fermentation process was stopped and the nutrient medium (fresh compost) from the central and peripheral zones was separated. The single sectors, as well as the cold finger condensate and filtrate, were worked up separately to determine the distribution of the xenobiotics during the decomposition processes.

Generally, extraction of the compost was carried out using polar solvents such as methanol, ethyl acetate, acetone, and diethyl ether; in fact, by this method, in contrast to aprotic unpolar solvents, the amount of co-extracted material increases; on the other hand, an estimate can be made that polar solvents give a more complete extrac-

tion of conversion products containing polar groups. Only in the case of buturon is benzene used as a extraction medium in order to prevent carbamate formation, as was observed in the extraction of monuron with ethanol.[16] Extraction required up to 48 hours.

Non-extractable residues were determined by means of burning and determination of the ^{14}C—CO_2 formed. The aqueous phase of the cold finger condensate was extracted with methylene chloride or diethylether. Activated charcoal and silica gel were extracted and measured directly in counting vials using toluene or dioxane.

Using the same procedures in the "working up process" (Fig. 6), the radioactive balance was controlled by measurement with a scintillation counter. ^{14}C-toluene was used as internal standard for the determination of yields. The recoveries by extraction ranged

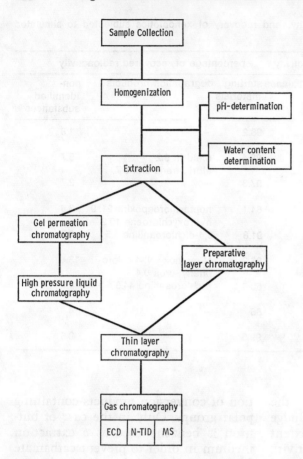

Fig. 6 *Analysis of xenobiotics in waste compost.*

from 76 to 96%. The total balance including non-extractable residues exceeded 95% in all cases. The extract was separated using chromatographic methods. An aliquot part of the solvent was purified by gel filtration. With this simple and rapid separation method, related compounds of higher molecular weight could be removed. A neutral styrene-divinylbenzene copolymerisate (Bio-Beads SX) with a molecular weight limit of 1000 or 2700 Dalton was used as stationary phase and benzene was used as the mobile phase. Analysis of a single fraction showed that a specific percentage of active compounds was absorbed on higher weight substances (Fig. 7). Presumably,

adsorption via hydrogen bonding or hydrophobic interaction plays a role; however, chemical bonding is not excluded.

For further analysis, the lower molecular weight fractions were chosen. After changing the solvent (hexane), the filtrate was separated further using high pressure liquid chromatography on silica gel (grain size 10 μm) with hexane/acetone 9:1 as mobile phase. A UV absorption instrument or a differential refractometer was used as detector. The distribution of radioactivity in the individual fraction was determined by liquid scintillation counting. By comparison with elution volumes of reference substances, an

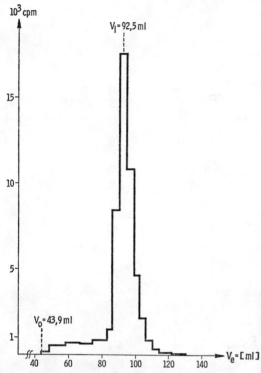

Fig. 7 Distribution of radioactively labeled Monolinuron and its degradation products from gel permeation chromatography.

initial qualitative determination of conversion products was possible (Fig. 8). The radioactive fractions obtained

from high pressure liquid chromatography were investigated individually by thin layer chromatography. By using suitable developing systems, a correlation of substances with comparison substances was already possible. For the purpose of identification, gas chromatography with N-TID and ECD (comparison of retention times) or coupling of gas chromatography/mass spectrometry with peak selector were used.

Alternatively to the above, the extract was also placed directly on a preparative thick layer plate, developed, and the distribution of radioactivity measured on a scanner. Because of the entrained accompanying materials of higher molecular weight, the adsorption phenomena led to unspecific distribution of a portion of the radioactive material over the plate; these peaks were especially apparent in the range of Rf values from 0 to 0.05.

The thick layer plates were divided into several zones corresponding to the distribution of radioactivity; these were removed and extracted with a polar solvent. The eluate of each individual zone was placed on a thin layer plate and developed with other developing

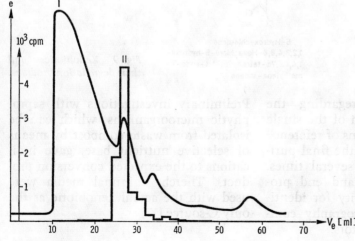

Fig. 8 Distribution of radioactively labeled Monolinuron and its degradation products from high pressure liquid chromatography.
Curve I: extinction scanned by UV detector.
Curve II: radioactivity scanned by liquid scintillation counter.

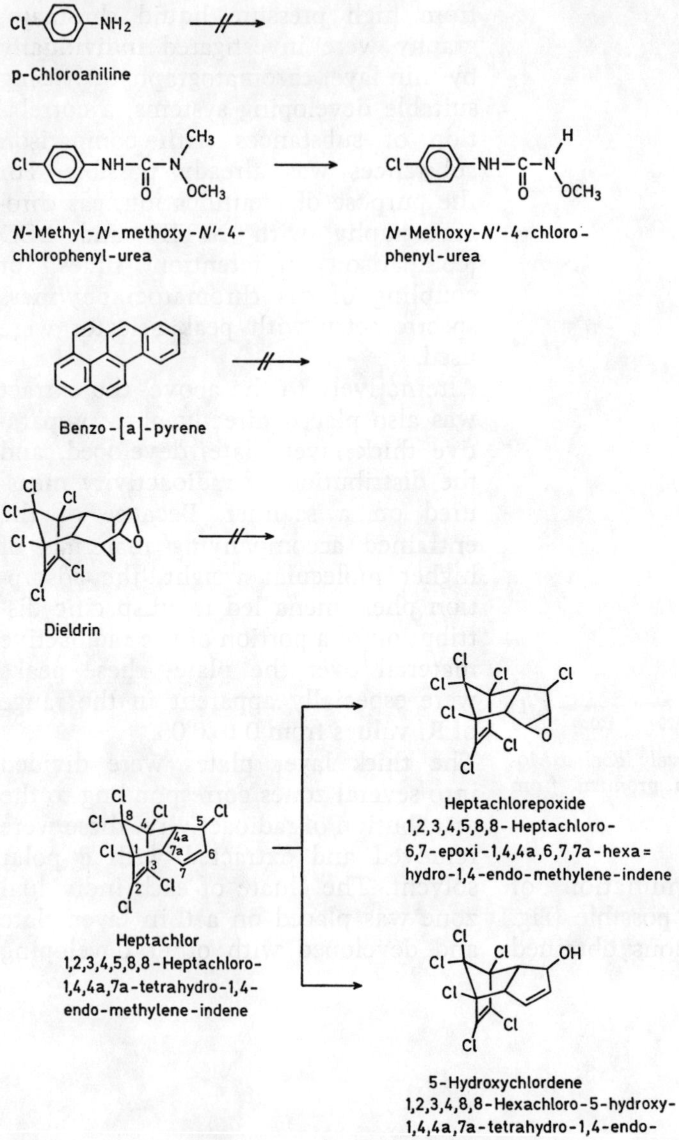

Cl—⟨O⟩—NH₂   —‖→

p-Chloroaniline

Cl—⟨O⟩—NH—C—N⟨CH₃ / OCH₃⟩   ⟶   Cl—⟨O⟩—NH—C—N⟨H / OCH₃⟩
                    ‖   ‖
                    O   O

N-Methyl-N-methoxy-N'-4-   N-Methoxy-N'-4-chloro-
chlorophenyl-urea   phenyl-urea

Benzo-[a]-pyrene   —‖→

Dieldrin   —‖→

Heptachlorepoxide
1,2,3,4,5,8,8-Heptachloro-
6,7-epoxi-1,4,4a,6,7,7a-hexa=
hydro-1,4-endo-methylene-indene

Heptachlor
1,2,3,4,5,8,8-Heptachloro-
1,4,4a,7a-tetrahydro-1,4-
endo-methylene-indene

5-Hydroxychlordene
1,2,3,4,8,8-Hexachloro-5-hydroxy-
1,4,4a,7a-tetrahydro-1,4-endo-
methylene-indene

Fig. 9 Xenobiotics and their degradation products.

systems. Conclusions regarding the qualitative determination of the single zones were made by means of reference substances. If necessary, the final purifications were repeated several times. In general, the initial and end products had sufficient purity for identification by gas chromatography combined with specific detectors.

Preliminary investigations with saprophytic microorganisms, which can be isolated from waste compost by means of selective nutrient base, gave indications to the expected conversion products. Thereby minimal media were used with the actual xenobiotic as the only C-source.(17)

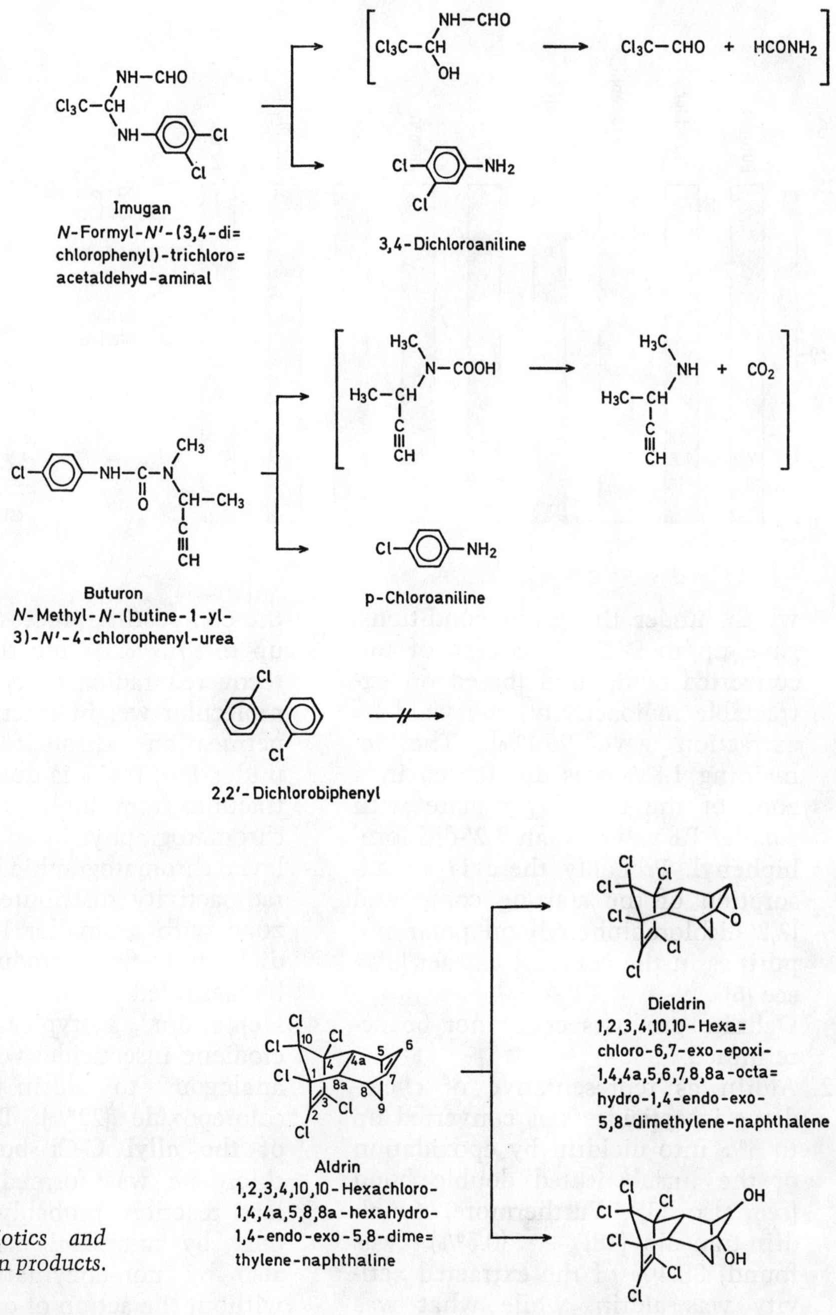

Fig. 9   Xenobiotics and their degradation products.

9. Behavior of Xenobiotics in Composting

The analytical investigation of residues gave the following pattern of the behavior of xenobiotics in waste composting (Fig. 9 and 10):

1. 2,2-Dichlorobiphenyl as representative of low chlorinated PCB's showed itself as a persistent compound

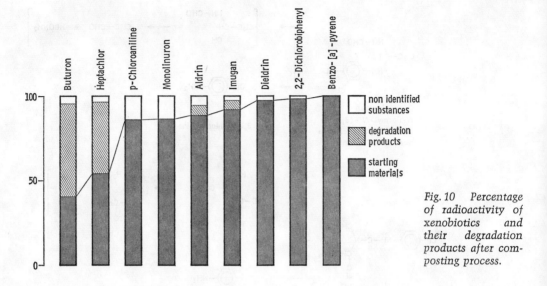

Fig. 10   Percentage of radioactivity of xenobiotics and their degradation products after composting process.

which, under the given conditions, gave up to 98.2% recovery of unconverted compound (based on extractable radioactivity; the yield by extraction gave 96.1%). The remaining 1.8% was distributed in a zone of the thin layer plate with smaller Rf values than 2,2'-dichlorobiphenyl. Probably there is an adsorption of the starting compound (2,2'-dichlorobiphenyl) on polar impurities in the compost extract (also see [6]).
Definite products could not be determined.

2. Aldrin as representative of cyclodiene insecticides was converted up to 5% into dieldrin by epoxidation of the unsubstituted double bond (see also [3]). Furthermore, 6,7-aldrin-transdiol (18)   (0.7%) was found; 88.4% of the extracted activity was aldrin, while what was stated in [1] holds for the remaining 5.7%.

3. Dieldrin, an additional cyclodiene insecticide, studied here as a reaction product from aldrin, withstood

the composting process unconverted up to 97.3%. While the portion of recovered radioactivity in the high molecular weight fraction from gel-permeation chromatography was under 1%, 0.4% is due to the polar fraction from high pressure liquid chromatography. In subsequent thin layer chromatographic analysis, this radioactivity distributed itself in a zone with a smaller Rf-value than dieldrin. Definite products could not be identified.

4. Heptachlor, a typical model cyclodiene insecticide, was epoxidized analogous to aldrin to a heptachlorepoxide (22%). By hydrolysis of the allyl C-Cl bond hydroxychlordene was formed (10%). The last reaction probably occurs not only by microbial conversion but also by non-enzymatic hydrolysis without the action of organisms.(19)

5. Buturon, a phenylurea herbicide, was hydrolyzed to p-chloroaniline (45%) and evidently to carbamic acid, the latter forming a primary amine. Since the isotopes of radio-

active buturon were localized exclusively in the phenyl ring, the suspected carbamide formation could not be followed. However, analogous investigations follow a similar degradation path.(20)

6. Monolinuron, an additional important compound in the class of phenylurea herbicides, showed the smallest recovery by extraction compared to the other applied xenobiotics: about 76.5% of the applied radioactivity was recovered with organic and inorganic solvents. Isolation was achieved with the help of gel permeation chromatography. By this method the higher molecular weight fractions which are recovered with small elution volumes contain up to 9% of the recovered radioactivity (Fig. 7). This phenomenon, as well as the inadequate recovery by extraction, indicates adsorption or chemical bonding of the polar starting substance and its conversion products with compounds of higher molecular weight (e.g., as conjugates). In the lower molecular weight fractions from gel permeation chromatography, up to 86.2% unconverted monolinuron was recovered; a demethylated compound, N-methoxy-N'-4-chlorophenyl-urea (0.4%) was isolated. Surprisingly, the expected hydrolysis product p-chloroaniline was not found.

7. Imugan, a substituted acetaldehyde aminal in the fungicide series, shows hydrolytic cleavage to 3,4-dichloroaniline (5.5%). The amount of further potential hydrolysis products (trichloroacetaldehyde and formamide, Schiff base) could not be investigated further for the reasons given in [5].

8. p-Chloroaniline, a typical conversion product of phenylurea herbicides, was recovered to at least 86% as the starting material. Here also, strong adsorption on the waste material was pronounced and led to a comparably small recovery by extraction. Extraction of the compost with acid or alkali increased the total amount of recovery; however, here also the greater portion of radioactive substance(es) was adsorbed on polar accompanying materials. From this the relatively greater portion (9.3%) of radioactive material, which has an Rf value below 0.05 by preparative layer chromatography, can be understood. Presumably it is starting material (p-chloroaniline) which is strongly bound to carrier material.

9. Benzo-[a]-pyrene, representative of carcinogenic anilated hydrocarbons whose occurrence in waste, sewage sludge, and compost has been proven (21), withstood the composting process without chemical conversion. Recovery by extraction was nearly 93%. The portion of extracted radioactivity in the higher molecular weight fractions by gel permeation chromatography was correspondingly small (less than 1%).

In the investigation of distribution tendency of the xenobiotics, it was shown that, in general, up to 16% of the applied radioactivity from the central zone was fond again in the peripheral compartment. Only 2,2'-dichlorobiphenyl was transported in large amounts and found in the peripheral zone up to 34%. The distribution within zones showed an expected transport preference in the region of the gas stream as can be demonstrated by several examples. Generally the loss by vaporization was unusually small: the total recovered radioactivity from the cold finger condensate and activated charcoal filter was less than 1% in all cases. In the case of monolinuron, it was noted that solid adsorption on accompanying compounds decreased the rate of vaporization to under ‰.

10. Summary and Interpretation

The results of the investigation on the behavior of organic xenobiotics in composting can be summarized as follows:

1. Municipal waste in the FRG (refuse, sewage sludge) and its composting products contain polychlorinated biphenyls; all investigated samples had PCB's with an average chlorine content between 54 and 60%. The PCB-content of the investigated samples ranged between 0.4 and 9.7 $\mu g/g$. In addition, chlorinated hydrocarbons like DDT, DDE, lindane, aldrin, dieldrin, and hexachlorobenzene could also be detected.

2. 2,2'-Dichlorobiphenyl remains unconverted in the composting process to at least 98%. From experience, a degradation of higher chlorinated biphenyls occurring in waste is improbable because they are generally even more persistent.

3. Cyclodiene insecticides such as aldrin and heptachlor are converted to a known extent to degradation products such as dieldrin, aldrin-trans-diol, heptachlorepoxide, and hydroxychlordene. In the example of dieldrin, it is seen that the "degradation" stops at the more persistent breakdown product.

4. Xenobiotics from the class of fungicides and herbicides have conversion rates between 0.4 and 50%. The degradation products (e.g., chlorinated aniline) again form stable compounds which, under the conditions of composting, were converted only in insufficient amounts.

5. Benzo-[a]-pyrene, a carcinogenic hydrocarbon, which occurs in waste, sewage sludge, and waste compost in concentrations between 0.4 and 1.7 $\mu g/g$, withstands the decay process unconverted to nearly 100%. A similar resistance is also expected for other hydrocarbons such as benzo-[e]-acephenanthrylene.

6. Thus a great number of the investigated xenobiotics were either not converted in the composting process or were converted only in negligible amounts. In no case could a complete degradation be established. On the other hand, it is known that some of the formed conversion products are persistent compounds which do not undergo further chemical conversion in the course of the decay process.

7. These or similar conclusions are also permissible with the view point that individual species of microorganisms are capable of conversion or metabolism of xenobiotics. These microorganisms as "specialists" under extreme restrictions (minimum medium with the respective pollutant as the only C-source, enzyme induction through biodegradable analogues) can achieve a degradation of xenobiotics. Under the restrictions which exist in the composting process and which were investigated in this work the possibility that these microorganisms are able to replace other populations is very small (rich nutrient supply, deficient specific enzyme induction).

8. Thus compasting of waste is no suitable treatment for decontamination of waste, at least for the investigated substances. On the con-

trary it must be feared that persistent pollutants such as fused-ring aromatic or chlorinated hydrocarbons, which already circulate in ecological cycles, are admitted into the environment especially to cultivated land by introduction of composting products. Thus the pollutants are distributed widely and can enter into biological cycles.

9. The often repeated hypothesis that, considering the preservation and restoration of organic materials in natural circulation, composting is termed the waste treatment method most "friendly" to the environment (22) is questioned because of consideration of the occurrence and behavior of xenobiotics in composting. On these grounds, the use of waste compost for agricultural purposes needs a critical check-up.

An evaluation under the aspect of the mentioned criteria for environmental conformity shows that waste composting can indeed be established as hygienically unobjectionable; however, under toxicological and, above all, ecological chemical aspects it can be shown to be doubtful.

11. Perspectives

From the results of this work, some new viewpoints develop for the ecological-chemical evaluation of waste treatment methods. Thereby, some of the arising questions can be answered through directed experiments:

1. Investigations of the behavior of other characteristic xenobiotics (e.g., plasticizers, chlorinated paraffins, organophosphorus compounds) in waste composting in order to make far reaching statements on the deg-

radability of xenobiotics.

2. "Kinetic" studies on the degradation rate in the various phases of microbial decomposition processes with the aim of optimizing the conversion rates

3. Investigations on plant resorption of xenobiotics which appear in waste compost; studies on vaporization and spreading of these xenobiotics in compost and the adjacent substrate with the aim of finding quantitatively the spreading of pollutants occurring in waste compost

4. Development of faster and possibly more universal analytical methods with good reproducibility for continual supervision of the amount of xenobiotics together with a quality control of the waste treatment products from an ecological-chemical view

The long range view involves a comparative evaluation of various waste treatment techniques (incineration, disposal, pyrolysis, etc.) under the criteria of environmental conformity in order to arrive finally at ecological-chemical preference lists. These priority listings should enable a choice of an optimal waste treatment method including geographic, economic, and social boundary conditions.

References:

1 Gesetz über die Beseitigung von Abfällen, Bundesgesetzblatt I, 49, 873, vom 10. 6. 1972.
2 Spendlove, M. J.: Umschau, 73, 364 (1973).
3 Coenen, R., et al: Naturwissenschaften, 59, 106 (1972).
4 Teeuwen, E.: "Environmental Quality and Safety", 2, 53 (1973).
5 Kumpf, W., K. Maas, H. Straub: "Handbuch der Müll- und Abfallbeseitigung", Band 2, Erich-Schmidt-Verlag, Berlin 1973.
6 Müller, G., G. Ritter: Zbl. Bakt. II, 127, 270 (1972).
7 Farkasdi, G.: Zbl. Bakt. II, 124, 334 (1970).

8 Knösel, D.: A. Rész, Städtehygiene, 24, 143 (1973).

9 Carnes, R. A., et al.: Arch. Environ. Cont. Tox., 1, 27 (1973).

10 Müller, W., F. Korte: Naturwissenschaften, 61, 326 (1974).

11 Müller, W., F. Korte: Chem. Zeit, 7, 112 (1973).

12 Müller, W., et al.: (Manuscript in preparation).

13 Roszinski, H., F. Herzel: Z. Kulturt. Flurber., 13, 304 (1972).

14 Borneff, J., et al.: Umschau, 73, 626 (1973).

15 Shamiyeh, N. B., L. F. Johnson: Soil, Biol. Biochem., 5, 309 (1973).

16 Lee, S., S. C. Fang: J. Ass. Off. Anal. Chem., 54, 1361 (1972).

17 Müller, W., et al,: GSF-Bericht Ö 104, Gesellschaft für Strahlen- und Umweltforschung, München (1974).

18 Korte, F., W. Kochen: Med. Pharmacol. Exp., 15, 28 (1966).

19 Miles, J. R., C. M. Tu, C. R. Harris: J. Econ. Entomol., 62, 1334 (1969).

20 Engelhardt, G., P. R. Wallnöfer, R. Plapp: Appl. Microbiol, 22, 284 (1971).

21 Wagner, K. H., I. Siddiqi: Naturwissenschaften, 60, 160 (1973).

22 Knoll, K. H.: Umschau, 72, 45 (1972).

PCBs and Environmental Contamination *)

By **W. Klein** and **I. Weisgerber**

Institut für ökologische Chemie der Gesellschaft für Strahlen- und Umweltforschung mbH, München, Western Germany

Summary

A survey is given on reported PCB-residues and accumulation in various environmental media including water and aquatic environment, flora, animals, food, and humans. Model studies with technical PCB, as well as with pure individual components, are presented from various working groups. As examples, metabolic studies and photochemical experiments are discussed. According to our present knowledge, a major metabolic pathway in animals and plants is hydroxylation, often followed by methylation or conjugation. By UV-irradiation, however, oxygenation, dechlorination and chlorination, polymerization and isomerization may occur.

Zusammenfassung

Es wird eine Übersicht über bisher gefundene PCB-Rückstände und die Akkumulierung in verschiedenen Umweltmedien einschließlich Wasser und wäßrigen Umweltmedien, Pflanzen, Tieren, Nahrungsmitteln und Menschen gegeben. Danach werden Modellversuche verschiedener Arbeitsgruppen mit technischem PCB sowie mit reinen Einzelkomponenten beschrieben. Als Beispiele dienen Metabolismusarbeiten und photochemische Versuche. Nach dem jetzigen Stand unserer Kenntnisse ist die Hydroxylierung, die oft gefolgt wird von Methylierung oder Konjugation, eine wesentliche Metabolismusroute in Tieren und Pflanzen, während durch UV-Bestrahlung Oxygenierung, Dechlorierung und Chlorierung, Polymerisation und Isomerisierung eintreten können.

I. Background, Use, and Trends

PCBs became a big issue during the last few years, and PCB-literature is increasing exponentially. Thus, in this review, only some selected aspects will be discussed.

PCBs have been recognized as an environmental hazard in two aspects; one is the interference in pesticide residue analysis, demonstrating the wide-spread occurrence of these compounds (1), and the other is the occurrence of yusho, an accidental poisoning caused by contaminated rice oil.(2, 3)

When discussing general environmental contamination apart from accidents, the question arises in regard to the sources since, apart from the use as plasticisers, which has never been more than 20% of total use, they have never been intentionally brought into the environment. The main use has been and is still today in capacitors and transformers. They are chemicals with unique technical properties — the ef-

*) This paper was presented in part at the 3rd International Symposium on "Chemical and Toxicological Aspects of Environmental Quality", Tokyo, November 19–22, 1973.

Pesticide	No. Composits	ppm Fat Basis
HCB	2	T, 0.01
2,4-DB	1	0.012
Diazinon	1	0.005
PCB (Cal' d as Aroclor 1254)	2	0.09, 0.13
PCB (Cal' d as Aroclor 1254)	1	0.12
PCB (Cal' d as Aroclor 1254)	1	0.05

Fig. 1   PCB and Pesticides in Food.(8)

ficiency of the technical product can easily be varied according to the technical requirements. PCBs have been used since 1929 and the world production so far is in the range of 1—2 million tons. The sources of PCBs recognized in the environment so far are: (2)

Open burning or incomplete combustion of solid wastes containing PCB; vaporization from plastics, paints, and coatings, municipal and industrial sewers, improper waste disposal;

dumping of sewage sludge and waste.

According to Nisbet and Sarofim (4), the greatest loss of PCBs into the environment (U.S.) has been to dumps. Thousands of environmental and biological samples have been analyzed for PCBs, and they have been found in many organisms or environmental media; they have been found in sea water, river water, and aquatic organisms, in sediments of rivers, in wildlife and even in human mother milk in urban and rural areas.(5—7)

Although, for instance in the United States, monitoring for PCB residues has been included in the national monitoring programs, PCBs definitely do not represent a frequent residue in food (Fig. 1). Fig. 2 demonstrates that in a number of food samples which contained other residues, PCB could not be detected.(8) Some delicate questions remain open in regard to the quantitative determination of these compounds; nevertheless it can be stated upon these results that now, after measures have been taken against contamination of food during storage, processing and in commerce, these compounds do not represent a direct general hazard to man as regards their dietary intake. The other measures taken so far, namely to ban their use totally in open systems (OECD 1973), could lead to the prospect that further loads will be kept away from the environment. On the other hand, the substitutes for open use present some problems and further questions arise in the context of using only lower chlorinated products.

The compounds detected in environmental samples contained more than four chlorine atoms. Thus, it had been

	Dieldrin	Heptachlor-epoxide	BHC	Lindane	DDT DDE TDE	Br-	Cd	Malathion	PCB	HCB
Cereals	0.014	0.002	trace	0.008	0.079	94	0.06	0.096	n. d.	n. d.
Potatoes	0.057	0.003	0.004	n. d.	0.046	52	0.008	n. d.	n. d.	n. d.
Leafy vegetables	0.009	n. d.	0.001	n. d.	0.044	23	0.14	n. d.	n. d.	n. d.
Root vegetables	0.010	n. d.	n. d.	0.072	0.037	19	0.08	n. d.	n. d.	n. d.
Fruits	0.003	n. d.	0.003	0.003	0.484	16	0.07	0.053	n. d.	n. d.

Fig. 2   Maximum Values of Environmental Chemicals in Vegetable Food, USA 1968–1970, ppm.(8)

Aroclor Compound	Zero Time	1 Week	2 Week	4 Week	8 Week	16 Week
1242	87	76	73	60	53	43
1248	80	60	50	78	54	38
1254	70	45	42	51	41	37
1260	–	39	40	42	42	51
1262	78	31	33	32	57	33
1268	75	18	–	27	65	37
X	82	49	52	50	50	40

Fig. 3   Disappearance of Residues of Aroclor from Raw River Water.(9)

concluded that the higher chlorinated PCBs represent the environmental hazard. Therefore, two measures have been taken: one is to restrict the use of commercial products containing high chlorinated compounds, and the other one is to find low chlorinated "PCBs" with the technical properties of the higher chlorinated ones. Low chlorinated commercial products, however, contain between 5 and 10% of compounds with more than 4 chlorine atoms; and the technological approach to replace higher chlorinated PCBs by lower chlorinated alkyl PCBs requires thorough scientific investigation of the consequences. Further progress along these lines should, however, represent a solution to the environmental problems with PCB.

In the following section, a few data on the occurrence and elimination of PCB-residues are reviewed.

Fig. 3 demonstrates that, indeed, lower chlorinated technical products disappear more rapidly from the environment than higher chlorinated ones.(9) The information given in Fig. 3, however, does not allow one to conclude that the compounds are broken down. They may evaporate into the atmosphere or may be adsorbed from the water by gravel, sand, and organic matter, including algae.(10)

The pattern of PCB-components in environmental samples as demonstrated by packed-column-GLC is constant. Differences can be seen only by capillary-column-GLC, but as expected, there is a wide range of residues reported in the literature from just detectable to a few hundred or even thousand ppm. Fig. 4 gives just a few examples.

In agricultural soils in remote areas, PCBs could not be detected; they could,

Place	Sample	Concentration	Ref.
Atlantic Ocean	plants and animals	1–100 ppb	CH. E. N., Dec. 13, 1971
Atlantic Ocean	zooplankton	1.5 ppm	
Chesapeake Bay	soft shelled clams	0.2 ppm	
Golf of Mexico	oysters	3 ppm	
Cornish Coast (Scotland)	blubber of seal pups	160 ppm	Chem. and Industry Nov. 20,
U. K.	sewage sludge (dry)	1–185 ppm	1971
Munich	sewage sludge	10 ppm	unpubl.

Fig. 4   PCB in Aquatic Environment.

however, be found in the ppb-range in generally contaminated soils. In this context, attention should be drawn to the fact that, for many published data, the determination of detection limit, recovery, relative standard deviation, or background is not reported. This may result in doubts on the reliability of these data. Particularly, neglect of the total laboratory background may lead to analytical results which are higher than the real PCB-content of the analyzed samples.

In the following studies on PCB in flora, which were carried out in our institute, a background of 5 ppb has been determined and taken into account for the analytical results.

Samples of wild growing plants were taken from densely populated areas near motor highways and from remote areas in the Alps. The samples taken near the motor highways contained between 0.09 and 0.95 ppm PCB of their dry weight. The samples from the Alps contained only an average of about one third of these values; however, also in these regions, a minimal concentration of 0.05 to 0.15 ppm was observed. Fig. 5 shows the results of these studies. Since the transport of PCB by water or by men can be excluded, it may be assumed that PCB is transported via the atmosphere from populated to remote regions. Eggs of predatory birds from the Alp region were found to contain up to 286 ppm of PCBs.[11]

The problem remains that PCBs in the global environment amounting to roughly 1 million tons (extrapolation of Sarofims data to world charge) when biologically available can be accumulated in food chains. Fig. 6 gives an example for the biological magnification in 3 aquatic organisms.[12]

The following section presents some findings demonstrating the conversion of lower chlorinated biphenyls to hydrophilic products in organisms and the

Sampling near Autobahn Pfaffenhofen or Irschenberg		Sampling Remote Area Alps, Tegelberg, or Hüttensee	
Plant	ppm dry weight	Plant	ppm dry weight
Moss	0.13	Moss	0.06
Heathen	0.28	Heathen	0.08
Raspberry leaves	0.15–0.27	Raspberry leaves	0.11
Larch leaves	0.95	Larch leaves	0.17
Rushes	0.12	Rushes	0.07
Hay	0.13	Fern	0.08–0.15
Grass	0.20–0.46	Juniper	0.07
Grass silage	0.40–0.42	Dwarf fir	0.09–0.10
Clover	0.33	Alder leaves	0.09
Mint	0.20		
Strawberry leaves	0.16–0.46		
Blackberry leaves	0.49		
Fir apex	0.31		
Oak foliage	0.50		
Pine leaves	0.11		
Water hemlock	0.23–0.57		

Fig. 5   PCB in Flora.(11)

Organism	Exposure Concentration	Time	Final Concentration
oysters	5.0 ppb	24 weeks	425.0 ppm
pink shrimp	2.50 ppb	22 days	510.0 ppm (hepatopancreas)
spot	1.0 ppb	28 days	37 ppm

Fig. 6   *Accumulation of Aroclor in Marine Organisms.(12)*

breakdown of PCB under the influence of sunlight.

II. Model Studies with Technical PCB and its Components to Study Environmental Alleviation of PCB-Pollution

1. Metabolism Studies

Since commercial PCB is a complex mixture of compounds, and since its individual components are not easily obtainable by synthesis, earlier metabolism studies were carried out with the technical mixtures. After administration of the technical mixtures to various organisms, the GLC-patterns of the residues in the organisms and excreta were compared to those of the administered mixture and quantified. The changes in the pattern showed that metabolic conversions had occurred, but with different rates for each component. In general, the lower chlorinated components disappeared to a greater degree than the highly chlorinated ones. In the following section, some examples for such work are given. For example, studies in rats have been carried out by Grant et al.(13) Male rats were orally dosed with Aroclor 1254, and residues were found in all tissues analyzed, with the greatest concentration in the fat. The GLC-EC-pattern of the residues was different from the standard mixture administered, indicating that all components

were not metabolized at the same rate, but that the lower chlorinated compounds had higher metabolism rates. Aroclor 1254 residues in the brain, spleen, blood, testes, heart, kidney, and fat were reduced by 90 to 33% within 20 days. Aroclor 1254 increased significantly the size of the liver and also the percent lipid in the liver.

Bird experiments were carried out by Bailey and Bunyan.(14) In the liver extract, a higher degradation of the lower chlorinated components of Aroclor 1242 and Aroclor 1254 was observed. Regular checks made on excreta revealed no gross excretion of unchanged material.

Preferential losses of single components of Aroclor 1254 from soils have been reported by Iwata et al. (15), indicating that soil microorganisms, too, have different metabolism capacities for each PCB-component.

For metabolism studies in which individual metabolites are to be identified, pure and chemically defined chlorobiphenyls were synthesized and applied to various organisms. All metabolites identified up to now were hydroxylated products with one, two, or three hydroxy groups, or the respective conjugates. Block and Cornish (16) have shown that 4-chlorobiphenyl was converted to 4-chloro-4'-hydroxybiphenyl and its conjugates by rabbits.

Metabolic behaviour of 4-chlorobiphenyl, 4,4'-dichlorobiphenyl, 2,2',5,5'-tetrachlorobiphenyl, and 2,2'4,4',5,5'-

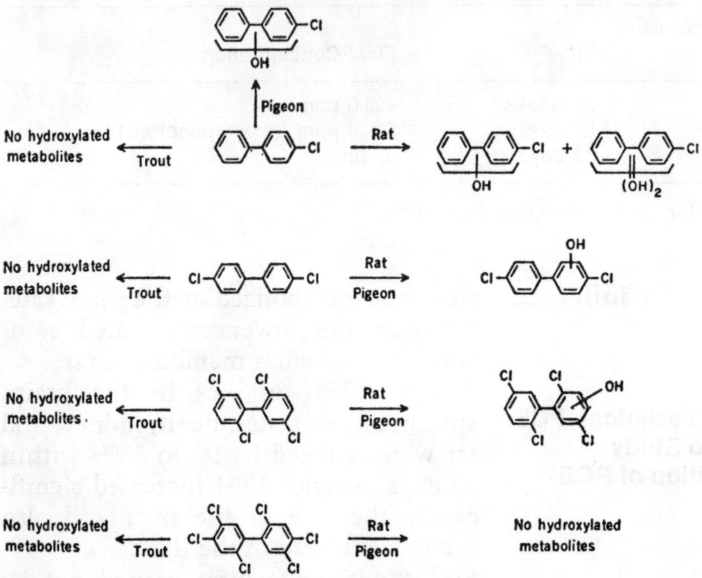

Fig. 7 Scheme for the Metabolism of Chlorobiphenyls in Three Animal Species.(17)

hexachlorobiphenyl have been studied by O. Hutzinger et al. (17) in pigeons, rats, and brook trout (Fig. 7). Their results show the conversion of 4-chloro-, 4,4'-dichloro-, and 2,2',5,5'-tetrachlorobiphenyl isomers to hydroxy derivatives by the rat and pigeon, whereas no hydroxy metabolites were detected in the excreta of the brook trout. The symmetric compound 2,2',4,4',5,5'-hexachlorobiphenyl was not metabolized to hydroxy products by the animals used in the experiments.

The metabolic fate of 3,3',4,4'-tetrachlorobiphenyl in rats was studied by Yoshimura and Yamamoto.(18) About 25 mg each was administered orally to adult male Wistar rats every third day for 9 days. The feces and urine collected for 12 days after the first administration were extracted and analyzed. The fecal extract only contained at least 3 metabolites along with unchanged 3,3'4,4'-TCB. A major metabolite was identified as either 2- or 5-hydroxy-3,3',4,4'-TCB; the other metabolites were phenolic in nature, too.

The excretion rate of unchanged 3,3', 4,4'-TCB and of the main metabolite in the rat feces was determined during 14 days after oral administration of 3,3',4,4'-TCB at a single dose of 25 mg. It was found that about 64% of the dose was excreted unchanged; the excretion of the metabolite in the feces accounted for only 3.3% of the dose during 14 days.

2,2',4,4'-Tetrachlorobiphenyl was injected intravenously to mice, and a mixture of 2,2'3',4-tetrachlorobiphenyl and 2,3',4,4'-tetrachlorobiphenyl to Chinese dwarf quails and Japanese quails.(19) Distribution in the body was studied in these species. Additionally, mice excreta were collected and analyzed. It was found that the amount excreted with the feces from mice reached a maximum a few days after the administration. The metabolites showed a very close resemblance with hydroxy biphenyls.

2,2',4,5,5'-Pentachlorobiphenyl-^{14}C was investigated by Berlin et al.(20) The compound was administered intravenously and orally to mice. The half-life of the PCB was about 6 days, and it was excreted almost entirely in feces. No information is yet available on whether the measured radioactivity is due to PCB or its metabolites.

Goto et al. (21—22) have studied the fate of ^{14}C-2,3-dichlorobiphenyl, 2,4,6-trichlorobiphenyl, 2,3,5,6-tetrachlorobiphenyl and 2,3,4,5,6-pentachlorobiphenyl in rats. In contrast to the above mentioned studies, these compounds contained all the chlorine atoms in one ring, and the other ring was free for hydroxylation. Indeed, hydroxy metabolites were identified for all four PCBs. In case of 2,3-dichlorobiphenyl, hydroxylation occurred in the 4'-position. In case of 2,4,6-trichlorobiphenyl, the 4'-hydroxyderivative, the 3',4'-dihydroxy-derivative, and the 3',4'-hydroxy-methoxy-derivative were detected. For both the tetra- and pentachlorobiphenyl, the 4'-monohydroxy-derivatives, the 3'-monohydroxy-derivatives, the 3',4'-hydroxy-methoxy-derivatives, and the 3',4'-dihydroxy-derivatives were identified.

The following studies related to disappearance of residues have been carried out in our institute, partly in cooperation with Albany Medical College of Union University, Albany, New York. 2,2'-Dichlorobiphenyl and 2,4'-dichlorobiphenyl, both ^{14}C-labeled, were synthesized starting from ^{14}C-benzene.(23) 2,2',5-Trichlorobiphenyl-^{14}C was commercially available. These three compounds were used for metabolic studies in rats, rat liver preparations, monkeys, microorganisms, and higher plants, as well as for a study of the behaviour of PCB during waste composting.

a) Fate of 2,2'-Dichlorobiphenyl-^{14}C in Rats (24)

2,2'-Dichlorobiphenyl-^{14}C was applied orally to rats for 42 days, with a daily dosage corresponding to about 1 ppm in the diet. Urine and feces were collected daily and checked for radioactivity. After 36 days, the daily excreted dose equalled the daily administered dose, and the excretion curve reached a plateau level which has been confirmed during 6 additional days. When the daily excreted radioactivity equalled about 99% of the daily administered radioactivity, the daily administration was discontinued. The stored radioactivity was excreted by the rats within 8 days. From feces extracts, two monohydroxy derivatives, three dihydroxy derivatives, one trihydroxy derivative, a small amount of unchanged PCB, and a small conjugate fraction were isolated; in urine extracts, two monohydroxy derivatives, three dihydroxy derivatives, one monomethoxy derivative, and a conjugate fraction were found.

b) Conversion of 2,2'-Dichlorobiphenyl-^{14}C, 2,4'-Dichlorobiphenyl-^{14}C, and 2,2',5-Trichlorobiphenyl-^{14}C by Rat Liver Enzymes(25)

2,2'-Dichlorobiphenyl-[14]C (0.66 μM/ml), 2,4'-dichlorobiphenyl-[14]C (0.12 μM/ml), and 2,2',5-trichlorobiphenyl-[14]C (0.23 μM/ml) were incubated with the microsomal fraction of rat liver homogenate for 1 hour at 37° C. During this time, 10% of the 2,2'-dichlorobiphenyl, 36% of the 2,4'-dichlorobiphenyl, and 31% of the 2,2',5-trichlorobiphenyl were converted to metabolites.

With the exception of small amounts of polar material which were not identified, all metabolites turned out to be hydroxylation products. From the experiment with 2,2'-dichlorobiphenyl, all four theoretically possible monohydroxy isomers were detected; additionally, four dihydroxy compounds were found. From 2,4'-dichlorobi-

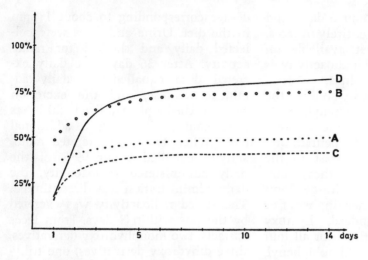

Excretion of radioactivity:

after 2,4'- dichlorobiphenyl appl., urinary A, total B

after 2,2,5 trichlorobiphenyl appl., urinary C, total D

Fig. 8 Excretion of Radio-activity by Rhesus Monkeys after Application of 2,4'-Dichlorobiphenyl and 2,2',5-Trichlorobiphenyl. (26)

phenyl, two monohydroxy- and two dihydroxy-metabolites were identified and from 2,2',5-trichlorobiphenyl, three monohydroxy- and two dihydroxy-metabolites.

c) Metabolism of 2,4'-Dichlorobiphenyl-¹⁴C and 2,2',5-Trichlorobiphenyl-¹⁴C in Monkeys (26, 27)

2,4'-Dichlorobiphenyl-^{14}C and 2,2',5-trichlorobiphenyl-^{14}C were injected intravenously into four female rhesus monkeys in doses between 16.8 and 566 ppb of body weight. After administration of 2,4'-dichlorobiphenyl, an average of 73.5% of the radioactivity had been excreted both in urine and feces within 14 days (about 51% in urine and 22% via feces). Although the dose level varied by a factor of 1 to 33, there was no significant change in the percentage eliminated. The radioactivity was excreted very rapidly, with the maximum amount in the first day after administration. The biological halflife differed within individual animals and with dose from 1.1 to 2.9

days. Continuation of the experiment for another 7 days until day 21 resulted in an increase of eliminated radioactivity of only 0.58% (0.35 in urine, 0.23 in feces). 14 days after application, about 7% of the administered radioactivity could be detected in the fatty tissue, while the blood level was negligible.

After application of 2,2',5-trichlorobiphenyl, 82% of the given radioactivity was recovered in the excreta after 14 days. The halflife was 2.1 days, in relation to 1.1 days for the comparable dose of the dichlorobiphenyl. The excretion of both compounds is presented in Fig. 8.

The excreted radioactivity consisted, for both compounds applied, only of metabolites; no parent material could be detected in excreta. For each PCB, a highly hydrophilic zone of 1% and 4%, respectively, could not be identified. The remaining radioactivity was found to be due to hydroxylated compounds, partly in conjugated form.

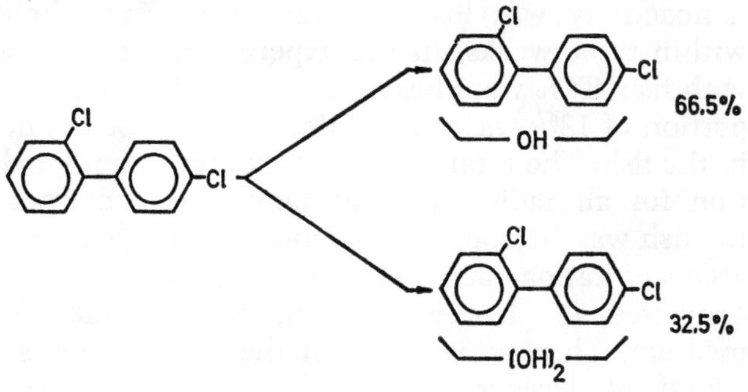

Fig. 9 Metabolites of 2,4'-Di-
chlorobiphenyl Formed by the
Rhesus Monkey, and Amounts
Eliminated in % of Totally Eliminated Radioactivity.(27)

From 2,4'-dichlorobiphenyl, three
monohydroxy- and three dihydroxy
derivatives were identified; the monohydroxy products constitute the major
portion of the excreted radioactivity.
The metabolites are presented in Fig. 9.
From 2,2',5-trichlorobiphenyl, three
monohydroxy products, two dihydroxy
products, and one trihydroxy product
were observed; contrary to the 2,4'-
dichlorobiphenyl experiment, the main
forms of excretion were the dihydroxy
derivatives. Fig. 10 shows a survey of
the metabolites formed from 2,2',5-
trichlorobiphenyl.
Total conversion of both compounds to
hydroxylated metabolites and the rapid
excretion might indicate that accumulation of these low chlorinated PCBs in
primates will not occur. As the dichlorobiphenyl is mainly excreted as
monohydroxy derivatives, and the trichlorobiphenyl as dihydroxy derivatives, it may be concluded that, for the
excretion of medium chlorinated PCBs,
a higher degree of hydroxylation is
needed, since they are less polar. A
higher degree of hydroxylation, however, will cause a longer retention time
of the compound in the body, resulting
in a slower excretion rate. Furthermore,
the chance of poly-hydroxylation will
be reduced with increasing chlorine
content; thus, for highly chlorinated

PCBs there exists the possibility of accumulation in the body.
d) Fate of 2,2'-Dichlorobiphenyl-
^{14}C in Goldfish (28)
In a laboratory aquarium experiment,
2,2'-dichlorobiphenyl-^{14}C was applied
to an aquatic ecosystem consisting of
water, sand, aquatic plants (Elodea
canadensis) and goldfish. In order to
check the role of fish in PCB metabolism, another experiment was carried out simultaneously with the same
ecosystem without fish. In this experiment without fish, 98% of the applied

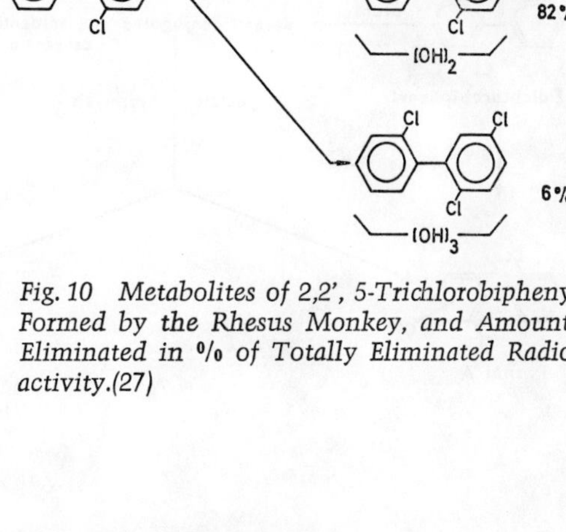

Fig. 10 Metabolites of 2,2', 5-Trichlorobiphenyl
Formed by the Rhesus Monkey, and Amounts
Eliminated in % of Totally Eliminated Radio-
activity.(27)

radioactivity was lost by evaporation within three weeks. In the experiment with fish, 87% were lost; the recovered portion of 13% was mostly (10%) stored in the fish. The total residue concentration for all radioactive substances in the fish was 4.15 ppm, corresponding to a concentration factor of about 500 as compared to the surrounding water medium. The radioactivity in the fish contained, besides unchanged PCB, a very small amount (1.33% of recovered radioactivity) of metabolites. They were found to be two monohydroxy derivatives of 2,2'-dichlorobiphenyl. Contrary to the findings of Hutzinger et al. (17) with other biphenyls and brook trouts, 2,2'-dichlorobiphenyl is metabolized by fish but only in small amounts.

e) Degradation of 2,2'-Dichlorobiphenyl-¹⁴C by Microorganisms (29–31)

When 2,2'-dichlorobiphenyl-¹⁴C was subjected to a waste composting process, only small amounts (about 2%) were lost by evaporation. The recovered portion was mostly due to the unchanged compound (about 98%). Equally, the compound is not metabolized by several soil microorganisms tested. Only Norcardia spec., which grows on 2,2'-dichlorobiphenyl as carbon source, formed about 4% of an unknown

metabolite. It should be emphasized that experimental results with isolated individual strains of microbes are scientifically of great interest but do not predict the real behaviour in the environment.

f) Metabolism of 2,2'-Dichlorobiphenyl-¹⁴C in Higher Plants

Since PCBs have been detected in water and aquatic plants, aquatic systems were used for metabolism studies, too. In order to determine only the plant metabolism of 2,2'-dichlorobiphenyl, and to exclude influences of water or soil microorganisms, the compound was first applied directly to the leaves of a marsh plant, Veronica beccabunga.(32) In order to get sufficient amounts of metabolites for identification, the dose applied was much higher than found in environmental samples (about 423 ppm of fresh plant weight at the time of working up). Within six weeks, most of the radioactivity was lost by evaporation. 6.8% of the applied amount was recovered. 4.6% had penetrated the plants, and 1.9% was found on the plant surface where it had been applied. Only traces of radioactive substances had been transferred to the soil and the water in which the plants were cultivated.

Whilst the radioactivity found on the leaf surface was due mostly to the

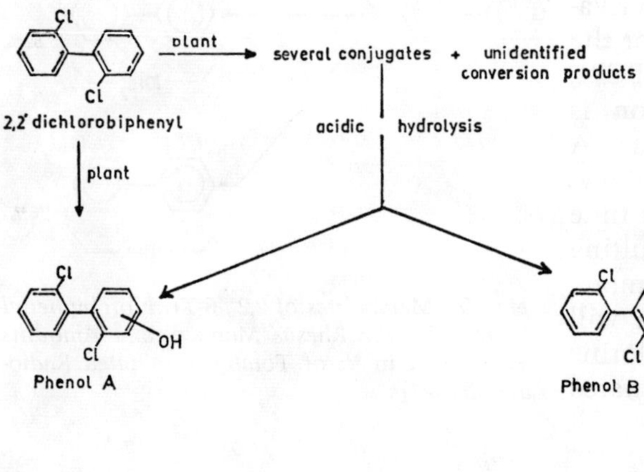

Fig. 11 Metabolism of 2,2'-Dichlorobiphenyl-¹⁴C in Veronica Beccabunga.(32)

Experiment	Water extract	Soil extract	Plant extract	Unextractable radioactivity	Total radioactivity
Experiment with Ranunculus	2.2	6.4	17.7	1.0	27.3
Experiment with Callitriche	1.9	15.6	7.6	1.3	26.4

Fig. 12 Distribution of Radioactivity in Plants, Soils, and Water, 4 Weeks after Application of 2,2'-Dichlorobiphenyl-^{14}C to Water (in % of Applied Radioactivity).(33)

unchanged parent compound (97.4%), the radioactivity extracted from the plants contained more than 30% of metabolites. Most of these metabolites were highly hydrophilic; after hydrolysis, two monohydroxy derivatives of the dichlorobiphenyl were identified. One of these was also found unconjugated in very small amounts. Fig. 11 shows a survey of the metabolism of 2,2'-dichlorobiphenyl in the marsh plant.

Two further plant metabolism studies were carried out with aquatic plants submerged in water and rooted in soil.(33) For these experiments, the species Ranunculus fluitans and Callitriche spec. were selected, and 2,2'-dichlorobiphenyl-^{14}C was applied to them indirectly via the water medium (13.7 ppm in Ranunculus water, and 14.5 ppm in Callitriche water). Four weeks after application, water, plants, and soils were worked up. Fig. 12 shows the distribution of the radioactivity in plants, soils, and water in both experiments.

Only 27.3% of the applied radioactivity could be accounted for in the experiment with Ranunculus and 26.4% in the experiment with Callitriche. Most of this radioactivity has been absorbed by the plants and soils, whereas only 2.2% and 1.9%, respectively, remained in the water medium.

The activity absorbed by the plants consisted mostly of unchanged parent compound. Additionally, unknown products were detected, as well as a group of conjugates, which, upon hydrolysis, yielded 2 isomers of monohydroxy-2,2'-dichlorobiphenyl.

The radioactivity in the water, however, contained more than half of the radioactivity as metabolites. These were isolated, and the main product was found to be a dechlorinated substance, the structure of which has not yet been decided. As by-products two monohydroxy derivatives were identified which were identical to those found in the plants and in the foliar application experiment with Veronica beccabunga. Additionally, small amounts of a dihydroxy compound as well as traces of unidentified conversion products were detected. Fig. 13 gives a survey of the metabolism of 2,2'-dichlorobiphenyl in these two ecosystems.

2. Studies on Abiotic Conversions

From the abiotic conversions of PCB, the degradation by irradiation is most important. Even the presence of the PCB in the environment might be due partly to photoreactions. It has been shown (34, 35) that part of the PCB in the environment does not arise from

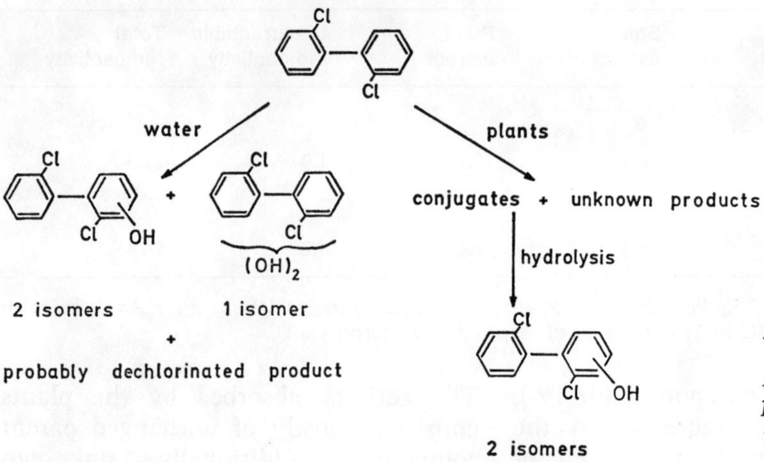

Fig. 13 Metabolism of 2,2'-Dichlorobiphenyl-^{14}C in Water and Plants after Application to Water.(33)

technical use of PCB but from DDT vapor. Dichloro-, trichloro-, and tetra-chlorobiphenyl are formed slowly from DDT via DDE and DDMU. The further fate of PCB during irradiation was investigated by several authors.

Hutzinger et al. (36–38) have studied the photochemical decomposition of individual PCBs as well as technical PCB mixtures. In all cases they obtained polar oxygenated products.

K. Andersson et al. (39) have irradiated 2,2',4,4',6,6'-hexachlorobiphenyl in methanol with a lamp (λ max. = 290–430 nm) for 30 minutes. They discovered that around 80 compounds were formed during this irradiation, and the majority of these were oxygenated chloro aromatics. Among these, a series of o-methoxy-PCBs were identified, and another group of compounds was considered to be methoxy substituted chlorodibenzofurans. Dechlorination was a main decomposition pathway. The authors could identify PCBs with 5, 4, 3, 2, and 1 chlorine atoms.

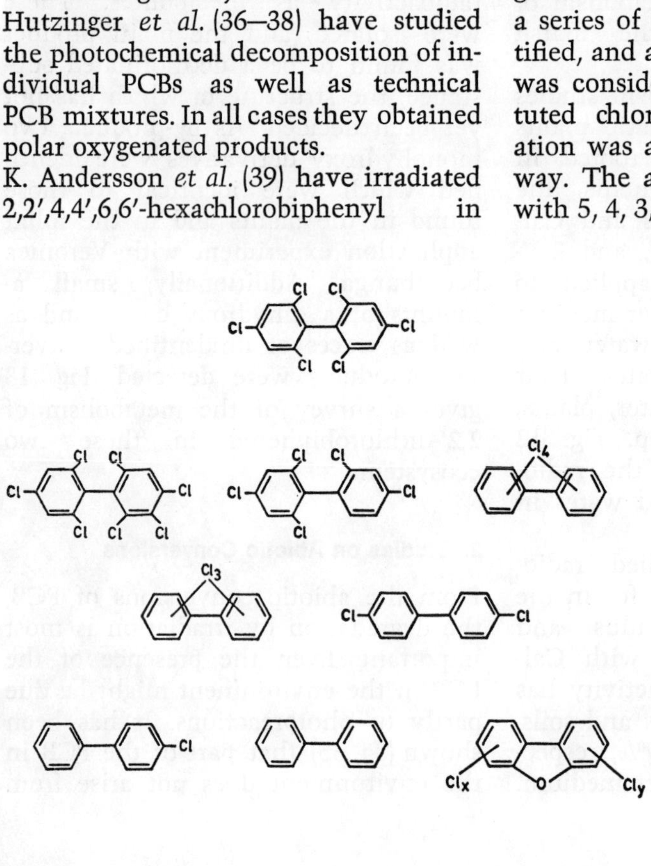

Fig. 14 Conversion of Hexa-chlorobiphenyl by UV-Irradiation.

The abiotic degradation of polychlorinated biphenyls was investigated in our institute using UV-light of various wavelengths.[40] In such experiments with 2,2',4,4',6,6'-hexachlorobiphenyl in non-polar solvents, a stepwise displacement of chlorine atoms, with the final production of unsubstituted biphenyl, could be demonstrated. Polymerizations and isomerizations are possible side reactions. In polar solvents, oxygen-containing compounds are also formed, for example, hydroxylated products. It is possible that the extremely toxic polychlorodibenzofurans, which have been detected as impurities in various industrial products, are also formed (Fig. 14). Irradiation in the solid or liquid phase without solvent produces higher chlorinated products such as hepta- or octachlorobiphenyl. Further irradiation experiments with hexa-, tetra- and dichlorobiphenyls in solvents, in the solid state, and in the gaseous phase resulted in dechlorinations and chlorinations, oxygenations, isomerizations and polymerizations.[41] Photochemically induced oxygen atoms cleaved, besides other reactions, the biphenyl system to form benzene derivatives. Under environmental conditions, however, such compounds have not been reported so far.

III. Conclusion

According to our present knowledge, PCBs do not represent a serious and general hazard — at least with the measures taken now — but they represent an excellent model for the investigation, under all environmental aspects, of persistent chemicals occurring unintentionally in the environment. We should learn from the PCB case that it is very important to know,

from production and use, the potential occurrence of chemicals in order to develop specific methods for them and not to rely on sensitive methods for some classes of chemicals only.

References:

1 Jensen, S.: New Scientist 32, 612 (1966).
2 Summarized in: Polychlorinated Biphenyls and the Environment, Interdepartmental Task Force on PCBs, Washington, D. C., May 1972. (Distributed by the National Technical Information Service, U. S. Department of Commerce, Springfield, Virginia 22151.)
3 Kuratsune, M., T. Yoshimura, J. Matsuzaka, and A. Yamaguchi: H. S. M. H. A. Health Reports 86, 1083 (1971).
4 Nisbet, I. C. T., and A. F. Sarofim: Environm. Health in Perspective, 1972.
5 Acker, L., and E. Schulte: Naturw. 57, 497 (1970).
6 Acker, L., and E. Schulte: Dtsch. Lebensm. Rdschau 66, 385 (1970).
7 Savage, E. P., J. D. Tessari, J. W. Malberg, H. W. Wheeler, and J. R. Bagby: Bull. Environm. Contam. Toxicol. 9, 222 (1973).
8 Corneliussen, P. E.: Pest. Monit. J. 5, 313 (1972).
9 Eichelberger, J.: PCB Newsletter, July 28, p. 8 (1971).
10 Bauer, U.: Dissertation Universität Münster (1972).
11 Müller, W., et al.: unpublished.
12 Duke, T. W.: PCB Newsletter, July 28, p. 10 (1971).
13 Grant, D. L., W. E. J. Phillips, and D. C. Villeneuve: Bull. Environm. Contam. Toxicol. 6, 102 (1971).
14 Bailey, S., and P. J. Bunyan: Nature 236, 34 (1972).
15 Iwata, Y., W. E. Westlake, and F. A. Gunther: Bull. Environm. Contam. Toxicol. 9, 204 (1973).
16 Block, W. D., and H. H. Cornish: J. Biol. Chem. 234, 3301 (1959).
17 Hutzinger, O., D. M. Nash, S. Safe, A. S. W. De Freitas, R. J. Norstrom, D. J. Wildish, and V. Zitko: Science 178, 312 (1972).
18 Yoshimura, H., and H. Yamamoto: Chem. Pharm. Bull. 21, 1168 (1973).
19 Melvås, B., and J. Brandt: PCB-Conference II, Stockholm, Dec. 1972.

20 Berlin, M., J. C. Gage, and S. Holm: PCB-Conference II, National Swedish Environment Protection Board, Publications 1973: 4 E.

21 Goto, M., K. Sugiura, M. Hattori, T. Miyagawa, and M. Okamura: Chemosphere 3, 227 (1974).

22 Goto, M., K. Sugiura, M. Hattori, T. Miyagawa, and M. Okamura: Chemosphere 3, 233 (1974).

23 Attar, A., R. Ismail, D. Bieniek, W. Klein, and F. Korte: Chemosphere 2, 261 (1973).

24 Kamal, M.: Dissertation Universität Bonn, unpublished.

25 Greb, W., W. Klein, F. Coulston, L. Golberg, and F. Korte: Bull. Environm. Contam. Toxicol., in press.

26 Greb, W., W. Klein, F. Coulston, L. Golberg, and F. Korte: Chemosphere 2, 143 (1973).

27 Greb, W., W. Klein, F. Coulston, L. Golberg, and F. Korte: Bull. Environm. Contam. Toxicol., in press.

28 Herbst, E.: Diplomarbeit Universität Bonn, 1974.

29 Müller, W., H. Rohleder, W. Klein und F. Korte: Gesellschaft für Strahlen- und Umweltforschung mbH, München, GSF-Bericht Ö 104 (1974).

30 Müller, W. P. und F. Korte: Naturwiss. 61, 326 (1974).

31 Müller, W.: Dissertation Universität Bonn, 1974.

32 Moza, P., I. Weisgerber, W. Klein, and F. Korte: Chemosphere 2, 217 (1973).

33 Moza, P., I. Weisgerber, W. Klein, and F. Korte: Bull. Environm. Contam. Toxicol. 12, 541 (1974).

34 Plimmer, J. R., and U. I. Klingebiel: Chem. Commun. 648 (1969).

35 Moilanen, K. W., and D. G. Crosby: 165th Meeting, Amer. Chem. Soc., Dallas, Texas, April 10, 1973; in: T. H. Maugh II, Science 180, 578 (1973).

36 Safe, S., and O. Hutzinger: Nature 232, 641 (1971).

37 Hutzinger, O., S. Safe, and V. Zitko: Environm. Health Perspectives 1, 15 (1972).

38 Hutzinger, O., W. D. Jamieson, S. Safe, and V. Zitko: ACS, 164th National Meeting, Div. Water, Air, and Waste Chemistry, p. 74 (1972).

39 Andersson, K., C.-A. Nilsson, Å. Norström, and C. Rappe: PCB Conference II, Stockholm 1972, General Discussion.

40 Hustert, K. und F. Korte: Chemosphere 1, 7 (1972).

41 Hustert, K. und F. Korte: Chemosphere 3, 153 (1974).

Index

Environmental Quality and Safety

Volume 4

Global Aspects of Chemistry, Toxicology and Technology as Applied to the Environment

Editors: Prof. Dr. F. COULSTON, Prof. Dr. F. KORTE
Assistant Editors: W. Klein, I. Rosenblum

1975. VIII, 277 pages, 115 illustrations, 48 tables 17 x 24 cm
‹Thieme Edition› cloth DM 64,—

ISBN 313516201 X

Environmental Quality and Safety-Supplements

Volume 1
Heavy Metal Toxicity, Safety and Hormology

Editors: Prof. Dr. F. COULSTON, Prof. Dr. F. KORTE
Assistant Editors: W. Klein, I. Rosenblum

By Prof. T. D. LUCKEY, Ph. D., B. VENUGOPAL and D. P. HUTCHESON, Columbia/Miss./USA

1975. VI, 120 pages, 12 figures, 44 tables, 17 x 24 cm
‹Thieme Edition› cloth DM 35,—

ISBN 3135163016

Joint edition with Academic Press New York · San Francisco · London

---

Volume 2
Lead

Guest Editors: Dr. T. B. GRIFFIN, ICES, Alamogordo N.M./USA and Dr. J. H. KNELSON, EPA, Res. Triangle, N.C./USA

1975. VIII, 299 pages, 108 ill., 166 tables, 17 x 24 cm
‹Thieme Edition› cloth DM 58,—

ISBN 3135164012

Joint edition with Academic Press New York · San Francisco · London

---

Synthesis

International Journal of Methods in Synthetic Organ Chemistry

Volume 8 (1976)

Editors: G. SCHILL, Freiburg/Br.; G. SOSNOVSKY, Milwaukee/Wis.; H. J. ZIEGLER, Basle

Editorial Office: R. E. Dunmur and W. Lürken, Stuttgart

Published monthly. Annual subscription price DM 198,— (**Special price** for undergraduate and postgraduate students **DM 128,40**)

Georg Thieme Publishers Stuttgart